T5-DGI-448

FUZZY LOGIC IN KNOWLEDGE-BASED SYSTEMS, DECISION AND CONTROL

edited by

Madan M. GUPTA
Intelligent Systems Research Laboratory
College of Engineering
University of Saskatchewan
Saskatoon, Saskatchewan, Canada

and

Takeshi YAMAKAWA
Department of Electrical Engineering
and Computer Science
Kumamoto University
Kumamoto, Japan

1988

NORTH-HOLLAND • AMSTERDAM • NEW YORK • OXFORD • TOKYO

ISBN: 0 444 70450 7

Publishers:
ELSEVIER SCIENCE PUBLISHERS B.V.
P.O. Box 1991
1000 BZ Amsterdam
The Netherlands

Sole distributors for the U.S.A. and Canada:
ELSEVIER SCIENCE PUBLISHING COMPANY, INC.
52 Vanderbilt Avenue
New York, N.Y. 10017
U.S.A.

LIBRARY OF CONGRESS
Library of Congress Cataloging-in-Publication Data

Fuzzy logic in knowledge-based systems, decision, and control / edited by Madan M. Gupta and Takeshi Yamakawa.
p. cm.
Bibliography: p.
ISBN 0-444-70450-7 (U.S.)
1. Expert systems (Computer science) 2. Automatic control.
3. Fuzzy systems. I. Gupta, Madan M. II. Yamakawa, Takeshi, 1946-
.
QA76.76.E95F885 1988
006.3'3--dc19
88-11161
CIP

PRINTED IN THE NETHERLANDS

DEDICATION

To:

Professor Lofti A. Zadeh
for providing us with a tool of graded membership
which has revolutionized the logic and computing methods
for analyzing the information arising from human cognitive processes.

इन्द्रियाणि मनो बुद्धिरस्याधिष्ठानमुच्यते ।
एतैर्विमोहयत्येष ज्ञानमावृत्य देहिनम् ॥४०॥

40. *indriyāṇi mano buddhir*
asyā 'dhiṣṭhānam ucyate
etair vimohayaty eṣa
jñānam āvṛtya dehinam

"The senses, the mind and the intelligence are the breeding grounds of desire and lust; they veil the real knowledge of the living entity and bewilder the embodied soul".

इन्द्रियाणि पराण्याहुरिन्द्रियेभ्यः परं मनः ।
मनसस्तु परा बुद्धिर्यो बुद्धेः परतस्तु सः ॥४२॥

42. *indriyāṇi parāṇy āhur*
indriyebhyaḥ paraṁ manaḥ
manasas tu parā buddhir
yo buddheḥ paratas tu saḥ

"The active senses are superior to the passive matter; mind is higher than the senses; intelligence is still superior than the mind, but the soul is the most superior".

"Bhagavad-Gita"

CONTENTS

APPENDICES

BIBLIOGRAPHIC INFORMATION

THE COMPANION VOLUME "FUZZY COMPUTING: THEORY, HARDWARE, AND APPLICATIONS"

FOREWORD

The collection of contributions which constitute this volume reflect the many important developments in the theory of fuzzy sets and its applications which have taken place since its inception in 1965. What one notices first is that the theory has become much richer and much more sophisticated. And second, its applications have grown at an accelerating rate both in number and variety.

One of the major branches of the theory of fuzzy sets which is the unifying theme of this volume is that of fuzzy logic. As its name implies, fuzzy logic is the logic underlying modes of reasoning which are approximate rather than exact. In compliance with the spirit of logic, it attempts to be precise. Thus, fuzzy logic is, perhaps somewhat paradoxically, a precise system for imprecise reasoning.

The need for a logic of approximate reasoning has become more obvious during the past decade as a result of the advent of knowledge-based systems – systems in which much of the expert knowledge is imprecise, incomplete, or not totally reliable. At present, most of the techniques which are used to deal with uncertain knowledge in such systems are still based on classical probability theory. What this demonstrates is not that classical probability theory is sufficient for the management of uncertainty in expert systems. Rather, it reflects the fact that all of us tend to employ the tools we know how to use. Thus, fuzzy logic will become a widely used tool for the conception and design of knowledge-based systems when its body of concepts and techniques becomes a familiar subject. This is not yet the case today, but the day when it might be true is brought closer by the many texts which are now in print and the collections of authoritative research contributions such as those which constitute the present volume.

An application area of fuzzy logic in which there is a great deal of activity at present, especially in Japan, is that of fuzzy logic control. Among the recent outstanding examples of this application are the new subway system in Sendai, Sugeno's Fuzzy Car, Hirota's Fuzzy Robot, Fuji's Universal Controller, and many other systems ranging from engine control to tunnel digging machinery and container-ship cranes. It may well turn out that these clearly visible applications will pave the way for a wider use of fuzzy logic in areas in which there is still some skepticism regarding its advantages over the more conventional methods for dealing with uncertainty and imprecision.

The articles in this volume cover a broad spectrum of topics ranging from classification structures and computer vision to team decision problems, forecasting and statistics. They clarify some of the outstanding problems, introduce new concepts and techniques, and point to novel applications. The editors, Professors Gupta and Yamakawa, deserve our thanks and congratulations for putting together a volume that is a source of many new ideas and an authoritative exposition of techniques which are central to the theory of fuzzy sets and its applications to knowledge-based systems, decision and control.

Lofti A. Zadeh
Berkeley, CA
February, 1988

PREFACE

To the expanding information-based technological society – we present another two volumes with the titles "*Fuzzy Computing: Theory, Hardware and Applications*", and "*Fuzzy Logic in Knowledge-Based Systems, Decision and Control*". These two volumes complement five previously published volumes with which one of the editors (M.M. Gupta) has been associated for the past fifteen years through the North-Holland/Elsevier Science Publishing Company.

FUZZY COMPUTING: THEORY, HARDWARE AND APPLICATIONS

The first of these two volumes presents thirty-two studies and is divided into three parts. The **first** part presents some thoughts on cognitive uncertainty, perception, and perspectives on cognitive computing. It also deals with the theoretical aspects of fuzzy computing including some important recent studies made in fuzzy logic, vision perception, associative pattern recognition, models of uncertainties, and the theoretical framework of prolog machines.

The **second** part of this volume is devoted to the hardware aspects of fuzzy computing. This part reports some recent development work in fuzzy electronic circuits that can be used for fuzzy computing, fuzzy memory modules, hardware for rule-based systems, multivalued CCD circuits and the organization of fuzzy micro-processor modules. Some of the seminal studies reported in this part may lead to the development of new fuzzy computing machines in the decade to come.

The **third** part of the volume deals with various applications of fuzzy logic and fuzzy computing. These studies range from fuzzy logic in expert systems to computerized electrocardiography, medical decision making, medical expert systems, occupational and psychosocial stress management, fuzzy knowledge networks, decision support systems, expert control for large scale industrial processes, rule-based self-organizing controllers, self-regulating PID controllers, image segmentation, knowledge engineering, multiobjective decision making, production management and soft clustering.

During the coming decade, this information based technological society is going to witness an explosion in many types of uncertainties and its cognate information. Therefore, we considered it essential to include some aspects of information classification and some basic philosophical questions that arise in the context of information processing. In particular, some preliminary studies of cognitive uncertainty, cognitive information, cognition, perception and uncertainty, and cognitive-neural computing based on fuzzy logic are reported. Hopefully, these new notions and studies will lead to some new areas of research.

The studies reported in this volume represent the work of some fifty authors from eleven countries*.

FUZZY LOGIC IN KNOWLEDGE-BASED SYSTEMS, DECISION AND CONTROL

This volume reports the work of some thirty-two authors from fifteen countries in twenty-three articles**. The work is divided into two parts.

The **first** part presents theoretical studies carried out in the field of cognition, perception and fuzzy logic. In particular, articles dealing with flexible classification structures, entropic models, ambiguity in human support systems, fuzzy probability measures, plausible reasoning, fuzzy numbers, expert systems, fuzzy modeling, multifactorial functions, systems with ambiguous states, graded consequence in multiple valued logic, multicriteria analysis, and formalization of quantitier phrases are reported. Also, this part puts forth some philosophical questions in the context of cognitive information and cognitive computing and their applications in the design of intelligent systems.

The **second** part of the volume contains studies on the application of fuzzy logic in knowledge-based systems, and decision and control problems. In particular, it presents, heuristic learning methods in expert systems, fuzzy control, human reliability, fuzzy analogical reasoning, ultra fuzzy information structure and fuzziness and rare items.

The studies reported in these two volumes clearly indicate an increasing interest in the theory of fuzzy sets with diversified and exciting applications not reported before. The new studies in fuzzy logic and fuzzy computing will help to develop new hardware, and this new hardware will help in the simulation of cognitive processes and in the development of cognitive-neural computing tools. These studies are important if we wish to emulate human intelligence and reasoning power on a silicon chip (machine). Some of the new emerging technologies, especially those based upon optoelectronics, gallium arsenide (GaAs), and molecular electronics, will help further in the development of fuzzy computing and cognitive-neural tools.

Although the editors have, to some extent, reviewed each article, these multi-authored volumes contain large variations in the style of writing, English and notations, which could not be unified.

* Algeria (1), Canada (5), Czechoslovakia (1), China (10), France (4), Greece (2), India (2), Italy (5), Japan (8), Poland (3), USA (9).

** Belgium (2), Brazil (2), Canada (3), China (5), France (1), FRG (1), GDR (1), India (1), Israel (1), Italy (1), Japan (3), Norway (1), Poland (3), Spain (2), USA (5).

The fields of cognitive information, fuzzy logic and fuzzy computing are currently expanding at an exponential rate, and the work presented in these two volumes is necessarily limited. It is hoped, however, that these studies will convey to the readers the excitement, advances, promises and challenges that the field offers to our expanding infomation-based technological society in the area of cognitive processes, knowledge-based systems, expert systems, and decision and control problems, of the future.

January 1, 1988

Editors:
Madan M. Gupta
University of Saskatchewan

Takeshi Yamakawa
Kumamoto University

ACKNOWLEDGEMENTS

These two volumes, entitled *"Fuzzy Computing: Theory, Hardware and Applications"* and *"Fuzzy Logic in Knowledge Based Systems, Decision and Control"*, represent the work of eighty-two individuals from eighteen countries*. It is gratifying to acknowledge the authors who have contributed to the exploration of these new theories, to the development of new fields of application, to the stimulation of the advancement of scientific information, and to the reinforcement of international co-operation in this important field of fuzzy logic, fuzzy computing, and computing tools dealing with cognitive information.

The University of Saskatchewan provides a stimulating intellectual environment, which has encouraged interaction with the international scientific community. It is, therefore, a pleasure to express once again gratitude to the University of Saskatchewan, to its College of Engineering, and to the research students and researchers in the Intelligent Systems Research Laboratory for their contribution to these newly developing theories and technology.

As always, the support from North-Holland has been generous. We are especially indebted to Drs. Gerard Wanrooy, our acquisition editor, who has provided us with much useful advice, and his production and marketing staff for their infectious enthusiasm for the series of volumes published over the last several years.

Finally, the editors record their appreciation to Miss Elizabeth Nikiforuk, the Research Assistant in the Intelligent System Research Laboratory, for her helpful and smiling attitude and for converting many uncertainties into useful information during the editorial phases of these two volumes.

January 1, 1988

Madan M. Gupta
University of Saskatchewan

Takeshi Yamakawa
Kumamoto University

*Algeria (1), Belgium (2), Brazil (2), Canada (8), Czechoslovakia (1), Republic of China (15), France (5), FRG (1), GDR (1), Greece (2), India (3), Israel (1), Italy (6), Japan (11), Norway (1), Poland (6), Spain (2), USA (14).

PART I

THEORY OF FUZZY LOGIC

Fuzzy Logic in Knowledge-Based Systems,
Decision and Control
M.M. Gupta and T. Yamakawa (Editors)

COGNITION, PERCEPTION AND UNCERTAINTY*

Madan M. Gupta

Intelligent Systems Research Laboratory, College of Engineering, University of Saskatchewan, Saskatoon, Saskatchewan, S7N OWO, Canada.

A beautiful blanket wraps our scientific environment and thoughts, but it is woven with fragile threads of two kinds of uncertainties. The first arises from the physical systems and the second from human thinking, reasoning, cognition and perception. A distainful act towards this wrapper may break the fragile threads and then our thoughts will just be naked, emotionless and impotent. Let us create more respect for this fragile uncertainty and generate intelligible rules for its morphology.

The theory of deterministic systems and stochastic processes has been an integral part of my graduate studies, teaching and research. In fact, I was, and I am still, a member of the school of *determinism* and *stochasticism*. It was in the summer of 1968 that I first had the opportunity to listen to Professor Lotfi A. Zadeh, the exponent and founder of *'Fuzzy-ism,'* at the Symposium on Adaptivity and Sensitivity held under the auspices of the International Federation of Automatic Control (IFAC) at Dubrovnik, Yugoslavia. His lucid exposition on Fuzzy Set Theory, the theory which deals with the information and uncertainty arising from human thinking, reasoning, cognition and perception, aroused my interest in this new class of mathematics and, since then, I have become also a student of the *School of Fuzzy-ism*.

During my studies, I soon realized that the 'lack of uncertainty' or the excess of 'precision and certainty' in our engineering design, decision and control problems are providing us with unrealizable solutions. The certainty (precision) has become an absolute standard in both our teaching and research, and far too often it is introduced into our scientific works without thought or feeling. This is the attribute which does not exist in our cognition, perception, reasoning and thinking. A baby of one week starts recognizing his mother by smiling at her, but the cognition, perception and recognition of his mother and his environment do not take place in the precise sense. This same imprecision or uncertainty will play an important role in thinking and reasoning throughout his life.

In our scientific decision and control studies, we have realized that the path of 'precision' is rapidly leading us towards unrealizable goals, but we still continue to move along that path. Of course, we know that any 'precise' design will be doomed to failure. Behind the curtain of 'precision' we have become impotent, and have lost scientific creativity.

*To: Professor Peter N. Nikiforuk

There are various classes of uncertainties, however, for the discussion purposes here we classify these uncertainties into two broad categories: *U-Type One,* and *U-Type Two,* where U-stands for uncertainty.

The *U-Type One* uncertainty deals with the information or phenomenon which arises from the random behaviour of physical systems. The pervasiveness of this type of uncertainty can be witnessed in the random vibrations of a machine, randomness of a message, random fluctuations of electrons in a magnetic field, diffusion of gases in a thermal field, random electrical activities of the cardiac muscles, uncertain fluctuations in the weather pattern and the turbulent blood flow through a damaged cardiac valve. The *U-Type One* uncertainty has been studied for centuries, and we have a very rich statistical theory to characterize such a random phenomenon. The calculus of mean and variance is very rich in this respect and is being used very widely.

The *U-Type Two* uncertainty, unlike the *U-Type One,* is the uncertainty that deals with phenomena arising from human thinking, reasoning, cognition and perception processes, or cognitive information in general. This is a subject which has been either neglected or taken very lightly. The cognition and perception of the physical environment through our natural sensors (eyes, ears, nose, etc.), the perception of pain and other similar biological events through our nervous system and neural networks deserve special attention. The 'perception phenomenon' associated with these processes are full of 'uncertainties' and cannot be characterized by conventional statistical theory. We can feel pain: 'the back is *very painful*', but this pain can be neither measured nor characterized using statistical theory. Similarly, we express our perception linguistically, 'this *red* flower is just *beautiful* and is full of *pleasing fragrance*'. This corresponds to the 'perception' of our physical environment where *'red'* and *'beautiful'* describe the visual perception, whereas *'pleasing fragrance'* describes the perception of smell. Again, we cannot characterize these perceptions using the strength of the statistical theory.

The *U-Type Two* uncertainty and its cognate, the *cognitive information,* involve the activities of the neural networks. To non-scientists, it may seem strange that such familiar 'notions' have recently become the focus of intense research. But it is the 'ignorance' of these notions, and their possible technological applications in intelligent man-made systems, and not 'familiarity' with them which has forced scientists to conduct research in the field of *U-Type Two* uncertainty and *cognitive information.*

The development of the human cognitive process and the perception of his environment starts taking shape with the development of imaginative power in a baby's brain. A baby in the cradle can recognize the human face long before it is conscious of any visual physical attributes of humans or its environment.

In spite of the richness of conventional mathematical methods, they are very often thought to be dry and cold. One reason lies in their inability to describe the beauty of white mountains, blue lakes, the rising sun, the full moon, or the richness of the fragrance of a spring flower. No doubt, one can estimate the volume of snow or the heights of the mountains, or the frequencies of vibrating musical strings, but the conventional mathematical methods cannot be used to narrate logically the feelings and the emotions associated with perception.

The study of such formless uncertainties provides us with a scientific challenge. Scientists have started now to think of giving a morphology to this amorphous uncertainty. In the past, mathematicians have distained this challenge and have increasingly chosen to flee from natural mentation by devising theories unrelated to human perception, feelings and emotions.

It was in 1965 when Lotfi A. Zadeh published his first celebrated paper on *Fuzzy Sets* and it is now almost two decades ago since he first introduced me to this new type of information and uncertainty. He showed me the path which leads to somewhat beautiful gardens full of immortal and ever increasing fragrance. Though I was taught the notions of cognition and perception at school, I was very ignorant about the uncertainty and its pervasiveness around these notions. Indeed this uncertainty has been distained by scientists and mathematicians.

No one had seen the beauty of these *Fuzzy Sets* before Professor Lotfi Zadeh, and it was he who showed promise of consolidating this beauty into an organized field with rich theories and promising applications.

Professor Zadeh coined the word *Fuzzy Sets*. *Fuzzy Sets* deal with sets of objects or phenomena which are vague and do not have sharp boundaries. The calculus of *Fuzzy Sets* is a very promising tool for dealing with the *U-Type Two* uncertainty (just as probability theory deals with the *U-Type One* uncertainty). Indeed, the applications of these fuzzy sets, which once were thought to be dull and dry, can be found in many scientific and scholarly works. It is true that Boole introduced the beautiful notion of binary sets which are so pervasive in our digital world, however, this beauty is naked and without any adornment. Boolean logic is unable to model the human cognition and thinking process. This is the very reason that no one today is indifferent to the *logic of Fuzzy Sets*. In fact, many view their first encounter with the fuzzy logic as a totally new and exciting experience in their scientific life.

From the purely mathematical view point, the evolution of the theory is very exciting but complex. Many scientific theories start by borrowing notions from the already developed areas of mathematics but, in this case, Professor Zadeh introduced the basic notion of 'vagueness' having no sharp morphology and which is so common in human thought processes.

Indeed, Professor Zadeh laid the foundation of fuzzy mathematics on a very robust rock. It now serves the needs of many existing scientific disciplines, but equally important is that many new disciplines, such as the study of neural networks, have started arising around this mathematics. Thus, this mathematics has united several noble (both old and new) narrow streams of scientific disciplines into one while, at the same time, instilling life into several other streams that have been dormant.

Since 1968, when the notion of fuzzy mathematics was first introduced to me, I have done many studies on this topic. These studies have brought many intellectual and scientific revelations to me. Presently, I am engaged in several scientific studies using fuzzy logic. For example, the research in my laboratory is heavily committed to the *U-Type Two* uncertainty through such studies as neural networks and neural computing, signal and image processing, cognitive information processing, cognitive vision fields, perception, etc., with applications to intelligent robotic systems and medical imaging. The newly developing technology based upon optoelectronics and molecular electronics computing will make it possible to process the *U-Type Two* uncertainty and cognitive information associated with signals and images the way our brains do.

Bibliography

1. R.E. Bellman, and L.A. Zadeh, "Decision Making in a Fuzzy Environment", *Management Science,* (1970), 17, B.141-B.164.
2. M. Black, "Vagueness: An Exercise in Logical Analysis", *Philosophy of Science,* 4, (1937), 427-455.
3. L. Brillouin, "Science and Information Theory", *Academic Press,* New York, (1956).
4. R.C. Conant, "Law of Information Which Govern Systems", *IEEE Trans. Systems, Man. and Cybernetics,* Vol. 6, (1976), 334-338.
5. M. Conard, "The Lure of Molecular Computing", *IEEE Spectrum,* Vol. 23, No. 10, Oct. 1986, pp. 55-60.
6. I.R. Goodman and H.T. Nguyen, "Uncertainty Models for Knowledge-Based Systems", *North-Holland,* New York, 1985.
7. M.M. Gupta, "Fuzzy Automata and Decision Processes: The First Decade", *Sixth Trienniel World IFAC Congress,* Boston, Cambridge, August 24-30, (1975).
8. M.M. Gupta, "On the Cognitive Computing: Perspective," in Fuzzy Computing: Theory, Hardware and Applications, North Holland, (1988).
9. A. Kaufmann, "Introduction to the Theory of Fuzzy Subsets", Vol. 1, *Academic Press,* New York, (1975).
10. A. Kaufmann and M.M. Gupta, "Introduction to Fuzzy Arithemtic: Theory and Applications", *Van Nostrand Reinhold,* New York, 1985.
11. G.J. Klir, "Where Do We Stand on Measures of Uncertainty, Ambiguity, Fuzziness and the Like", *Fuzzy Sets and Systems,* Special Issue on Measure of Uncertainty, Vol. 24, No. 2, November 1987, pp. 141-160.
12. K. Kornwachs and W. von Lucadou, "Pragmatic Information as a Nonclassical Concept to Describe Cognitive Processes", *Cognive Systems,* 1, (1985), 79-84.
13. F.M. Reza, "An Introduction to Information Theory", *McGraw-Hill,* New York, (1961).
14. L.A. Zadeh, "Fuzzy Sets", *Information and Control,* 8, (1965), 338-353.
15. Proceedings of the International Conference on "Bioelectronics and Molecular Electronics Devices," R and D Association for Future Electronics Devices, Tokyo, Nov. 1985.

Fuzzy Logic in Knowledge-Based Systems,
Decision and Control
M.M. Gupta and T. Yamakawa (Editors)

ON THE COGNITIVE COMPUTING: PERSPECTIVES*

Madan M. Gupta, Intelligent Systems Research Laboratory, College of Engineering, University of Saskatchewan, Saskatoon, Saskatchewan,

1. Introduction

It was a cold New Year's eve. After cleaning the clutters of 1987 from my office and arranging certain tasks for 1988, I was driving home. Light snow was falling and the roads were icy, but not too busy. Suddenly, my hands steered the car towards the side lane and my foot worked the brake rendering the car motionless. I soon realized that a speeding car, chased by the police, was coming towards me in my lane and would have collided with my car had I not steered it to the side lane. I took a long breath of relief and thanked the Almighty for giving us such a wonderful sensory system and cognitive computational organ, the brain, which can perform many urgent tasks, such as this collision avoidance, at an ultra high speed. The *natural sensors* (in this case my eyes and ears) were acquiring the information continuously and relaying it through the sensory neurons to the cognitive processor (my brain) which perceived the environment and the danger associated with it just in time. The *cognitive processor* then synthesized the control signals and transmitted them through the motor neurons to my hands and foot, which activated the steering and braking mechanisms to avoid the collision. I was left alone on the road and the experience created a number of emotional feelings in my mind: thoughts of drunk drivers on the road, on modern technology, on a government which encourages the selling of liquor in order to collect revenue and then spends the same revenue to curb drunk drivers, and so on.

This incident upset me so greatly that a magnitude of thoughts appeared momentarily keeping the *cognitive processor* busy in various types of *reasoning* and *perception*. After having supper and sitting down in my study, I recalled a thought: the Almighty has given many wonderful attributes to humans. The attributes of 3 **h** which I described in one of the earlier papers need special mention here.

The attributes of 3 **h** are associated with the **h***and*, the **h***ead* and the **h***eart*. We have sucessfully mimiced the functions of the **h***and*, and have created many mighty machines. Now we are working on the emulation of some of the important attributes of the **h***ead* (brain) such as thinking, reasoning, or in general cognition and perception, on a silicon chip and the creation of an intelligent machine. The **h***eart* is associated with the attribute 'emotions', and this is the attribute which is mainly responsible for the 'survival' of humanity. Perhaps this is the subject which must be considered very seriously for future research because a *mighty thinking* machine

*To Professor Arnold Kaufmann

without *emotions* (**heart**) may become the cause for the destruction of humanity.

2. Mentation, Cognition and Fuzzy Logic

In biomedical engineering we apply the principles of the natural sciences and engineering to the benefit of the health sciences. Here, we shall take an inverse biomedical engineering [(biomedical engineering $)^{-1}$] approach and shall try to apply the biological principles to the solution of some engineering problems. In particular, engineers are investigating the problem of creating *intelligence* in a robotic system. The thought of the creation of intelligence on a silicon chip (machine) creates some strange feelings in our minds.

Intelligence implies the ability to think, reason, learn and memorize, or, in general, it refers to the human mentation and cognition process. One of the last frontiers of science, perhaps its ultimate challenge, is understanding the biological basis of mentation and cognition: how we think, reason, learn, remember, perceive and act. I still cannot understand how the brain perceived the dangerous driving situation and acted instantaneously while it might take several seconds to multiply two three-digit numbers. How do the genes contribute to the process of mentation and cognition and how do they develop with the environment?

Here we have two computational tools: the carbon based organic brain which has existed in humans and animals for several billions of years, and the silicon based modern computers which have evolved only over the last three decades. Recent technological advances in computer hardware have made it possible to carry a very powerful computer in a brief case which is ultra fast and efficient in numerical computations. However, the *'cognitive information'* the information which our natural sensors acquire, is *not numerical,* but the *'mentation process'* can process such information very efficiently and act upon it accordingly. The modern day computers fail to process such cognitive information.

The fact that the human mentation and cognition process is so marvelously efficient and effective poses a question for scientists and engineers: *Can some of the functions and attributes of the human sensory system, mentation and the cognitive processor, and motor neurons be emulated in a robotic system?*

For such an emulation process, it is necessary to understand the biological and physiological functions of the brain. It is a difficult question to answer. However, it is felt that if we examine some of the 'mathematical aspects' of our thinking process and 'hardware aspects' of the 'neurons,' the principle element of the brain, we may succeed to some extent in our emulation process.

The mentation and cognitive activity of the brain, unlike the computational function of the binary computer, is based upon the *relative grades* of the information acquired by the natural sensory system. The conventional mathematical tools, whether deterministic or probabilistic, are based upon some absolute measure of the information. Our natural sensors acquire information in the form of *relative grades* rather than absolute numbers. The perception and 'action' of the cognitive process also appear in the form of *relative grades.* While driving on an icy road, for example, we perceive the driving environment in a *relatively graded* sense and act accordingly.

The mentation and cognitive process thus acts upon the graded information. Information may appear in a numerical form (temperature of the body is $38.4^0 C$), however, during the process of cognition we perceive this temperature as *near normal,* in the form of a *relative grade*. Thus, the cognitive process acts upon the different forms of information and this leads to 'formless' uncertainty: *temperature is near normal*. This 'formless uncertainty' provides us with a new scientific challenge. Recently, scientists have started to think of giving a morophology to this amorphous uncertainty. In the past, mathematicians have distained this challenge and have chosen to avoid natural mentation by devising theories unrelated to human perception and the cognitive process.

It was in 1965 when Lotfi A. Zadeh published his first celebrated paper on *fuzzy sets*. It is also now almost two decades since he first introduced to me this new type of information and uncertainty and showed me the path which leads to somewhat beautiful gardens full of immortal and ever increasing fragrance. Though I was taught the notion of cognition and perception at school, I was very ignorant about this uncertainty and its pervasiveness around these notions.

No one had seen the beauty of *fuzzy sets* before Lotfi Zadeh and it was he who first showed the promise of consolidating this beauty into an organized field full of rich theories and promising applications.

The theory of fuzzy logic is based upon the notion of *relative graded membership* and so is the function of the mentation and cognitive process. In the past, studies of cognitive uncertainty and its cognate, the cognitive information, were hindered by the lack of suitable tools for modeling such information. However, with the introduction of the theory of fuzzy logic, it is possible now to expand studies in this important field of cognitive information, neural networks, and cognitive-neural computing tools.

My own laboratory is heavily committed to studies in the field of cognitive information processing, cognitive vision fields, vision perception, cognitive-neural computing tools, and cognitive feedback controllers with promising applications to intelligent robotic systems and medical image processing. Most of these studies make use of the theory of fuzzy logic.

3. Perspectives

Recent progress in information-based technology has significantly broadened the capabilities and application of computers. Today's computers are merely being used for the storage and processing of numerical data (information). Should we not re-examine the functions of these computing tools in view of the increasing interests in subjects such as knowledge-based systems, expert systems and intelligent robotic systems and for solving problems related to decision and control? Human mentation acts upon cognitive information and the cognitive information is characterized by using relative grades: *"Although it is snowing, it is not very cold"*. Human mentation and cognition function by using fresh information (acquired from the environment by our natural sensors) and the information (experience, knowlege-base) stored in the memory.

Shannon's definition of *'information'* was based upon certain physical measurements of random activities in systems, in particular, in communication channels. This definition of information was restricted only to a class of information arising from physical systems.

If we want to emulate some of the cognitive functions (learning, remembering, reasoning, intelligence and perceiving, etc.) of humans in a machine, we have to generalize the definition of information and to develop new mathematical tools and hardware. These new mathematical tools and hardware must deal with the simulation and processing of cognitive information. Many new notions, although still at a primitive stage, are springing up around the mathematics of fuzzy logic and, hopefully, we will be able to nurture some interesting studies in the not too distance future.

Bibliography

1. R.E. Bellman, and L.A. Zadeh, ''Decision Making in a Fuzzy Environment'', *Management Science,* (1970), 17, B.141-B.164.
2. M. Black, ''Vagueness: An Exercise in Logical Analysis'', *Philosophy of Science,* 4, (1937), 427-455.
3. L. Brillouin, ''Science and Information Theory'', *Academic Press,* New York, (1956).
4. R.C. Conant, ''Law of Information Which Govern Systems'', *IEEE Trans. Systems, Man. and Cybernetics,* Vol. 6, (1976), 334-338.
5. I.R. Goodman and H.T. Nguyen, ''Uncertainty Models for Knowledge-Based Systems'', *North-Holland,* New York, 1985.
6. M.M. Gupta, ''Fuzzy Automata and Decision Processes: The First Decade'', *Sixth Trienniel World IFAC Congress,* Boston, Cambridge, August 24-30, (1975).
7. M.M. Gupta, ''Cognition, Perception and Uncertainty'', in Fuzzy Logic in Knowledge-Based Systems, Decision and Control, North Holland, (1988).
8. A. Kaufmann, ''Introduction to the Theory of Fuzzy Subsets'', Vol. 1, *Academic Press,* New York, (1975).
9. A. Kaufmann and M.M. Gupta, ''Introduction to Fuzzy Arithemtic: Theory and Applications'', *Van Nostrand Reinhold,* New York, 1985.
10. G.J. Klir, ''Where Do We Stand on Measures of Uncertainty, Ambiguity, Fuzziness and the Like'', *Fuzzy Sets and Systems,* Special Issue on Measure of Uncertainty, Vol. 24, No. 2, November 1987, 141-160.
11. K. Kornwachs and W. von Lucadou, ''Pragmatic Information as a Nonclassical Concept to Describe Cognitive Processes'', *Cognive Systems,* 1, (1985), 79-84.
12. F.M. Reza, ''An Introduction to Information Theory'', *McGraw-Hill,* New York, (1961).
13. L.A. Zadeh, ''Fuzzy Sets'', *Information and Control,* 8, (1965), 338-353.

Fuzzy Logic in Knowledge-Based Systems,
Decision and Control
M.M. Gupta and T. Yamakawa (Editors)

A FLEXIBLE CLASSIFICATION STRUCTURE

Ellen HISDAL

Institute of Informatics, University of Oslo, Box 1080 Blindern, 0316 Oslo 3, Norway.

The main theme of this paper is that an exact, nonfuzzy tree structure can be used as a starting point for the construction of membership and of multiple partition structures. This applies both to the semantic classification structures themselves, and to their storage structure in a lexicon.

In the majority of significant applied classification problems one meets (A) The fuzzy set difficulty of borderline cases, and (B) The difficulty of choosing a suitable partitioning of a given class. E.g., the partitioning of 'organism' into the set {animal, plant} , or, alternatively into {sea-organism, land-organism} . In the absence of these two problems, a classification system can be represented as a pure tree structure. This paper describes the necessary modifications in the above 'semantic tree' in the presence of problems (A) and (B). Both modifications lead to new 'semantic structures'. The 'multiple partition structure' for the treatment of problem (B) allows more than one partition of a node into a set of subnodes. The membership modification for the treatment of problem (A) is based on the definition of first- and higher-order parent families of a node in a pure tree (def. 6.1). Structurally, the membership modification is a network consisting of the 'originator tree', with additional upwards or 'is a' pointers from any node belonging to the i-th order parent family of a leaf to an element of the (i+1)-st order parent family of this leaf. The values of the upward or 'is a' pointers in a membership structure (from a node nj to a node nk) are identified semantically with Zadeh's possibility measure of nk. They are shown to depend strongly on the situation to which the assignment of the nodes nj, nk refers (SIM versus RR situation, see section 6). In contrast, the grade of membership of each node is independent of these two reference situations, as well as of the prior distribution of the objects over the attribute values.

Semantically, the membership modification is an expectation over the different substructure trees contained in a special multiple partition structure. The surface-structure of the membership graph, although not its conceptual deep-structure, is simpler than the multiple partition structure. The latter is not a simple graph at all, but a structure of nested trees in the sense that one node of the outer tree can contain several alternative subtrees.

The paper also describes the structure of a lexicon which is part of a natural language system called Alex, and which implicitly contains the semantic structures. Each lexicon entry has also basically a pure tree structure. No modification of this structure is required for the storage of a membership structure. In contrast, the semantic multiple partition structure is stored in a multiple partitioned lexicon-entry tree; in the sense that a whole entry-subtree is wrapped up into a leaf of the outer tree. Similar modifications of pure entry trees can be used for the representation of lexicon terms with several meanings, or of information supplied by different informants or at different times.

Keywords: Multiple partition classification structure, Membership classification structure, Interlaced IS-trees, Inferences with probabilities and memberships, Associative memory.

1. Introduction

1.1 Overview

The theory of fuzzy sets is concerned with partial grades of membership of objects or attribute values in classes. Its main concern is thus to find a good solution to the classification problem.

The first step in every type of human information processing, as well as its equivalent in artificial intelligence, is always a pattern recognition and classification step. When we meet another person, we classify her/him as to sex, age etc., and as to whether the person is a previous acquaintance or not. When we drive a car, and see another car approaching, we classify it as to direction, speed, and as to whether it comes from the right or left. In speech recognition we classify sounds according to the words which they symbolize, and combinations of words as to the sentence which they make up. The next information processing step is then the one of drawing inferences from the classification, and possibly acting on the basis of these inferences. The structure of our knowledge base must, as far as possible, facilitate the logical inference procedure, as well as the construction and updating of the knowledge base itself.

The classification or taxonomy problem is thus a very important, as well as a difficult one. In this paper we shall often illustrate it by examples from the biological classification of organisms, of which there exist about 5 million different species. Curtis (p. 371, 379) says that taxonomy has become the focus of some of the most heated arguments of modern biology. The reason for this is that different groups of researchers use different classification criteria. Thus 'pheneticists' classify organisms only according to observable characteristics, while 'cladicists' ignore overall similarities, and classify them solely on the basis of their branching in the course of evolutionary history. For some organisms these two systems result in completely different grouping. For example, crocodiles belong to the class of reptiles according to the pheneticists. According to the cladicists, they are not reptiles but belong to the same class as birds.

Some years ago, a group of students (Haug and Johansen, 1983; Ohren, 1986), together with the author, started work on the construction and programming of the 'Alex' computer system in Oslo. This work is now being continued by Andreassen (1988) and Utne (1988). 'Alex' is an interactive natural language system with emphasis on the representation and retrieval of the meaning of the incoming information, independently of the particular words used by the informant. For example, we require that the information 'Margy is a woman' and 'Margy is a female, adult human' must either have the same representation in the knowledge base or 'lexicon' of the Alex system. Or there must exist automatic procedures which can convert the two pieces of information to each other such that the system answers 'yes' to the question 'Is Margy a woman?', even if the incoming information was the second sentence.

The result of this requirement has been that the focus of the Alex system has shifted more and more in the direction of building up a good classification system. The requirement also resulted in the recognition that there does not exist one absolute, 'kernel' classification system which stands above all others. All classification structures which conform to the requirement of nonfuzzy partitions in the downward direction (but not necessarily in the upward one, see fig. 4.3) must be acceptable as equally valuable systems; and they must be able to coexist simultaneously in the lexical knowledge base. The classification of humans into {children, men, women} , before they are classified into instances, is not inferior to a classification in which instances (with attribute values) are the direct descendants of the 'human' node. And the pheneticists' classification of organisms is neither better nor worse than the cladicists' one. The two classifications use different partitioning criteria, and our knowledge base should be able to handle both of them. If the pheneticists' and cladicists' classifications are, later on, followed by still newer ones, then our knowledge base should be able to store the new system without destroying the old ones because much of the literature will still refer to the old systems.

There exist many more biological classification systems than the two modern ones that we have mentioned. The classification of many of the 5 million species may be the same in several systems, while only the higher nodes of the classification tree differ. For example, the subclasses of 'bacteria' may be the same in the the two partitioning systems of 'organism' depicted in fig. 3.1; although the first system classifies the 'bacteria' class itself as being a subclass of 'plant', while the second system classifies it as a subclass of 'protista'. The structure of our knowledge base should be such that the identical 'bacteria'-subtrees of the two classification systems need not be represented twice. For this reason, and for an even more cogent one mentioned in section 3.1, we have ruled out the device of numbering the different pure classification trees.

A nonfuzzy classification of a class into disjoint subclasses, and of the latter into subsubclasses etc., can be visualized as a tree structure. In section 2 we consider such nonfuzzy 'semantic trees', and their representation in a knowledge base which has the form of a lexicon. The name of each node in the semantic tree is a term in the lexicon, and the structure of the semantic tree is implied by *child* (or 'may be a') and *parent* (or 'is a') 'pointers' from one term to another. In addition to these two pointers the lexicon entry contains syntactic information. It may also contain other types of information such as attribute values pertaining to the term.

The choice of a good structure for the lexicon entries themselves is essential for the retrieval of the information contained in this entry. In the Alex system, the entry of each lexicon term has basically also a pure tree structure (see sect. 2.2 and 7).

Our previous example from the biological classification systems demonstrates the insufficiency of a single, nonfuzzy semantic classification tree. Several examples of other types of insufficiencies of such a tree are presented in section 3.

Two main modifications of the semantic trees are suggested to remedy these insufficiencies. The first of these is presented in subsection 3.1. This 'multiple partition' modification, allows for the possibility of two or more partitions of a node in the semantic tree into two or more child families.

The second, or membership modification of the semantic tree is described in sections 4-6. It allows for the possibility of partial belongingness of one class or node to more than one parent class. In this case two or more parents may share a given child. Or, said in another way, a given node in the semantic 'tree' may have more than one parent *although a subject who performs an LB (labeling) or YN (yes-no) experiment now operates with a single, nonfuzzy partition of each node, using his estimate* u *of the object's attribute value as a classification criterion.*

The retrieval of information concerning the multiple partition modification necessitates modifications of the lexicon entries from their original pure tree structure (see figs. 2.2, 3.2). There are also additional reasons why a lexicon, which aims at being a flexible knowledge base, must make allowance for deviations from the pure tree structure. Modifications of the pure entry tree structure for the purpose of retrieving and processing information concerning multiple partitions of a node in a semantic tree are discussed in sections 3 and 7. Essentially the same techniques can be used for the representation of ambiguous terms, or of information supplied by different informants, or at different times, about the same term.

Subsection 1.2 discusses shortly the classification structures of the Alex system from an expert-system point of view. The most important terminology, and some definitions, are listed in subsection 1.3.

In summary, we consider in this paper two tree structures and their modifications: The semantic trees, and the entry trees of a lexicon. The lexicon implicitly stores the semantic structures, as well as other pertinent information, about each term.

1.2 The Expert-System Point of View

If we take the expert-system point of view, then the lexicon of the Alex system is our dynamic knowledge base acquired by the system through input from the user in the dialogue mode.

The present paper treats mainly the static knowledge base put in by the expert. This knowledge consists of the structural relations which the Alex system imposes upon the dynamic knowledge base. Namely the basic tree structure of the lexicon entries, and the semantic structures (tree, multiple partition or membership structure) stored implicitly in the lexicon.

In addition the paper treats the inferences to be drawn from these structures when the user asks questions of the system. In a typical expert-system, the interesting inferences are those concerning the real or 'exact' attribute value u^{ex} of an object which has been classified by a subject in a 'nonexact experiment'.

The inferences are drawn by following the 'may be a' or downwards pointers of the semantic structure. They are presented in the form of the probability $P(u^{ex}|n)$ that an object which has been assigned the label of a node n of the structure has an exact attribute value u^{ex}. In this connection, the term 'exact attribute value' should be understood in the wide sense defined in requirement 2 et seq. of sect. 6.2. It may be a numerical or a linguistic value. The $P(u^{ex}|n)$ probabilities are treated in sect. 6.2, def. 6.4 et seq. .

The present paper makes no use of surface or ad hoc knowledge in the static knowledge base, but only of deep-knowledge with a firm theoretical foundation. However, the paper does not yet treat in detail the case in which the information supplied by the user about a given object consists of the values of two or more attributes, or of the grades of membership of the object in the classes described by such collections of values. This subject will be treated in Hisdal (1987a, part 1.6).

1.3 Terminology

We list here some of the most important terminology, notation and definitions used in this paper.

Cousins: Two nodes nj, nk of a tree are cousins iff neither of them is an element of the rootpath of the other. Siblings are a special case of cousins.

Δu_n : Quantization-interval for the label of node n. Nonfuzzy interval (or closed hyper-volume) in U for which the subject assigns the label $n \in \{n_l\}$ in an LB or YN experiment, where $\{n_l\}$ is a generalized partition.

est: Superscript for the estimate by the subject of a probability distribution relating to his fuzziness #1a, 2a or 3a.

Exact experiment: A semantic experiment in which the subject measures (or is told) the exact attribute value of each object. Consequently $u = u^{ex}$ for an exact experiment.

Exact attribute value: See u^{ex}.

Families: The collection of nodes $\{nj_{l_j}\}$ lying at the arrowheads of the downwards (or 'may be a') pointers from a node nk in a tree or membership structure is called the child family of nk; in the case of a tree, the child family of nk is also called the sibling family of any of the elements of $\{nj_{l_j}\}$. The parent family $\{nk_{l_k}\}$ of a node nj in a pure tree is a set consisting of 1) The parent nk of nj. 2) All nodes of the tree whose parent belongs to the rootpath of nk. See also defs. 6.1, 6.3.

Generalized family or generalized partition of the root of a tree, or of a substructure tree contained in a multiple partition structure: A generalized family is a subset of the set of nodes of the structure defined recursively as follows. The singleton set whose element is the root is a generalized family. If $\Lambda 1 = \{n1_{l1}\}$, $l1 = 1, \ldots, L1$ is a generalized family, then the set $\Lambda 2 = \{n2_{l2}\}$, $l2 = 1, \ldots, L2$, as defined below, is also a generalized family. $\Lambda 2$ is obtained from $\Lambda 1$ by replacing every element of $\Lambda 1$ either by itself, of by all the elements of its child family. Any node of a tree (considered as the root of a subtree or substructure) has its generalized familie(s) unless it is a leaf. A generalized family of the root of a tree is also called a generalized family of the tree. In a semantic structure, the elements of a generalized family represent a nonfuzzy partition of the root as used in an LB experiment by a given subject according to his estimate u of the object's attribute value. The child family of a node nk is a generalized family of nk. The parent family of a node nk is a generalized family of the whole tree.

Instance: Synonym for 'object' on which the subject performs the semantic experiment.

$\Lambda = \{n_l\}$ =label set (see Hisdal 1987a, part 1.3). The set of labels of the elements of a generalized family (see above) of the semantic tree or membership structure.

LB experiment: A Labeling experiment in which a subject assigns a label from a label-set to an object (see Hisdal 1987a, part 1.3).

MU experiment: A grade of membership experiment in which the subject is asked to assign a grade of membership value $\mu \in [0, 1]$ to an object concerning the label n (see Hisdal, 1987a, part 1.3).

Originator Tree of a membership structure: The pure tree from which a membership structure is derived (sect. 6.2).

$P(n|u^{ex})$: Labeling probability or likelihood distribution of n over u^{ex}. Probability that an object with attribute value u^{ex} (as measured by the experimenter) will be labeled $n \in \{n_i\}$ in a nonexact YN or LB experiment performed by a subject. Superficially stated, it is identified in the TEE model with $\mu_n(u^{ex})$ as elicited from that subject in an exact MU experiment (Hisdal 1986a,b, 1987a,b, LB,YN-MU assumption of equivalence).

$P(u^{ex}|n)$: n-qualified probability distribution over u^{ex} = probability that an object which has been labeled $n \in \{n_i\}$ by the subject has the attribute value u^{ex} (see def. 6.4 et seq.).

$P(u^{ex})$: unqualified or prior probability distribution over u^{ex} (see sect. 6.5).

$P(u|u^{ex})$: Error curve for a given subject, and a given set of conditions of observation.

Rootpath: ROOTPATH(n), a rootpath (with respect to u^{ex}) from a node n in a membership structure. Let the node n belong to the i-th order parent family of u^{ex} (defs. 6.1, 6.3). ROOTPATH(n) is any sequence whose j-th element, $j = 1, 2, \ldots,$ belongs to the $(i+j)$-th parent family of u^{ex}. The last element of ROOTPATH(n) is the root of the structure.

Semantic experiment: An LB or YN or MU experiment.

Semantic Structure: A graph with a single root at the top, leaves at the bottom and directed arcs. We talk about an upward and a downward structure depending on whether all the arcs are directed upwards or downwards. The upward arcs are also called 'is a' pointers, the downward arcs 'may be a' pointers. (A semantic multiple partition structure is a nested graph (tree). A semantic membership structure is a nonnested graph.)

Subnode or Descendant, Supernode: Let Λ_i, Λ_j be the i-th and j-th order parent families of a leaf u^{ex} of a membership structure (defs. 6.1, 6.3), where $i = 0, 1 \ldots,$ $j = 1, 2 \ldots,$ and $j > i$. Then any element of Λ_j is a supernode of any element of Λ_i, and any element of Λ_i is a subnode or descendant of any element of Λ_j with respect to u^{ex}. 'Parent' and 'child' (sometimes called 'direct' or 'first-order' parent and child) are special cases of 'supernode' and 'subnode' respectively.

Substructure Tree of a multiple partition structure: One of the pure trees contained in this structure.

Tree Parent of a node n in a membership structure: The parent of n in the originator tree of the structure.

u: Estimate of the object's attribute value by the subject in a nonexact experiment.

u^{ex}: Exact attribute value of object as measured by the experimenter. It is often multi-dimensional. See also sect. 6.1, requirement 2 et seq.

YN (yes-no) experiment: An experiment in which a subject answers 'yes' or 'no' to the question of whether an object is $n \in \{n_i\}$ (see Hisdal 1987a, part 1.3).

2. The Two Pure Tree Structures

2.1 Semantic Trees

A nonfuzzy classification scheme can be visualized in the form of a semantic tree, such as that of fig. 2.1 in which each node (oval) represents the name of a class; and the elements of its 'child family' represent a partition of this class.

In all our structures we require that this partition be nonfuzzy; i.e., the elements of a child family must be disjoint, and their union must be equal to the parent class. In the sense that when a subject S has assigned to an object the label of a node n of the tree in an LB or YN experiment, then he is also able to assign to it simultaneously exactly one of the labels of the child family of n. We shall see in sections 4 and 5.2 that, rather surprisingly, there is no contradiction between this requirement on the one hand, and, on the other hand, the elicitation in a MU experiment of a partial belongingness of an element of a class to several parent classes.

The tree structure of fig. 2.1 is called a semantic IS-tree. Its arcs, considered as upward pointers, are called 'is a' pointers because their meaning is

$$\text{a } < \text{n2} > \text{ is a } < \text{n1} > \; , \tag{2.1}$$

when n1 is the parent of n2; for example, 'a dog is a mammal'. The arcs, considered as downwards pointers, are called 'may be a' pointers because they signify that

$$\text{a } < \text{n1} > \text{ may be a } < \text{n2} > \; , \tag{2.2}$$

for example, 'a mammal may be a dog'.

The rectangular nodes of the tree of fig. 2.1 signify no longer names of classes, but names of instances belonging to that class. In a natural language sentence, the first indefinite 'a' article in (2.1), and the second one in (2.2) must be deleted when n2 is an instance. (Sentences (2.1), (2.2) must also be modified when $< n1 >$ or $< n2 >$ is a non-countnoun such as 'flour'.)

We can draw 'positive syllogistic inferences' (or 'transitive inferences') from the tree structure by following the rootpath of a node; e.g., 'a dog is a vertebrate'. And we can draw negative inferences, such as 'a fly is not a mammal', when neither of the two nodes lies on the rootpath of the other.

Semantic trees replace the quantifiers of first-order mathematical logic. Thus the 'is a' relation of eqn (2.1) would be expressed in mathematical logic as the universal quantification: 'For all instances x (if x is a $< \text{n2} >$ then x is a $< \text{n1} >$'); while (2.2) is a typical existential quantification: 'There exist instances x of $< \text{n1} >$'s that are instances of $< \text{n2} >$'s'.

In addition to the parent and child pointers, a node in a semantic tree can be supplied with attributes and with information concerning the allowed intervals for their values. For example, 'height' is an attribute which applies to the 'human' node with the allowed interval (0, 250] cm; or with the allowed interval in the form of a complete linguistic label set (see Hisdal 1987a part 1.3; also figs. 6.1a,b,c here) such as { small, medium, tall }.

Attributes are inherited downwards in a tree, and their allowed value-intervals must be successive subsets when we proceed in the downwards direction. In subsection 3.3 we

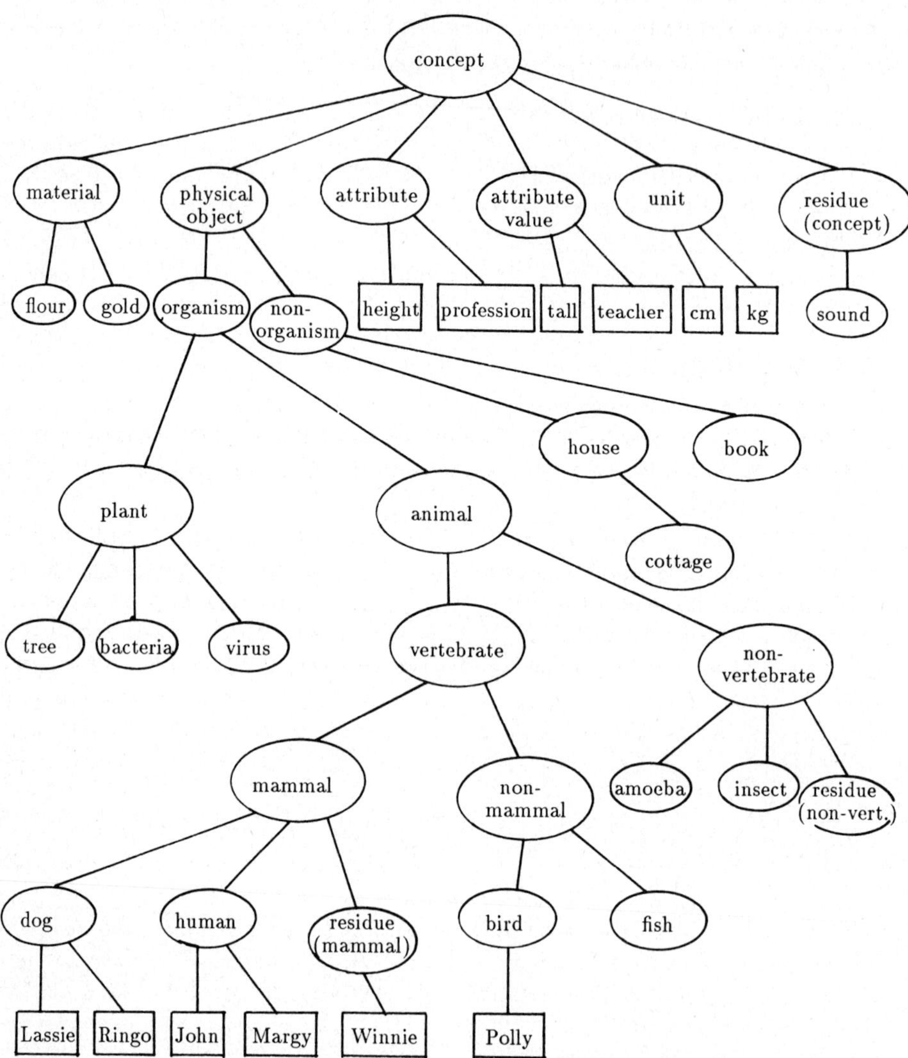

Fig. 2.1. The default semantic tree of the Alex3 system. Each arc of the tree is an 'is a' pointer in the upwards direction (e.g., 'a dog *is a* mammal') and a 'may be a' pointer in the downwards direction (a mammal *may be a* dog). In addition, attributes can be associated with each node; e.g., 'weight' for 'physical object'. Attributes are inherited downwards. The square nodes represent instances. When the child family of a node n is not complete, then a node 'residue(n)' should be added to it because every child family represents a partition of the parent node. In the figure most of the residue nodes have been left out for visual clarity. When a node n has only one child m and its residue, then the name of the latter can be replaced by 'non-m <parent name>; e.g., 'non-mammal vertebrate'. In the figure we have left out the <parent name> part. It is however needed because otherwise we would get such inferences as 'an insect is not a non-mammal'.

discuss why attribute values (e.g. 'female' or 'tall') cannot generally be used as node names of additional nodes in the semantic tree, even in the nonfuzzy case. However, attribute values and node names can be mutually exchanged in two informationally-equivalent semantic trees. This situation is summed up at the end of subsection 3.3.

In addition to semantic IS trees we have also semantic IN trees with inferences such as [(Paris is in France) $\wedge$ (France is in Europe)] $\rightarrow$ (Paris is in Europe); and semantic PART trees with inferences such as [(a finger is a part of a hand) $\wedge$ (a hand is a part of a body)] $\rightarrow$ (a finger is a part of a body).

The common denominator of IS, IN and PART trees is a nonfuzzy subset relation between a child and its parent; such that the child family of a node is a nonfuzzy partition of the parent node.

2.2 Entry Trees

Every semantic tree which belongs to the classification system is stored implicitly in the lexicon of the Alex system, such that the name of each node of the semantic tree is a term in the lexicon. A lexicon term, together with the information relating to it, will be called a 'lexicon entry'.

Fig. 2.2 shows a possible lexicon entry for 'vertebrate' in a *pretty-print* form. We see that 'vertebrate' has three subentries on the first level, 'syntactic category', 'IS', and 'PART'. The first of these is a necessary part of the lexicon of the more extended Alex system which contains also verbs and other syntactic categories. Even when we deal solely with the classification structure, we need the syntactic category 'adjective' in addition to 'noun'. Furthermore, we saw in connection with eqns (2.1), (2.2), that we must distinguish between proper nouns (used for instances) and common nouns (used for classes). Again, among common nouns we must distinguish between count nouns and non-count nouns (see e.g. Quirk and Greenbaum, p. 60).

The 'is a' and 'may be a' pointers of the 'vertebrate' node in the semantic tree, as well as the attributes which apply to 'vertebrate' and their values, are gathered together under the heading 'IS'. For terms which are also nodes in a semantic IN tree, we would have an additional 'IN' heading. For example, if our term had been 'California', then we might have 'proper noun' as the subentry of 'syntactic category', and 'state' as the subentry of 'is a'. On the same level as IS we would have an additional 'IN' subentry. This might have the subentries 'is in' with subentry 'United States', and 'contains' with subentries 'Los Angeles', 'San Francisco', 'residue [[empty(no)]]'.

The *pretty print* structure of fig. 2.2 is equivalent to a tree structure which we will call an 'entry tree'. 'vertebrate' is the root of the entry tree of fig. 2.2. Its children are 'syntactic category', 'IS', and PART. And the 'IS' node has the children 'is a', 'may be a', and 'attributes', etc. . Thus every line of fig. 2.2 is a node in the entry tree. (The structure in connection with the 'special information term' in the double square brackets under 'residue' is explained in section 7.3)

We see that the entry tree is a completely different tree from the semantic one. This remark seems superfluous here, but our experience in the Alex group has shown us that one has constantly the tendency to create even more difficulties than the many real ones, and to get involved in endless discussions, because one forgets to clarify which of the two trees one is talking about.

The precise storage structure of the entry trees in the Alex lexicon is described in section 7.

2.3 The Alex System as an Alternative Logical Reasoner with Quantifications and Implications

We touch in this subsection very shortly and superficially on the subject that the semantic tree structure, combined with its storage structure in the lexicon (see subsections 2.1, 2.2) can be used as an alternative to the structures and inference procedures of predicate calculus in connection with quantification and more general implications.

It is easy to build up an automatic reasoning system for positive syllogistic inferences with quantifiers. Thus P. Schefe described a version of the FUZZY computer system in a talk in the Fuzzy Set workshop at Queen Mary College, London, in 1981. This system could infer 'Socrates is mortal' from the two sentences 'All men are mortal' and 'Socrates is a man'. But it could not infer

$$\text{God is not a man} \tag{2.3}$$

from the sentences

$$\text{All men are mortal ,} \tag{2.4}$$

and

$$\text{God is not a mortal .} \tag{2.5}$$

The possibility of negative inferences depends on the disjointness of the elements of every child family in the semantic tree. Hendrix attacks this problem by using two types of '(is a)' pointers. 'ds' pointers belong to a set of mutually disjoint subsets of the parent node, while a node with an 's' pointer to the same parent may not be disjoint from each of the above subsets (referred to in Harris, 1985, p. 229). A somewhat different partial solution of the disjointness problem was used in the Alex2 system (Ohren, 1986). In this system the '(may be a)' pointers of the parent node were grouped into a 'disjoint' set and into a set called 'others'.

The Alex3 system pays particular attention to the possibility of drawing correct inferences in the case of negated statements. This is achieved by two means. (1) The multiple partition structure described in sections 3 and 7.4 which does not allow categorical negative inferences between two nodes, each of which belongs to a different child family, and thus to a different partition, of the same parent. (2) By the requirement that the child family of every term which is a non-leaf node in a single pure tree (with no multiple partitions) must represent a nonfuzzy partition of this node. This requirement agrees well with the TEE model according to which the attribute values of each element of the partiton are elicited in an LB or YN experiment performed with a single subject in the absence of fuzziness. We show in sect. 5.2 that the assignment to an object in a YN or LB experiment of a single element of a generalized nonfuzzy partition of the tree (see sect. 1.3) does not contradict the assignment to the object *in a MU experiment* of partial grades of membership in the different elements of such a partition.

The lexicon, and the implied semantic structure are built up by the Alex system via a man-machine dialogue which creates one semantic node at a time and inserts it as a term in the lexicon. The device by which the storage of a complete partition of the term (under its '(may be a)' pointers) is achieved is to store initially the singleton child family '(residue)' under the '(may be a)' entry of the term. When a semantic child nj of the term nk is later on created in the dialogue, it is inserted into the lexicon (1) as a new lexicon term, and (2) as an additional element of the child family of nk, provided that a question concerning the disjointness of nj from the already-existing non-residue elements of the child family of nk is answered in the affirmative. At each insertion, a further question is asked concerning the emptyness of the residue set until the latter is specified by the user to be empty (see the last subentry of '(may be a)' in fig. 2.2).

This system can readily be extended to a system which performs syllogistic inferences. One starts out with a semantic tree consisting solely of the root 'concept'. The statements on which the inferences are based are used to insert further nodes into this tree. For example, in connection with the syllogism of equations (2.3)-(2.5), 'All men are mortal' results in the insertion of 'mortal' as a child of 'concept', and the insertion of 'man' as a child if 'mortal'. 'God is not mortal' results in the insertion of 'not mortal' as a sibling of 'mortal', and the insertion of 'God' as a child of 'not mortal'. The question 'Is God a man?' is now answered by 'No' because 'man' is not in the rootpath of 'God', and 'God' is not in the rootpath of 'man'.

The semantic tree structure, together with its representation in the lexicon, can also be used for general implications, not necessarily connected with quantification. The following example is taken from Kleene (1968, p. 60):

1. If he doesn't tell her, she will never find out.
2. If she doesn't ask, he won't tell.
3. She did find out.
4. So she must have asked.

The reasoning connected with these items is performed in the following way in the Alex system. A semantic tree is initialized with the node 'root'. The first of the above items results in the creation of the mutually successive subnodes 'she will not find out' and 'he does not tell her' of 'root'. The last subnode is assigned the child 'she does not ask' by item 2. All these nodes are automatically inserted into the lexicon with their correct upwards and downwards pointers.

We now ask the question 'Did she ask?'. This results in the system looking for the node (i.e., the lexicon term) 'she asked'. (Just like Kleene, we ignore here the extra complication due to the past, present and future tenses.) Since it does not find this term, it looks for its negation, namely 'she did not ask'. This is found to be a node in the semantic tree which implies the node 'she did not find out'. Since this contradicts item 3, we find that 'she did not ask' is false and consequently that 'she did ask' is true.

The automatic building up in the Alex system of a lexicon with its implied semantic tree can thus replace the implication, and the universal and existential quantification of mathematical logic; as well as the procedures for drawing inferences from these structures.

```
vertebrate
  (syntactic category)
    (count noun)
  IS
    (may be a)
      bird
      reptile
      mammal
      residue     [[empty(yes)]]
    (is a)
      animal
    attributes
      length
        (0,40] meter
      weight
        (0,200 000] kg
  PART
    (has as part)
      backbone
      (chambered heart)
    (is a part of)
      nothing
```

Fig. 2.2 A possible lexicon entry for 'vertebrate' in 'pretty-print' form. Part of the precise, nested-list and numeral storage forms of the entry are shown in eqns (7.4) and (7.1) respectively. The term in double square brackets is the optional special information term of subsection 7.3. In the present case it serves to indicate that 'residue' is not the name of the English language for a vertebrate animal called 'residue', but the class of all vertebrates that have not been specified under 'may be a'. Its subentry indicates that the residue class is empty. (The double square brackets are used in the figure only for purposes of visualization. They do not occur in the final storage structure, see subsection 7.3.)

```
⟨term⟩
  syntactic category
    ...
  IS
    partition   [[true]]
     ⎧(may be a)
     ⎪   ...
     ⎪   :..
     ⎨attributes
     ⎪   ...
     ⎩   :..
     ⎧(may be a)
     ⎪   ...
     ⎪   :..
     ⎨attributes
     ⎪   ...
     ⎩   :..
    (partition tree)    [[((...)...(...))]]
     ⎧(is a)
     ⎪   ...
     ⎨attributes
     ⎪   ...
     ⎩   :..
     ⎧(is a)
     ⎪   ...
     ⎨attributes
     ⎪   ...
     ⎩   :..
```

Fig. 3.2. The approximate 'pretty print' form of a lexicon entry which is a node in a semantic structure that permits multiple partitions. The entry tree may have two additional nodes, 'partition' and '(partition tree)', as compared with the entry tree of fig. 2.2. The former node is needed when ⟨term⟩ is a multiply partitioned node in the semantic structure, the latter when a supernode of ⟨term⟩ is multiply partitioned. The purpose of these nodes is to signal to the system that a grouping of the 'may be a' (or 'is a') and 'attribute' values, referring to one of the semantic trees contained in the semantic structure, follows. Each group refers to one pure tree structure contained in the semantic structure. The entry of the special information term of '(partition tree)' is a list. Each element of this list refers to one of the braces underneath. It specifies the particular semantic tree (i.e. the subpart of the semantic structure) to which that brace refers. For more details, see sections 3 and 7.4.

3. The Multiple Partition Modification of the Semantic Tree

3.1 The Multiple Partition Modification

This subsection discusses one of the two modifications of the pure tree structure of subsect. 2.1. Examples of the use of the modified structure are then given in subsections 3.2-3.5.

In order to be able to retrieve the information concerning a semantic structure with one or more multiply partitioned nodes we must also modify the pure tree structure of the lexicon entries. An outline of this modification is illustrated in fig. 3.2. A more exact description of the storage and processing of the entry structures is given in section 7.

The modification of the semantic tree described here concerns the possibility of two or more partitions of a given parent node in the tree. Fig. 3.1 illustrates the case of an additional partition of the 'organism' node of fig. 2.1.

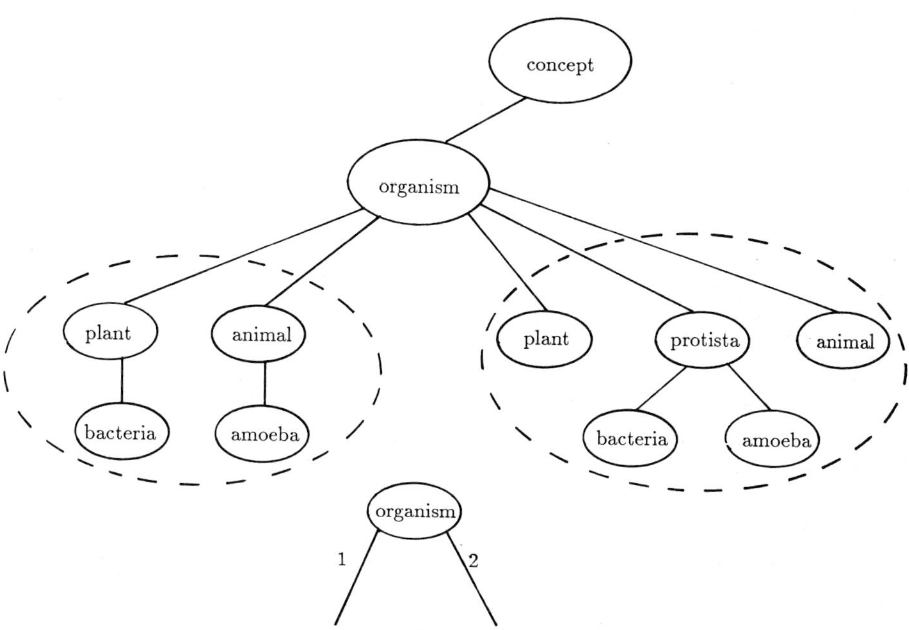

Fig. 3.1. The multiple partition semantic structure of a modification of the semantic tree of fig. 2.1. The present structure contains an additional partition of 'organism'. The bottom figure shows the partition tree of the semantic structure. It consists of a single node named 'organism' because this is the only node in the semantic structure which has a multiple partition. The node has two child pointers, corresponding to the two partitions, but no children. Note that 'plant' has somewhat different meanings in the two partitions because the sets of instances belonging to the two 'plant' classes are not identical; similar considerations hold for 'animal'. In contrast, the two 'bacteria' and 'amoeba' classes can have the same meaning in the two partitions.

The multiple partition structure of fig. 3.1 is equivalent to two pure tree structures; namely that of fig. 2.1 (many of the nodes of that figure have been left out in fig. 3.1), and another tree in which the subtree of organism has been replaced by that of the right hand partition of fig. 3.1. All the nodes which do not belong to the two alternative organism-subtrees are common to both of the two pure tree structures. Furthermore, the subtrees of 'bacteria', 'amoeba', 'vertebrate' and many other classes may again be common to both of the pure trees.

We may also have a multiple partition structure with more than one multiply-partitioned node. If nodes $n_1, \ldots, n_N$ have $m_1, \ldots, m_N$ partitions each, then the multiple partition structure is equivalent to $m_1 \times \ldots \times m_N$ pure trees. To store each of these separately in the lexicon, with a duplication of all the nodes which are common to some of the different trees, may be a waste of storage space of several orders of magnitude. Furthermore such a procedure would hide information as to which of the nodes have only one partition, common to all the trees. Finally we would have to have a system of numbering for the different pure trees in order to be able to refer to them. This problem is far from trivial, especially when one does not know in advance all the different partitions which may crop up in the course of the use of the system, and which may necessitate a renumbering of the trees. This would again require the updating of all references in the lexicon to the different trees.

As far as the storage in the lexicon is concerned, it is thus better to look upon the multiple partition structure as a single one which contains information concerning both the nodes which have multiple partitions, and the tree structure implied by each partition. We will call this structure the 'multiple partition semantic structure', or the 'semantic structure' for short.

The information about nodes that have multiple partitions can be thought of as being contained in a pure tree stucture which we will call the 'partition tree'. The name of a node in the partiton tree is identical with the name of a multilply partitioned node in the multiple partition semantic structure. And the number of nodes in the partition tree is equal to the number of nodes with multiple partitions in the semantic structure. If the latter is a pure tree, with no multiply-partitioned nodes, then the partition tree has 0 nodes.

The partition tree of the semantic structure of fig. 3.1 is shown in the lower part of that figure. It consists of a single node named 'organism'. The node has two child pointers but no children. Each of these two pointers has its individual name, namely the number of the partition of 'organism' to which it refers. If the subtree of one of the two partitions had contained another multiply-partitioned node n' in the semantic structure, then the corresponding child pointer of the partition tree would have had a child node labeled n' attached to it.

A possible way of storing and retrieving information concerning a multiply partitioned semantic structure is outlined in fig. 3.2. If the node 'term1' of a semantic structure has m different partitions, then the lexicon entry < term1 > contains m groups of 'may be a' and 'attribute' entries, one for each partition. Each such group is enclosed in a brace in the figure. A new node 'partition', with the special information term [[true]], is introduced as a supernode to these groups in order to warn the program that a grouping is about to take place. We shall see in sect. 7 that the m groups are stored in a list. Each group, i.e. each partition of 'term1', is identified by its number in the list.

Suppose now that 'term2' is a node belonging to one or more of the *semantic* subtrees corresponding to one or more of the partitions of 'term1'. The lexicon entry of 'term2' must now contain information which identifies the particular partition of 'term1' to which the 'is a' pointer of 'term2', as well as its collection of attribute values, refers. Again we may have a grouping, this time of 'is a', 'attributes', and possibly 'may be a' values. To warn the system that such a grouping follows, the additional node '(partition tree)' is introduced into the entry structure. This node has a special information term containing a list. The n-th element of this list identifies the particular 'substructure tree' to which the n-th group under '(partition tree)' refers.

The new 'partition' and '(partition tree)' nodes can be introduced as obligatory ones into the lexicon entry of every term which is a node of a semantic tree. Alternatively one can use the original structure of figure 2.2 in cases in which there would be only one group under 'partition' or '(partition tree)'. The subentries of '(partitition tree)' must now refer to the default semantic tree. (The default semantic tree uses the first partition of every multiply partitioned node=term of the semantic structure; corresponding to the first group under 'partition' of the lexicon entry of < term >.) In those cases in which a 'partition' or '(partition tree)' node is needed we add this node, together with a special information term, to the entry structure.

The examples of the following subsections illustrate the above description. They also show that there is a mutual dependency between the choice of the partition for a particular node of the semantic tree and the relevant attributes. This is why some attributes are grouped together with 'may be a' or 'is a' values in fig. 3.2. The attributes that are common to all groups are listed directly under 'IS'.

3.2 Different Spheres of Interest

In this and the following subsections we present different examples which illustrate the insufficiency of the pure tree structures, and the usefulness of the multiple partition device for the modification of the semantic trees. We start with an example given by Ohren (1986).

If our focus of interest is a biological one, then it is most natural to use the partition { vertebrate, non-vertebrate } of 'animal' in fig. 2.1. If our interest focuses on the geographical location of different animals, then the partition { sea animal, land animal } is more pertinent. These two partitions cannot be reconciled with each other to form a single semantic tree. For example, 'fish' and 'jelly fish' belong to the two different elements of the first partition respectively; while both belong to the first element of the second partition.

As another illustration, suppose that our focus of interest is a classification of people according to professions. The 'human' node of fig. 2.1 could then have an additional partition such as { carpenter, doctor, gardener, teacher, residue(human) }. The lexicon entry of < term > =human in fig. 3.2 will now have an 'IS' subentry with subentries '(is a)', 'attributes' and 'partition'. The subentries of 'attributes' are those attributes which are common to both partitions, together with their interval values; for example the attributes 'height' and 'weight', but not the attribute 'profession'. The subentry of 'partition' consists of two braces. The first of these lists the instances of 'human' under 'may be a'. In addition it has the subentry 'attributes' with subentry 'profession'. The latter may have a subentry consisting of the 'interval value' which 'profession' can assume for humans, namely a list of all professions. The second brace under 'partition' has the list of all professions as its 'may be a' value and no 'attributes' subentry.

The lexicon entries of instances of 'human', e.g. < term >=Margy, will now have the additional '(partition tree)' node of fig. 3.2 with the special information term which specifies the partition tree (see figure 3.1) for each of the two braces underneath. In our example, the special information term of '(partition tree)' would be ((human 1) (human 2)) in order to indicate that the first brace refers to the first partition of 'human', and the second brace to the second partition. The first brace has the '(is a)' entry 'human', and the 'attributes' entry 'profession' with subentry, e.g., 'doctor. While the second brace has 'doctor' as the value of 'is a' because 'doctor' is a child of 'human' in the semantic tree corresponding to the second partition of 'human'. We see that for < term >=Margy, the first and the second brace-group under the '(partition tree)' subentry of fig. 3.2 (together with the inference 'A doctor is a human') contain the same information. We can, of course, delete one of them if we wish. However, both from a theoretical point of view, and for applications to data base problems, it is important to understand the conversion process of the nodes of one tree to the attribute values of the other. We come back to this subject in subsection 3.3.

The entries in the lexicon representing profession values, e.g. < term >=policeman, will now also have a '(partition tree)' subentry under 'IS' whose subentry consists of two brace-groups. The first of these refers to the first partition of 'human'. Its has a subentry 'is a', with successive subentries 'attribute value', 'refers to attribute' and 'profession'; and a subentry 'may be a' which has the empty list as its subentry because 'policeman' is a leaf in the default semantic tree of fig. 2.1 (a sibling of 'teacher').

The second brace under '(partition tree)' of 'policeman' has also 'is a' and 'may be a', and possibly 'attributes' subentries. 'is a' has the subentry 'human'. 'Attributes' lists those attributes which come in addition to those listed under 'human'; e.g., '(carries a gun)', with subentry 'yes'. Furthermore it may list attributes of 'human' whose interval-values are narrowed down. For example, the interval-value for 'height' may be [180,250] cm, assuming that a minimum height of 180 cm is required for policemen. The 'may be a' subentry lists the instances of policemen.

Note that in contrast to the lexicon entry of 'human', 'may be a' is a subentry of '(partition tree)' for the lexicon entry 'policeman', not a subentry of 'partition'. The reason is that the term 'policeman' is not a multiply partitioned node in the semantic structure.

As a different example, consider concepts such as 'food' or 'clothing'. These do not fit anywhere into the tree of fig. 2.1. However, in an economist's world we could have {goods, residue(concept)} as a partition of 'concept', and {consumers' goods, producers' goods} as a partition of 'goods'.

In general, when we are not acquainted with the subject of the field in which a class, such as 'food', finds its place, then we can always use the expedient of creating a new partition of 'concept' into {food, residue(concept)} . Even a nonexpert can now easily classify bread, meat, apples, butter, milk etc. as subnodes of 'food'.

Multiple partitions can, of course, also occur for IN and PART semantic trees. For example, consider the partitioning of the surface of the earth into {land, sea} , then partitioning of the 'land' node into continents (defined as 'large and extensive land masses' in my Webster dictionary), and finally partitioning of each continent into countries. Most countries fit into this scheme. However, consider the Soviet Union. A considerable part of this country belongs to Asia, and another considerable part to Europe. This is not due to any fuzziness. When the latitude and longitude of a point in the Soviet Union are given, then we also know whether it belongs to Asia or Europe, provided that we have a specified, nonfuzzy boundary between these two continents (according to the Webster dictionary, both Europe and Asia are continents, the separating boundary being the Ural mountains). Similarly, the United States are mostly in North America. But a small part of it (Hawai) belongs to the Pacific Islands. We therefore have both a geographical partition into continents and a political partition into countries.

We can now speak of probabilities such as

$$P(\text{Europe} \mid \text{Soviet Union}) = P(\text{Europe, Soviet Union})/P(\text{Soviet Union}) \,. \qquad (3.1)$$

Their value depends on the criterion (experiment) for determining probabilities. If the criterion is the choice of a random point of latitude and longitude, then the numerator on the right hand side of (3.1) is the area of the European part of the Soviet Union, and the denominator is the area of the Soviet Union. If the criterion is the number of inhabitants, then we must operate instead with the number of inhabitants of these two areas.

Note that probabilities such as those of (3.1) have no connection with fuzziness. They are generally numbers between 0 and 1 even when the geographical and political boundaries are completely nonfuzzy. We should therefore not denote a probability such as that of the left hand side of (3.1) by $\mu_{Europe}(\text{Soviet Union})$. Similarly, we have probabilties such as $P(\text{vertebrate} \mid \text{sea animal})$ which should not be denoted by $\mu_{vertebrate}(\text{sea animal})$.

3.3 Exchange of Attributes and Nodes

In the last subsection we already mentioned an example in which the subnodes of a second partition of a node n (=human) of the semantic structure were named after the values of an attribute (profession) of the first partition. In the second partition we had then subnodes of 'human' such as 'carpenter', 'doctor', etc. . We note that the 'is a' and 'may be a' relations of the second semantic tree skip over the name of the attribute itself ('profession' in our example). This skipping over the name of the attribute is a linguistic property of the word 'is', in contrast to the word 'has'. (In contrast to 'John *is a* carpenter', we say 'John *has the profession* 'carpenter' ', not 'John *has a* carpenter' '.)

The automatic skipping over the name of the attribute when a node in a semantic tree is partitioned into subnodes can be a big advantage in those cases in which the attribute names are uncertain, and/or known only by experts. For example, the nodes 'dog', 'human' etc. in fig. 2.1 are values of the attribute 'species' according to biologists. Higher up in the tree, the corresponding attribute names are 'class', 'order', 'family', 'genus', 'kingdom' according to Curtis (1968, pp. 374, 375). These names are only known to experts, and even among them the terminology varies. The word 'is', and the semantic tree itself, skip elegantly over this difficulty.

In the following we consider a slightly more sophisticated example in which the exchange of attribute values against node names is mutual in the two subtrees of a doubly partitioned node.

Suppose that the lexicon contains the semantic tree of fig. 2.1, and that each instance of animal in the lexicon also lists the sex of the animal as an attribute. For some reason we are now interested in the collection of all instances of female and of male animals. At the same time we would like to retain information concerning the species of the animal, and as to whether it is a mammal, a vertebrate, etc. . This is again a typical data base problem. We shall assume in this subsection that no borderline cases of hermaphrodites exist, so that every animal can be unambiguously classified as being either female or male.

The first 'solution' to the

$$\{\text{female animal, male animal}\} \tag{3.2}$$

classification which suggests itself is to retain, on the whole, the pure tree structure of fig. 2.1, but to introduce a partition of 'animal' into the two elements of eqn (3.2). Each of the two new nodes, 'female animal' and 'male animal' respectively, would then be partitioned into

$$\{\text{vertebrate, non-vertebrate}\} \tag{3.3}$$

with their subtrees of fig. 2.1. We would then have a semantic tree in which the subtree of 'animal' of fig. 2.1 is duplicated, once as a subtree of 'female animal', and once as a subtree of 'male animal'. There are, however, two drawbacks to this 'solution'.

1) The first drawback is that we would not find the instances of female and male animals listed as 'may be a' pointers under the lexical entries of 'female animal' and 'male animal'. Instead we would find the 'may be a' pointers 'vertebrate' and 'non vertebrate'. We would have to look up each of these in the lexicon, follow their 'may be a' pointers etc., until we come to the instances. We would thus still be without an explicit list of pointers to male and female animals; it was just this list which was our objective with the new partition.

2) The second drawback of the semantic structure of duplicated subtrees is even more serious. When we follow the 'may be a' pointers in order to find the male and female instances, we come to nodes such as 'mammal'. But there are now two 'mammal' entries in the lexicon, one of them belonging to the semantic subtree of 'female animal', and one to that of 'male animal'. There is nothing in the two 'mammal' entries which tells us to which subtree it refers. Furthermore, we would now have such negative inferences as 'a mammal is not a mammal' because the two mammal nodes are cousins in the semantic tree.

A possible solution to drawback 2 is to concatenate the words 'female' and 'male' respectively with the names of each node in the two subtrees; resulting, e.g., in the lexical entries 'female vertebrate', 'male vertebrate', 'male mammal' etc. . This would lead to the creation of many new lexical entries, and would still leave us with drawback 1. A solution to these difficulties which makes use of the multiple partition structure of figs. 3.1, 3.2 is described below.

The 'animal' node of the semantic tree is supplied with two partitions, the first one being that of fig. 2.1 and eqn (3.3), and the second one that of eqn (3.2). The < term > = animal entry of the lexicon will now have two brace-groups under 'partition' in fig. 3.2. The 'may be a entries' of the two groups consist of the two elements of (3.3) and (3.2) respectively.

In the following we describe a way of retaining the information concerning, e.g., the 'humanness' of the different instances of animals in the { female animal, male animal } partition of 'animal' without having to specify a name for the attribute that can have 'human' as one of its values.

As an illustration, consider the lexicon entry of the instance < term > =Margy, assuming that we retain both of the above mentioned partitions of 'animal' in the lexicon. The '(partition tree)' node of Margy's lexicon entry tree will now have two groups of subentries referring to the first and second partition of 'animal' respectively. The first group refers to the semantic tree of fig. 2.1. It has the subentry 'human' under 'is a', and 'sex' (with subentry 'female') under 'attributes'.

The second brace-group under '(partition tree)' of < term > =Margy has the subentry 'female animal' under the subentry 'is a'. The trick of avoiding the name of the attribute to which the attribute-value 'human' refers is achieved by letting the 'attributes' subentry of the second group have a subentry called 'is a' whose subentry is 'human'. The 'is a' subentry of 'attributes' is now a substitute for the missing attribute name, namely 'species'. It can thus be considered to be an attribute on par with, e.g., 'height'.

The 'is a' name of the attribute has two advantages. 1) It tells us that even though we work with the '(partition tree)' group referring to the second partition of 'animal, we can use the 'human' node of the first partition of 'animal' for inferences, just as if we had worked with that partition. The 'is a' subentry of 'attributes' is supplied with a special information term [[((refers to) (partition of (animal (1))))]] in order to let the system know that its subentry 'human' is a subnode of the first partition of 'animal'.

2) The second advantage of the 'is a' attribute is that there is now no *necessity* to have a name for the attribute whose value is 'human'. And therefore there is also no *necessity* to store the word 'species' in the lexicon as being an attribute with possible values 'human', 'dog' etc. .

If we do wish to store the term 'species' in the lexicon, then it must be a child of 'attribute' in the default semantic tree (see second node level of fig. 2.1); i.e., the < term > =species entry in the lexicon would have a subentry 'attribute' of 'is a'. The 'is a' subentry itself would have a special information term [[((refers to) (partition of (animal (2))))]] because 'species' is already *implied* by the node *names* (e.g. 'dog') of the default partition (1) of 'animal' and should therefore not be used as an *attribute* of the nodes of the default semantic tree.

'Margy' would have two 'partition tree' braces as described before, the only difference being that the 'is a' subentry of 'attributes' in the second brace is replaced by 'species'.

To complete the description, we note that the lexicon entry for < term > ='(female animal)' will have a '(partition tree)' subentry with one brace group. The special information term of '(partition tree)' has the entry (animal (2)) to indicate that it refers to the second partition of 'animal'. The 'is a' subentry of '(partition tree)' itself has the subentry 'animal'; and its 'attributes' subentry has the subentry 'sex' with subentry 'female'. From the point of view of an understanding of natural language, the information that a female animal has the sex 'female' is, of course, redundant. However, it is the Alex system which must be able to infer this information. To avoid the double specification of 'female' and, at the same time, enable the system to infer the sex value, we can use the following device.

The term-name 'female animal' in the lexicon is replaced by the miniature entry tree

$$\text{female animal} \quad \Rightarrow \quad (\text{animal (IS (attributes (sex (female)))))} \,. \tag{3.4}$$

This whole subtree is thus wrapped up into the root of the entry tree, instead of letting the root be 'female animal'. A special information term [[(nested root)]] on the same level warns the system to expect such a structure. The specification of the sex value under the main 'attributes' entry is now unnecessary. The 'nested root' structure does not save space, but it results in a more consistent internal representation in the context of the Alex system. If we wish, we can also have the term 'female animal' in the lexicon with a 'see' pointer to the right hand side of (3.4).

We sum up this subsection by noting that a pure semantic tree structure, in which nodes have names that are values of the different *attribute values* which an instance may have, cannot be maintained when the different attribute values lead to different partitions of the set of instances. Such a structure leads to a duplication of node names, and consequently to difficulties in identifying the correct node, and to wrong negative inferences.

In order to retain a structure that allows correct inferences, the nodes of the tree must be supplied with attributes and their value intervals, as indicated in fig. 2.2. Which partition we choose for a given node depends on the application.

The multiple partition structure allows the storage of several semantic trees in one semantic structure. In our last example, the attribute values of the attribute 'sex' of 'animal' of the first partition, eqn (3.3), are used as node names in the second partition, eqn (3.2). And the attribute values (e.g. 'human', of the implied 'species' attribute) of the second partition are node names in the first partition. Nameless attributes can be stored as values of the *attribute* 'is-a', which can then refer to the other partition.

The Alex system allows us to have '(is a)' pointers from the instances to both partitions of 'animal'. Alternatively, it allows us to have '(is a)' pointers to one partiton only, e.g. the second. The information contained in the other partition can, however, be reached via the '(is a)' pointer under 'attributes'. Expressed in another way, the semantic structure of the subtree of 'animal' in fig. 2.1 corresponds to a tree structure of the attribute values of female and male animals in the semantic tree of the second partiton of 'animal'. Thus the multiple partition device allows us, in effect, to operate in our lexicon with a tree structure of attribute values.

3.4 Borderline Cases and Insertion of an Intermediate Term

In the last subsection we considered cases of exchange between node names and the names of attribute values in two subtrees of a doubly partitioned node. The meaning of a given name which was an attribute value in one subtree and a node name in another (such as 'female' or 'human') was, however, the same in both subtrees; the test for the identity of meaning being that the set of instances that have 'is a' pointers to the node 'human' and attribute value 'female' in the first subtree, is the same as the set of instances that have 'is a' pointers to the node 'female animal' and the attribute value 'human' in the second subtree.

In this and the next subsection we consider the case of a multiple partition in which a given node name can appear in two or more subtrees of a multiply partitioned node. However, the meaning of such a node may be somewhat different in the two partitions.

In the lexicon such nodes must then have separate entry trees, one for each meaning. This can be achieved by a structure analogous to that of fig. 3.2, except that the grouping starts directly under < term > . For example, if the term 'animal' has two different meanings, then it is supplied with a subentry 'meaning', directly under 'animal', which has a special information term [[true]]. This node, together with the presence of its special information term, warn the program that several groups may follow, one for each meaning. A given meaning-entry can have a 'partition' or '(partition tree)' subentry, just as in fig. 3.2.

In many applied classification systems one encounters the problem of borderline cases both in connection with classifications according to unidimensional attributes such as 'age' with values that are elements of the partition {young, old} ; and in connection with classifications according to multidimensional attributes such as 'organism' with the partition {plant, animal} . Although many objects fit easily into one of the elements of the partition (or of the 'legal label set' according to the terminology of Hisdal (1987a, part 1.3; 1987b)), the attribute values of others lie near the boundary of two or more elements. The exact fixation of the boundary hypersurface can vary in the opinion of different people. Furthermore, a slight error in the measurement or estimation of the attribute value of the object can easily change its classification. These two sources of uncertainty have been named 'fuzziness #3' and 'fuzziness #1' respectively in Hisdal (1986b).

The grade of membership treatment of this problem is given in sections 4-6 below. Another, often-used expedient in everyday language and in scientific applications is to decrease the 'linguistic unit of measurement' by creating a new quantization interval, with a new name, around the boundary region. In most cases the names of the elements of the first partition are also used in the second partition. But their meaning in the new partition is modified because the quantization intervals of the concepts which they represent are narrowed down. For example, in connection with our linguistic set of age values, it is usual to create a quantization interval labeled 'middle-aged' in the numerical attribute universe of 'age in years'. In biology the insertion of an intermediate term has occurred for the classification of organisms, where the concept 'protista' (microorganism like 'bacteria') was added as an element to the original bipartition of organisms into plants and animals (see fig. 3.1).

In contrast to the case of the previous subsection, the insertion of one or more intermediate terms does not change the universe of attribute values in which the partition is performed. However, a concept with a given name, e.g. 'plant', may now have two different meanings, depending on the partition to which it refers. In addition we have again two or more partitions of its supernode 'organism'.

3.5 The Center of the Universe Problem

Although Aristarchus of Samos already suggested the heliocentric system in the year 310 B.C. (Koestler, 1968 p. 51), the extremely complex Ptolemeian astronomy, which assumes that the earth is the center of the universe, prevailed for almost another two millenia before an obscure canon in Varmia, a remote outpost of Christendom, picked up the thread where Aristarchus had left off. His name was Copernicus.

Human beings have a hard time in accepting that they are not the most important, and very special objects in the universe. As a result, they do not consider themselves to be animals although they fit into this class according to their attributes. The everyday meaning of the class 'animal' is thus the biological class of animals minus the class of humans (see, e.g., the Webster dictionary for the two meanings of 'animal').

If we wish to allow for such an additional meaning of 'animal' in our lexicon, we must provide for the possibility of still another partition of 'organism', such as {plant, animal, human} . Every attempt to insert nodes such as 'mammal', 'non-mammal' into the semantic tree will now lead to inconsistencies, just as in subsection 3.3. Consequently the subnodes of 'animal' in the semantic tree of fig. 2.1 must be used as attribute values of both the 'human and the'animal' node in the semantic tree with the new partition of 'organism'. The attribute itself can be called 'is a' as before.

We have here a combination of the cases of subsections 3.3, 3.4. Just as in subsection 3.3, some of the nodes of fig. 2.1 are changed to attribute values in the semantic substructure tree of the new partition. But, at the same time, the meaning of a given term, in our case 'animal', is changed in the sense that the set of instances which belong to it is restricted to a smaller one as in subsection 3.4.

3.6 Summary

Two or more different partitions of a single node 'term1' in a semantic structure give rise to two or more 'substructure trees' (fig. 3.1). In the lexicon, the multiply partitioned 'term1' node gives rise to two or more groups of 'may be' pointers in the 'term1' entry (fig. 3.2). Each of these groups may have its particular set of attributes attached to it. 'Is a' pointers of a node 'term2', which is a subnode of 'term1' in the semantic structure, must make it clear to which partition they refer. The attributes that apply to 'term2', as well as its 'may be a' pointers, may be specific to the particular partition of term1 to which the 'is a' pointer refers.

The multiple partition structure can lead to redundant storage of information if the instances have '(is a)' pointers to two or more partitions of a given node which do not result in ambiguous node-names (see sect. 3.4 for the latter). The redundant storage of information is, howver, not necessary. We have shown in sect. 3.3 that in a well-constructed multiple partition structure all the '(is a)' classifications and attributes of the instances in the different semantic substructure trees can be inferred through a single '(is a)' pointer of each instance to a leaf of one semantic substructure tree only.

Finally we add that this paper does not describe the complete storage stucture for attributes. E.g., it is easy to store the *context dependent* numerical interval-values of linguistic attribute values (such as the year interval of 'old' for 'human' vs. 'old' for 'dog') in the Alex lexicon.

4. The Grade of Membership Modification of the Semantic Tree

4.1 The Multiple Partition versus the Membership Case

In the previous section we treated the case of two or more partitions of a single node $n1$ of the semantic tree into subnodes, thereby creating a semantic structure which is equivalent to two or more semantic trees. A node $n3$ of the semantic structure could then have more than one 'is a' pointer, where each pointer referred to an element of a different partition,

$$\{n2^p\} = \{n2^p_{l^p}\}, \qquad l^p = 1, \dots, L^p \qquad p = 1, \dots, P \tag{4.1}$$

of some supernode $n1$. The superscript p in (4.1) denotes the number of the particular partition. Thus each 'is a' pointer of a single node $n3$ referred to a different semantic tree (see fig. 4.1a for which P=2).

In the present subsection we treat solely the case of a single partition of each node of the semantic structure into subnodes. The structure is then a pure semantic tree as long as its arcs are considered to be downward, i.e. 'may be a', pointers.

However, a node $n3$ may now have two or more parent or 'is a' pointers to two or more elements of the single partition $\{n2\} = \{n2_l\}$, $l = 1, \dots, L$ of some node $n1$ (see fig. 4.1b).

The membership treatment is pertinent to the case of objects whose attribute value lies on the boundary of two or more adjacent siblings of the $n2$ family. An alternative treatment of this case was given in subsection 3.4 by creating a new partition of $n1$ with an additional sibling for attribute values near the boundary region. The treatment of the present section creates no new partition. Instead it assigns partial grades of membership to the object belonging to the class $n3$ in two or more sibling nodes of the $n2$ family. Fig. 4.1 illustrates the difference between the two treatments.

The above-mentioned two distinct treatments can thus be put to the same use, namely the classification of objects whose (generally multidimensional) attribute value lies near the boundary of two or more classes. If we wish, we can, of course, superpose the membership treatment also on the additional partition of the multiple partition treatment such that the right hand partition of $n1$ is also included in fig. 4.1b, and $n3$ may have more than one parent in this partition.

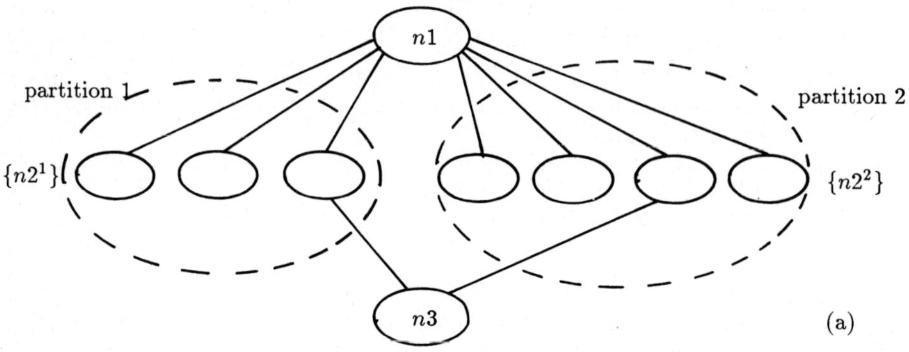

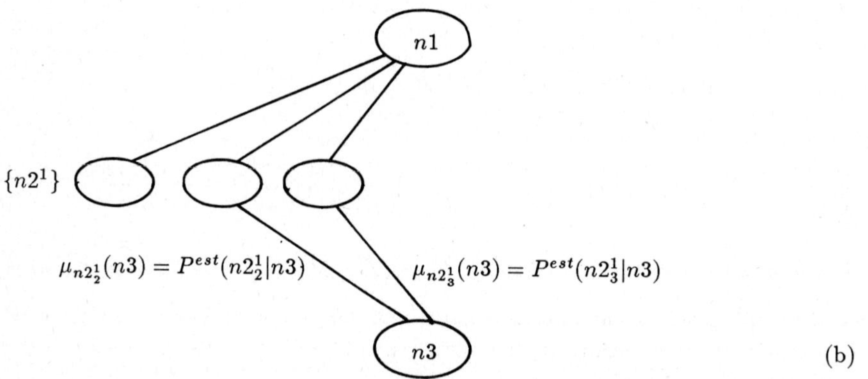

Fig. 4.1. A multiple partition structure (a) versus a grade of membership structure (b). (a) Part of a semantic structure with two partitions of $n1$. $n3$ has an 'is a' pointer to only one element of a given partition. (b) Part of a semantic structure in the case of a class $n3$ with partial grades of membership in two different elements of a single partition of $n1$. A node can now have two or more parent pointers to elements of the same family. The grade of membership values are interpreted in the TEE model as the subject's estimate of the probability that an object labeled $n3$ would be labeled $n2^1_l$ in a YN (yes-no) or LB (labeling) experiment performed in the presence of fuzziness #1, 2 or 3, and referring to the partition (nonredundant label set) $\{n2^1\}$. The membership values of $n3$ in the elements of $\{n2^1\}$ must add up to 1.

4.2 The Surface Structure of the Membership Modification

In this subsection we give a purely formal description of the membership structure and a superficial description of its meaning. The deep structure and meaning of the membership modification are discussed in sections 5 and 6.

Just as in the case of the multiple partition structure, the description of the membership structure starts out from a pure tree which is then modified. The membership modification is applied only to semantic trees in this paper. To make the discussion more concrete, we shall therefore refer it to the semantic case. This does not detract from the generality of the purely formal description.

Consider a pure tree structure. The unique role of the root, named 'concept', of this structure is not affected by the membership modification.

Let

$$\{n2_{l2}\}, \quad l2 = 1, \ldots, L2 \tag{4.2}$$

be a generalized partition of a node $n1$ in the pure tree structure. And let $n3$ be a child of $n2$, where $n2$ is one of the elements of $\{n2_{l2}\}$. The 'is a' or upwards pointers of the pure tree then imply that

$$P(n2 \mid n3) = 1 \ , \tag{4.3}$$

$$P(n2_{l2} \mid n3) = 0 \qquad \text{for } n2_{l2} \neq n2 \ , \tag{4.4}$$

$$P(n1 \mid n2_{l2}) = 1 \qquad \text{for } l2 = 1, \ldots, L2 \tag{4.5}$$

$$P(n1 \mid n3) = P(n2 \mid n3) P(n1 \mid n2) = 1 \ . \tag{4.6}$$

For a pure, nonfuzzy tree structure the right hand sides of (4.3)-(4.6) can assume only the values of either 0 or 1. In the case in which $P(nk|nj) = 1$, the relation

$$\text{Every } nj \text{ is an } nk \tag{4.7}$$

is generally valid. It implies that nk is a supernode of nj.

In the fuzzy or membership case the right hand sides of (4.3)-(4.6) may assume any value in [0,1]. (4.7) is then changed to

$$\text{An } nj \text{ is an } nk_{lk} \text{ to the degree } \mu_{nk_{lk}}(nj) \ , \qquad lk = 1, \ldots, Lk \tag{4.8}$$

Assuming that $n3$ is an exact node of a membership structure, and $\{n2_{l2}\}$ its parent family (see subsection 6.2), equations (4.3), (4.4) are replaced by the condition

$$\sum_{l2=1}^{L2} P(n2_{l2} \mid n3) = 1 \ , \tag{4.9}$$

in the membership structure, and (4.6) is replaced by

$$P(n1 \mid n3) = \sum_{l2=1}^{L2} P(n2_{l2} \mid n3) P(n1 | (n2_{l2}, n3)) \ . \tag{4.10}$$

Fig. 4.2 illustrates these relations for the case of $n3 =$virus.

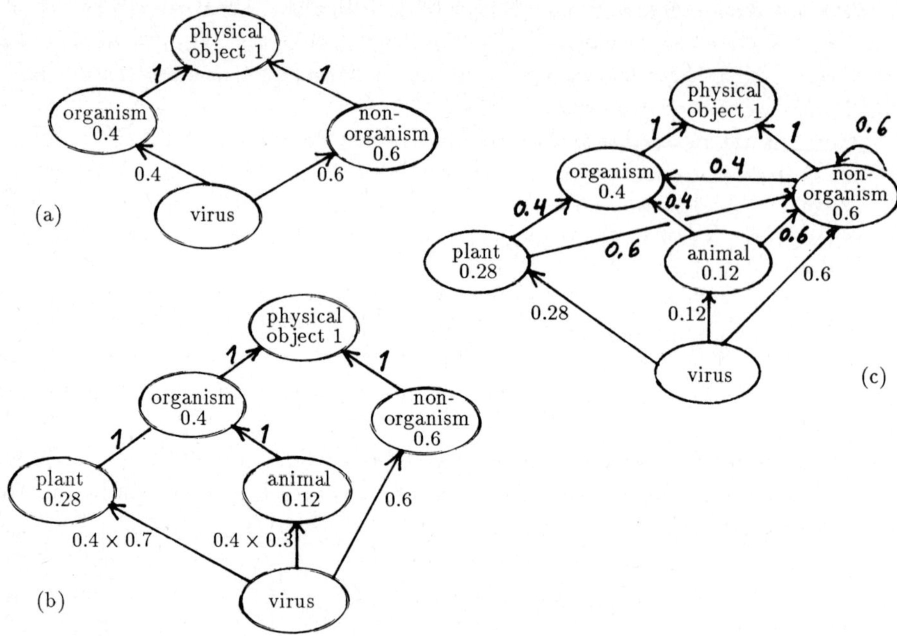

Fig. 4.2. Part of the membership structure derived from the originator tree of fig. 2.1, the exact node 'virus' being fuzzily classified. It is assumed to be a child of 'plant' or 'animal' in the originator tree. (a) is an abbreviated version of (b) and (c).

The numerical value inside a node nk is the membership value $\mu_{nk}(\text{virus})$. These values are part of the definition of the membership structure (see def. 6.5), and are thus the same in figs. (a), (b) and (c). The values of the 'is a' or upwards pointers of fig. (b) refer to a SIM situation (see sects. 1.3 and 6.3), those of fig. (c) to an RR situation (see sections 1.3 and 6.4). Note the consistent summing-up to 1 of the grades of membership of virus in any generalized family (partition) of the membership structure in all three figures.

The number along any 'is a' pointer from nj to nk is $P(nk|(nj, \text{virus}))$, see eqn (6.7) et seq.. E.g., consider the case when nj is an element of the parent family $\Lambda j = \{\text{plant, animal, non organism}\}$ of 'virus', and nk an element of its grandparent family $\Lambda k = \{\text{organism, non-organism}\}$ in figs. (b) and (c). The probabilities $P(nk|(nj, \text{virus}))$ refer to the space $\Lambda j \times \Lambda k \times U^{ex}$, where virus $\in U^{ex}$. The sum of the products of the pointer values along any path from 'virus' to a given $nk \in \Lambda k$ via a node $nj \in \Lambda j$ add up to $\mu_{nk}(\text{virus})$ in both figs. (b) and (c).

In the SIM case (b), the $P(nk|(nj, \text{virus}))$ values of the pointers are equal to $P(nk|nj)$, without the qualification by u^{ex} =virus. In the RR case of fig. (c), the pointer values are equal to $\mu_{nk}(\text{virus})$, see eqn (6.7b). In the same figure we have two 'upward' pointers which do not have an upward direction in the figure, namely from 'non-organism' to 'organism', and from 'non-organism' to itself. The pointers are, however, 'upward' in the sense that they point from an element of the parent family Λj of 'virus' to an element of its grandparent family Λk. 'non-organism' belongs to both of these families. These two pointers, as well as a third one (from 'animal' to 'non-organism') are absent in the SIM- figure (b) because the subject's etimate of the object's attribute value (in connection with fuzziness #1) is, for SIM reference, the same in both component experiments referring to Λj and Λk respectively. The pointer from 'non-organism' to itself has therefore the value 1 in the SIM case. The other two pointers have the value 0 because the nodes at the tail and at the head of each pointer are also elements of a single, nonredundant label set (Λk and Λj respectively for each of the two pointers). In contrast to the SIM case, the RR 'is a' pointers from an element of Λj to an element of Λk depend on the value of u^{ex} (where u^{ex} ='virus' in our figures). They can be averaged over u^{ex} (see fig. 6.1(c)).

As an example of the summing up to MU relation of eqns (6.7), (6.7b), we follow all paths from 'virus' to 'non-organism' in fig. (c) and obtain $P(\text{non-organism}|\text{virus}) = 0.28 \cdot 0.6 + 0.12 \cdot 0.6 + 0.6 \cdot 0.6 = 0.6 = \mu_{non-organism}(\text{virus})$.

The conditional probabilities $P(nk|nj)$ of this subsection (where nk is a supernode of nj) are interpreted as the probabilities that a subject will assign the label $nk \in \{nk_{l_k}\}$ to an object that has been labeled $nj \in \{nj_{l_j}\}$ (in a labeling or LB situation, see Hisdal (1987a, part 1.3)). According to the TEE model (see sect. 5), the estimate by the subject of such a probability is equal to the membership value in the class nk which he assigns in a MU experiment to an object that has been labeled nj ,

$$\mu_{nk}(nj) = P^{est}(nk|nj) \ . \tag{4.11}$$

The most prominent difference between a multiple partition modification of a tree and the surface structure of a membership modification is that when a node has more than one parent, then these parents are elements of the same generalized partition of some supernode in the membership structure. In the multiple partition structure they are elements of different partitions.

We summarize this section by noting that the structure of eqns (4.9),(4.10) is an acceptable and consistent description of the case in which an object may have partial grades of membership in several classes $n2_{l_2}$ which are elements of a single partition of some superclass $n1$. The membership modification of a pure tree consists in allowing two or more parents to share a given child; or, said in another way, a node may have more than one parent or 'is a' pointer, all of which are elements of one generalized partition in the originator tree. To illustrate this summary we show in fig. 4.3 a greater portion of the structure of which fig. 4.2b is a part.

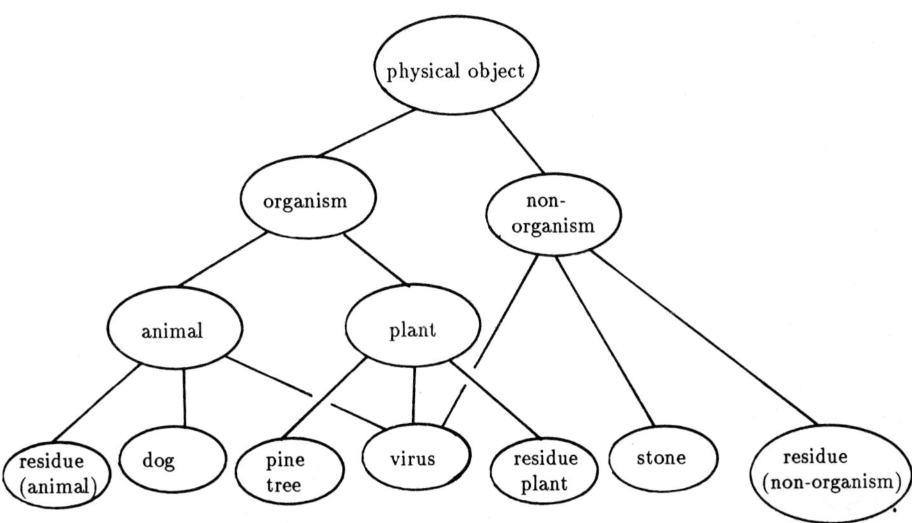

Fig. 4.3. This figure shows a larger part of the membership structure of fig. 4.2(b). The node 'virus' is common to the children families of both 'animal' and 'plant' and 'non-organism'.

5. The Deep-Structure Connection between the Membership and the Multiple Partition Modification of Trees According to the TEE Model.

5.1 The Structural Similarities and Differences between Pure Trees, Multiple Partition and Membership Structures

In the previous two sections we dealt with two modifications of pure tree structures, the multiple partition modification and the membership modification. As we have shown in subsection 4.1 and in fig. 4.1, the surface structure of these two modifications is completely different. The common properties of the two modified structures are some of the typical tree properties. Both structures consist of nodes and directed arcs which connect them. To every structure there corresponds an inverted one obtained by inverting the direction of all the arcs. We talk about an upwards and a downwards structure when the arcs point in the upward and the downward direction respectively. Both the multiple partition and the membership structure have one unique node called 'root' that has no upwards pointer in the upwards structure. And they have one or more nodes called 'leaves' that have no downwards pointers. By convention the root node is placed at the top of the structure and the leaves at the bottom.

The tree-structure concept of a sequence of nodes called the rootpath from a node n applies to both of the modified structures. However, for both of them there may exist more than one ROOTPATH(n). For the multiple partition structure, a rootpath from n is constructed by following the rootpath of one of the pure trees contained in the structure. For the membership structure, the possible ROOTPATH(n) sequences may depend on u^{ex} , the attribute value of the object to which they refer (see sects. 1.3 and 6.2).

At this point the similarity between the multiple partition and the membership structures stops. The membership structure (upward or downward) is a directed graph referring to the originator tree (see sect. 6.2). [1] In contrast, the multiple partition structure is not an ordinary graph at all but a nested one, more specifically a nested tree. This situation is illustrated by the broken, subtree-embracing nodes in fig. 3.1.

The surface structure of the membership modification, although not the conceptual deep-structure, is thus simpler than the multiple partition modification. This difference in simplicity is reflected by the representations of the two structures in the lexicon. A membership structure can be represented in the lexicon without destroying the pure tree structure of the lexicon entries. We need only store the originator tree and the membership functions of its nodes in the lexicon. The additional '(is a)' pointers of the membership structure can then be inferred by a straightforward algorithm (see sect. 6.6).

---

[1] A network without loops, i.e. with the possibility of several 'is a' pointers from a single node, has previously been suggested by Ohren (1986) at the end of her thesis on the Alex2 system. Her 'is a' pointers refer to one 'kernel node' and one or more 'non-kernel nodes'. In the context of the present paper, each of these nodes would belong to a different partition.

In contrast, the representation of a multiple partition structure in the lexicon (see sect. 7.4) necessitates a modification of the pure tree structure of the lexicon entries to the nested tree structure of fig. 3.2 which is structurally similar to the multiple partition modification of the semantic tree.

5.2 The TEE Model Interpretation of the Membership Structure as an Expectation over the Pure Trees Contained in a Multiple Partition Structure

The present section has two purposes. The first of these is to uncover a deeper-lying connection between the multiple partition structure and the membership structure, based on the meaning of grades of membership in the TEE model. The membership structure turns out to be an expectation over the different trees contained in a special multiple partition structure.

The second purpose is to clear up an apparent paradox in the surface description of the membership structure. We saw in section 4 that the membership structure leaves the pure downwards tree structure intact as far as the nonfuzzy partitioning of a node into a set of subnodes is concerned. On the other hand we find that a given instance may belong to two or more elements of a single partition when we follow the membership structure in the upwards direction. However, the classes to which these elements belong should have been (nonfuzzily) disjoint according to the downwards structure. In the following we describe the resolution of this apparent paradox in the TEE model (Hisdal, 1986a,b; 1987a,b).

According to this model, the nonfuzzy partitioning of a node into a set of subnodes implied by the downwards structure of the membership modification refers to the situation of a single YN (yes-no) or LB (labeling) experiment performed by a subject on a set of objects or instances. In contrast, the fuzzy partitioning implied by the upwards structure refers to the situation of a MU experiment in which the subject takes the possible presence of fuzziness #1,2 or 3 into account.

YN and LB experiments correspond to the situation in everyday discourse in which a subject describes an an object as being $n2$ (or λ according to the notation used in the TEE model papers), where $n2$ is chosen from a complete, nonredundant label set $\Lambda = \{n2_{l2}\}$, $l2 = 1, \ldots, L2$ (e.g., $\Lambda =$ { organism, non-organism}, or $\Lambda =$ {small, medium, tall} for figs. 4.2, 6.1 respectively).

The subject's criterion for the choice of label is based on his estimate u of the object's (often multidimensional) attribute value, and a partitioning of the universe U of estimated attribute values into L2 disjoint 'quantization intervals Δu_{l2} ', one for each element of Λ. In a given YN or LB experiment, in which the subject is instructed to choose one label from $\Lambda = \{n2_{l2}\}$ for each object, we have then a nonfuzzy partitioning of the set of objects used in the particular experiment, based on the value of u.

When the YN or LB experiment is an *exact* one, in which the subject S measures the attribute values of the objects exactly, then $u = u^{ex}$, where u^{ex} is the object's exact attribute value. Consequently, all objects with the same value of u^{ex} are assigned the same label by the subject. Equations (4.3)-(4.7) refer to this case.

When the experiment is nonexact, then objects with the same value of u^{ex} may be assigned different estimated attribute values u by the subject. Consequently two objects with the same u^{ex} value may be assigned different labels by the subject when u^{ex} lies near the boundary of the quantization interval of two neighboring labels. Furthermore, one and the same object may be assigned different labels by the subject in two different experiments in which it is observed under a different condition of observation (fuzziness #1). Alternatively, the given object may be assigned different labels by two different subjects, both of whom perform an exact experiment, because the two subjects operate with somewhat different threshold values for the boundaries between two labels (fuzziness #3). An analogous statement hold for fuzziness #2 (Hisdal, 1986b). For each of the three sources of fuzziness we can thus talk about

$$P(n2_{l'} \mid u^{ex}) \qquad l' = 1, \ldots, L2 \;, \tag{5.1}$$

the probability that an object with attribute value u^{ex} will be assigned the label $n2_{l'} \in \Lambda = \{n2_{l2}\}$ by the subject in the presence of fuzziness.

The situation of an LB or YN experiment is contrasted in the TEE model with that of a MU experiment in which S is told to assign a grade of membership to the object in the different elements of $\Lambda = \{n2_{l2}\}$ (Hisdal 1987a, parts 1.3,1.4; 1987b). Assuming that the MU experiment is an exact one, then S will, according to the TEE model, estimate the effects of errors of estimation of the attribute value in the case of a nonexact YN or LB experiment performed under a set of conditions of observation which corresponds, according to S's judgement, to the variable conditions of observation of everyday life to which he refers his fuzziness #1a. And he will assign to the object a membership value in the class $n2_{l'}$ that is equal to his estimate of the probability that the object would be assigned this label under these variable conditions,

$$\mu_{n2_{l'}}(u^{ex}) = P^{est}(n2_{l'} \mid u^{ex}) \;. \tag{5.2}$$

A similar effect occurs in connection with fuzziness #3a when the subject takes into account that different persons belonging to a population of subjects may operate with slightly different thresholds in U for a given label $n2_{l'}$. In an exact MU experiment, the right hand side of Eqn (5.2) then refers to S's estimate of the proportion of subjects whose quantization interval $\Delta u_{n2_{l'}}$ for $n2_{l'} \in \{n2_{l2}\}$ is such that it contains the object's $u = u^{ex}$ value. The case of a nonexact MU experiment, is discussed in Hisdal (1986b, section 3, fuzziness #1b).

The differentiation between LB or YN versus MU experiments thus clears up the apparent paradox of the nonfuzzy and fuzzy partitioning of a node into its child family implied by the downward and upward membership structure respectively.

To explain the deep-structure connection between the membership and the multiple partition structure we start with the example of intersubject fuzziness (fuzziness #3);

namely the fact that different subjects use somewhat different partitions in U for the same liguistic label set Λ. When S operates with fuzziness #3a, he then takes this variabilty in thresholds, with a consequent variability in labeling, into account. He thus establishes many multiple partitions of the supernode $n1$ of $n2$. Each partition is assigned a probability equal to S's estimate of the proportion of persons (in the population of subjects) who use this particular partition of U (or of U^{ex} in the case of an exact experiment).

In connection with fuzziness #1a, S operates with only a single partition in U, and consequently with a single quantization interval Δu_{l2} for each label $n2_{l2}$ in the universe U of *estimated* attribute values. This being his own subjective quantization interval for $n2_{l2}$. But he takes into account that the objects' u-value depends on the condition of observation. This results in a *multiple* partition of $n1$, the parent of the $\{n2_{l2}\}$ family, *with respect to the universe U^{ex} of exact attribute values.* Each partition has a probability that depends on S's estimate of $P(u|u^{ex})$, the distribution of errors of observation under the set of conditions of observation to which he refers his fuzziness #1a.

Irrespective of which of the three sources of fuzziness S takes into account, he may even work with a superposition of all three, he thus operates with a multiple partition of $n1$ into many child families $\{n2^1\}$, $\{n2^2\}, \ldots$. The number $L2$ of elements in the different partitions of the node $n1$, as well as their names, are the same. However, the quantization interval in U^{ex} of an element $n2_{l'} \in \Lambda = \{n2_{l2}\}$ with a given name varies somewhat from partition to partition. Consequently, identically named elements of Λ have somewhat different meanings in the different partitions.

The membership structure is obtained by computing for each subnode $n3$ of the $n2$ nodes the expectation value for the assignment of the *linguistic* label $n2_{l2}$ over the different occurrences of the node with this label in the different multiple partitions of n1. All nodes with a given name in the different partitions are then combined into a single node. This results in a formal structure without multiple partitions. An upwards pointer to $n2_{l2}$ is then drawn from every node $n3$ for which the subject's estimate of the above expectation value is bigger than 0; and the numerical estimated expectation value is attached to the pointer. We thus end up with a structure that is quite different from the multiple partition structure although the latter is used as an intermediate step in the construction of the membership structure.

In connection with fig. 4.1b, we said that the membership structure refers to the elements of a single partition of $n1$. We now modify this statement to say that from the point of view of the intermediate, multiple-partition step of the MU experiment, the membership structure refers to the elements of a single *linguistic* partition of $n1$ because the elements of the different partitions have the same labels. The grade of membership of a node $n3$ in $n2_{l2}$ is the expectation value of the node with the name $n2_{l2}$ over the different partitions of a multiply partitioned node $n1$ for objects that have been labeled $n3$

6. Semantically and Numerically Correct Membership Structures

6.1 Overview

In the last section we discussed shortly the meaning of the membership values in the TEE model. There are, however, still left some unanswered questions concerning the precise meaning of the relation between two nodes of a membership structure and of the numerical values of the upwards (is a) and downwards (may be a) pointers.

In subsection 6.2 we precisate this meaning by requiring that the construction of a membership structure must start out from a pure 'originator' tree (requirement 1 and def. 6.2). However, in contrast to the originator tree, the leaves of a membership structure are considered to be exact nodes, the other nodes being nonexact ones (requirements 2, 3). As a consequence, a node in a membership structure may have addtional 'is a' pointers to that of the originator tree. To which additional nodes nk a given node nj may potentially be connected by an 'is a' pointer is determined by the first and higher order parent families (def. 6.1) of one of the exact leaves. The numerical value of a pointer from a node $n1$ to a node $n2$ is interpreted as $P(n2|n1)$, the probability of assignment of the node at the head of the pointer to an object that has been assigned the node at the tail.

We have a potential 'is a' pointer from a node nj to a node nk, and a potential 'may be a' pointer from nk to nj whenever nj and nk belong to the i-th and $(i+1)$-st parent family respectively of a leaf (def. 6.2). The numerical values of these two pointers are defined to be $P(nk|nj)$ and $P(nj|nk)$ respectively. The numerical values of the upward and of the downward pointer between two nodes are thus, in general, quite different. This is in contrast to the present-day theory of fuzzy sets and possibility in which the two pointers are assumed to be numerically equal (see Hisdal, 1986a, difficulty 6 and eqn (2.2); also Hisdal, 1987a, part 1.4). The $P(nk|u^{ex})$ numerical values of the lowest 'is a' pointers from a leaf u^{ex} are equal to $\mu_{nk}(u^{ex})$, the value of the membership function of the node at the head of the pointer for the exact attribute value u^{ex}.

The numerical values of the higher 'is a' pointers, and those of all the 'may be a' pointers, are not unique for a given membership structure (with given grade of membership functions for each node, see def. 6.5). The values of the higher 'is a' pointers are shown to depend strongly on the reference situation to which the assignment of the node at the tail and at the head of the pointer refers. There are two categories of such reference situations, SIM and RR. These are treated in subsections 6.3 and 6.4 respectively. The membership structure for SIM reference turns out to be a pure tree structure except for the relation between the two lowest node levels (see figs. 4.2(b), 6.1a).

In the more complicated case of an RR reference situation it turns out that also the higher-up nodes are fuzzily classified in the sense that a given node may have more than one 'is a' pointer. Furthermore we may have one set of numerical values of the RR 'is a' pointers for each u^{ex} (figs. 4.2c and 6.1b). Averaging the numerical value of each 'is a' pointer over u^{ex} results in a set of values for these 'expected 'is a' pointers' which depends on $P(u^{ex})$, the prior distribution of the set of objects over u^{ex} (fig. 6.1c). The values of both the SIM and the RR expected 'may be a' pointers depend always on the prior distribution $P(u^{ex})$. This distribution is discussed in subsection 6.5.

In contrast to the above-mentioned $P(u^{ex})$-dependent pointer values, the membership functions of the nodes, the values of the lowest 'is a' pointers and the values of all the SIM 'is a' pointers are independent of $P(u^{ex})$.

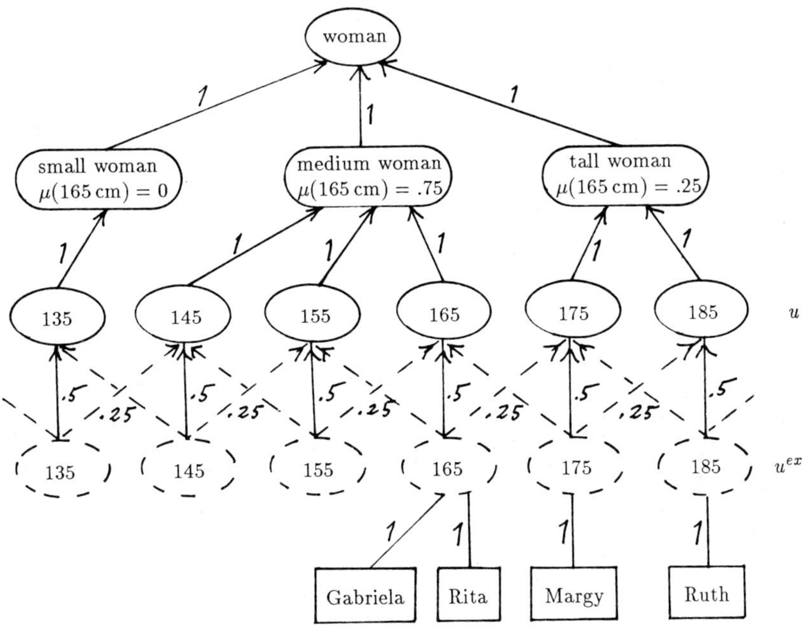

Fig. 6.1a. A membership structure with SIM pointers for the label set {small woman, medium woman, tall woman}, for a subject who assumes the following quantization intervals in the universe U of estimated attribute values in an LB (labeling) experiment:
$\Delta u_{tall} = \{175 \text{ cm}, 185 \text{ cm}, \ldots\}$;
$\Delta u_{medium} = \{145 \text{ cm}, 155 \text{ cm}, 165 \text{ cm}\}$;
$\Delta u_{small} = \{\ldots, 125 \text{ cm}, 135 \text{ cm}\}$.
These intervals, and the $P(u|u^{ex})$ error function (indicated by the numbers on the 'is a' pointers from the u^{ex} nodes) are the same as those used in the TEE model papers (Hisdal, 1986a,b, 1987a,b).

The fully drawn nodes belong to the originator tree of the structure (sect. 6.2, requirement 1), and the fully drawn arcs point towards the tree parent. These nodes and pointers refer to a situation in which all nodes are assigned to an object under the same point of the set of conditions of observation in connection with fuzziness #1.

The broken oval nodes are exact nodes of the membership structure (def. 6.3 and requirements 2, 3). They are assigned to the instances by the experimenter who is assumed to have a perfect measuring apparatus. All the nodes above the exact ones, except the root, are nonexact ones assigned by the subject either in a nonexact YN (yes-no) or LB (labeling) experiment, or in an exact MU (grade of membership) experiment. In the latter case, the grade of membership which the subject assigns to an instance in one of the classes whose label is that of a nonexact node (the two node levels under the root) is, according to the TEE model, equal to the subject's estimate of the effects of fuzziness #1, 2 or 3. (Concerning nonexact MU experiments, see Hisdal (1986b, source of fuzziness #1b). A node such as 175 cm actually means the interval 175±5 cm.)

The membership structure has additional 'is a' pointers and an additional linguistic-duplicate lowest node level (broken arcs and nodes) as compared with the originator tree. The values of the 'is a' pointers (except those from the broken, exact nodes) are for SIM reference concerning the assignment of the node at the tail and that at the head of the pointer respectively (see sections 6.2, 6.3). The summing-up to MU law, eqn (6.7a), holds for the figure. E.g., $\mu_{medium}(u^{ex} = 165) = 0.75$ is obtained by following all 'is a' pointers from u^{ex}=165 cm to 'medium woman'.

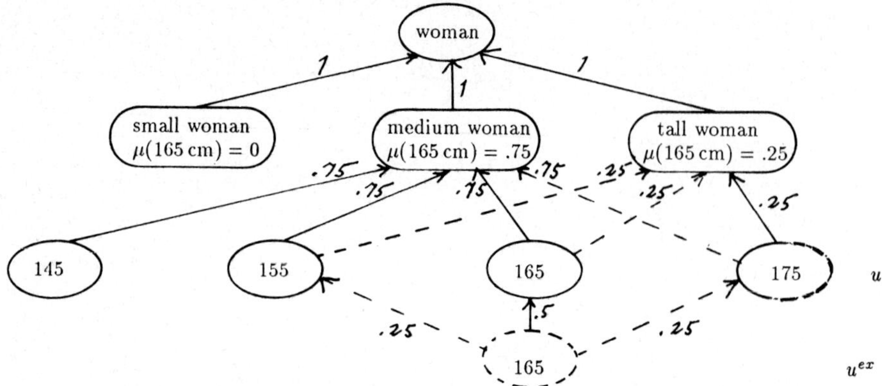

Fig. 6.1b. A membership structure with the same originator tree as that of fig. 6.1a and with RR 'is a' pointers. The values of these pointers, eqn (6.11), refer to u^{ex}=165 cm. The parent and grandparent families of this node (as well as of every other u^{ex} node, see def. 6.3) are $\Lambda1 = U = \{135\,\text{cm}, 145\ \text{cm}, \ldots\}$ and $\Lambda2$ = {small woman, medium woman, tall woman} respectively. The numbers inside the nodes of the $\Lambda2$ family are their grades of membership for u^{ex}=165 cm. The summing-up to MU law on upwards paths from u^{ex}=165 cm (see eqns (6.7a), (6.7b)) holds for these. (The fully drawn 'is a' pointers point to the tree parent, defs. 6.1, 6.3.)

In the RR case the 'is a' pointers from $\Lambda1$ to $\Lambda2$ depend on u^{ex}. We have additional such pointers as compared with the SIM case because, e.g., an instance with u^{ex}=165 cm could be assigned an estimated height value of 175 cm $\in \Lambda1$ in the first component experiment, and the label medium $\in \Lambda2$ in the second (based on a height estimate of 155 or 165 cm in the second experiment). This results in a nonzero RR pointer from u=175 cm to 'medium woman'.

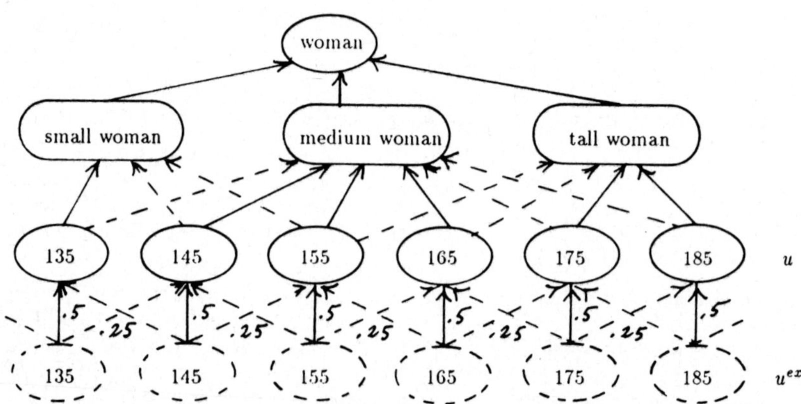

Fig. 6.1c. A membership structure with the same originator tree as that of figs. 6.1a, 6.1b, but with expected RR 'is a' pointers. Their values depend on the prior distribution $P(u^{ex})$ of the population of women over height, see eqns (6.12)-(6.14). All the indicated pointers have nonzero values, assuming that $P(u^{ex})$ is nonzero for all the u^{ex} nodes. (E.g., u=155 in component experiment 1 can be due to an object with u^{ex}=165; which may give rise to u=175, and therefore to the label 'tall' in component experiment 2. This results in an expected nonzero 'is a' pointer from 'u=155 cm' to 'tall' whose meaning is: 'A woman whose height is estimated to be 155 cm in one experiment has a nonzero probability of being labeled 'tall' in another (see caption to fig. 6.1a for assumed quantization interval for 'tall'). The 'summing-up to MU' law does not hold for the expected RR 'is a' pointers.

6.2 Definitions and Requirements Concerning a Membership Structure

This subsection lists the different requirements and definitions pertaining to a membership structure.

Requirement 1. *The construction of a membership structure starts out from the pure-tree classification structure* of a given subject in LB (labeling) experiments. The experiments refer to the assignment of the label of a given node n of the tree to an instance or object; n being an element of a 'label set' or 'generalized family' of the tree, $\Lambda = \{n_l\}, \quad l = 1, \ldots, L$. From now on we will not always repeat the requirement of the reference of the semantic LB, YN or MU experiment to a generalized family (partition) of the tree. A pure tree always refers to the case that the assignment of the labels nj, nk to a given object (where nk is a parent of nj) is performed in exact experiments or, more generally, under the same point of the space of fuzziness. (For example, in connection with fuzziness #3, this means that the labeling experiments concerning all the different partitions of the tree are assumed by the subject to be performed on all objects by the same person).

The pure tree from which a membership structure is constructed is called the originator tree of the membership structure.

Definition 6.1 of the i-th order parent family of a node $n0$ in a tree or in a membership structure. Let $n0$ be a non-root node in a pure tree structure. And let $n1$ be the parent of $n0$ and $n2$ the parent of $n1$, if it exists. Then the first-order parent family of $n0$ contains the following elements: 1) All the elements of $\{n1_{l1}\}$, the sibling family of $n1$ in the tree. 2) All the elements of $\{n2_{l2}\}$, the sibling family of $n2$, except $n2$. 3) If the parent $n3$ of the $\{n2_{l2}\}$ family has siblings, then these siblings are also elements of the parent family of $n0$. This process is repeated until we reach the root. It follows from this construction that all elements of a sibling family have the same parent family. (The node-level numbering of fig. 4.1 is reversed here.)

As an illustration, the first-order parent family of 'bacteria' in fig. 2.1 is {plant, animal, non-organism, material, attribute, attribute value, unit, residue(concept)}. The elements of this set are the parents of 'bacteria' in the membership structure, the element 'plant' being the 'tree parent'.

The (i+1)-st order *parent family* of $n0$ is the *parent family* of the i-th order *parent* of $n0$ in the tree, $i = 0, 1, \ldots$; the i-th parent of $n0$ being defined as the i-th element of the rootpath of $n0$. The 0-th parent of $n0$ is $n0$ itself, and the 0-th order parent family of $n0$ is the singleton $\{n0\}$.

Note that the i-th order *parent family* of a node $n0$ in a tree is a set consisting of the i-th order *parent* of $n0$ in the tree and of all the i-th and higher-order aunts of $n0$ (siblings of the i-th and higher order parents); but it does not contain the (i+1)-st and higher-order parents of $n0$. The purpose of this definition is to ensure that the i-th order parent family of any node of the tree, $i = 1, 2, \ldots,$ always represents a partition of the root of the tree.

The i-th order parent family of a node in a membership structure is defined to be identical with the parent family of this node in the originator tree (see requirement 1). The elements of the i-th order parent family of $n0$ are called the potential i-th order membership parents of $n0$. The i-th order parent of $n0$ in the originator tree plays a special role in the i-th order parent family and is called the i-th order 'tree parent' of $n0$.

It is important to notice that although it is easy to determine the first and higher-order parent families of a node from a glance at a pure tree or a membership structure with SIM 'is a' pointers, this is not the case when all the RR 'is a' pointers have been added to the structure. The parent families should therefore always be referred to the originator tree. (See also def. 6.3.)

end def. 6.1

Definition 6.2 The structure of the membership modification of a tree with its potential additional 'is a' and 'may be a' pointers. A membership modification of a pure tree is a structure consisting of the originator tree. In addition, the structure has potential 'is a' pointers from any element of the i-th order parent family of a leaf u^{ex} (or u) of the tree to any element of the (i+1)-st order parent family of that leaf, i=0,1,. . . . These additional pointers are 'potential' in the sense that their numerical values may be 0 in specific cases. Note that this additional set of arcs always includes potential 'is a' pointers from every i-th order tree parent n of u^{ex} to any element of the parent family of n *in the originator tree.*

From a membership structure with potential 'upward' or 'is a' pointers, we can always derive a corresponding structure with potential 'downwards' or 'may be a' pointers by reversing the direction of all the pointers. (For the numerical values of the 'is a' and 'may be a' pointers, see defs. 6.4, 6.6 and subsections 6.3, 6.4.)

end def. 6.2

Definition 6.3 of a membership structure with a linguistic-duplicate lowest node level (LDL) and of the parent and parent-family of the nodes of this level. It is often expedient to add an extra lowest node level to the membership structure, this level being a linguistic copy of the leaves of the originator tree. As in every membership structure (see requirements 2, 3 below), the labels of the lowest level are assumed to be assigned in an exact experiment, those of the next higher level in the presence of fuzziness #1, 2 or 3. Since there is no duplication of nodes in the originator tree, we must define separately the parent family of the additional nodes. We will denote one of these lowest nodes by u_i^{ex} (also when it has a nonnumerical label, see example below), and its linguistic duplicate in the next higher level by u_i. The set of nodes in each of the two levels is denoted by U^{ex} and U respectively.

The first-order parent family of any u_i^{ex} node is defined to be U, the set of nodes in the duplicate, next higher-level. u_i is considered to be the tree parent of u_i^{ex} for the purpose of constructing higher-order parent families of u_i^{ex} (see def. 6.1).

end def. 6.3

Figs. 6.1, 6.2 illustrate membership structures with a linguistic-duplicate lowest node level. The nodes of the two lowest levels of figs. 6.1a-c have labels designating intervals in the numerical attribute universe. Those of fig. 6.2 have nonnumerical linguistic labels. In principle, there is, however, no deeper-going difference between these two cases. According to the TEE model, nonnumerical labels of the two levels correspond, in the subject's mind, to intervals in the estimated and exact attribute universes respectively. (When the attribute universes are multidimensional, then these 'intervals' are hypervolumes.)

The next two requirements concern the precise meanings of the nodes of a membership structure.

Requirement 2. *The non-instance leaves of a membership structure must be exact, possibly multidimensional, attribute values* u^{ex} of the instances (more precisely, the u^{ex} values represent intervals in the universe U^{ex}). These leaves are called the 'exact nodes' of the membership structure. Instances can be hung on the exact nodes. Since every instance has a unique exact attribute value (assumed to be measured by the experimenter), it follows that each instance has only a single parent so that there is no fuzziness involved in the arcs connecting the instances with the exact nodes.

In this connection we allow a wide definition of the concept of an exact attribute value u^{ex} . The label of a node represents an exact attribute value when it refers to an LB experiment that always results, for all instances, either in the assignment to a given instance of this label, no matter how often the experiment is repeated; or to the assignment to the instance of the label another node belonging to the generalized family to which the experiment refers, no matter how often the experiment is repeated. An exact attribute value can be an interval in the universe U^{ex} of exact numerical attribute values. It can also be a linguistic value, such as 'virus' and the other leaves in fig. 4.3, provided that the assignment of these labels always satisfies the above condition for an exact value.

Each of figs. 6.1, 6.2 show membership structures in which the label sets of the two lowest, non-instance node levels are identical. However, the labels of the lowest levels are assumed to be assigned by the experimenter in an exact experiment (or by the subject S in the case of an exact LB experiment). The labels of the next higher node level are assumed to be assigned by S in a nonexact LB experiment. In figs. 6.1a-c, these exact and nonexact nodes have labels which are values of numerical intervals in U^{ex} and U respectively. The values of the 'is a' pointers between these two levels are equal to the error function $P(u|u^{ex})$. (The real error function of an LB or YN experiment; or the error function estimated by the subject in a MU experiment.)

Requirement 3. *A membership structure contains exact and nonexact nodes.* All the nodes of a membership structure except the root and the exact nodes described under requirement 2 are nonexact nodes. The assignment of the label of a nonexact node must be performed in a nonexact LB experiment in which fuzziness #1, 2 or 3 are present.

Definition 6.4. The meaning of the $\mu_{nk}(u^{ex})$ membership values of the nodes nk, of $P(nk|u^{ex})$, of $P(u^{ex}|nk)$, and of the values of the lowest 'is a' pointers. Let nk be an element of $\{nk_{ik}\}$, the first or higher-order, parent family of one of the exact nodes u^{ex} of the membership structure. The semantic experiments described in the sequel use this family as their reference label set (see Hisdal 1987a, part 1.3).

To define the membership value of u^{ex} in nk, we define first the conditional probability $P(nk|u^{ex})$. This quantity is measured in an LB experiment performed on one or more instances with the exact attribute value u^{ex}, the u^{ex} value being measured by the experimenter. In contrast to the pure-tree case of requirement 1, the classification is repeated many times in the presence of fuzziness #1, 2 or 3. For each classification we have a random reassignment of the instance to the points of the set pertaining to the source of fuzziness. E.g., for fuzziness #1, the instance is reassigned at random to a point of the set of conditions of observation, resulting in a probability distribution $P(u|u^{ex})$ for the estimated atribute value of an instance whose exact atribute value is u^{ex}. For fuzziness #3, the classification of the instance with attribute value u^{ex} is performed by randomly chosen subjects from the population of subjects (see Hisdal 1986b, fuzziness #3). The conditional probability $P(nk|u^{ex})$ is equal to the number of experiments that result in the assignment of the label nk divided by the total number of experiments. The $\mu_{nk}(u^{ex})$ membership value of u^{ex} in nk, as found from an exact MU experiment performed on an instance with attribute value u^{ex}, is interpreted as the subject's estimate of $P(nk|u^{ex})$, assuming the presence of fuzziness. In the sequel we shall assume that S's estimate is correct so that we can write,

$$\mu_{nk}(u^{ex}) = P(nk \mid u^{ex}) \ . \tag{6.1}$$

The left hand side of (6.1) represents the basic membership concept of fuzzy set theory. The right hand side is a well defined quantity computed, according to the TEE model (see Hisdal (1987b, eqns (5) and (9); (1987a, part 1.4)) from the formula

$$\mu_{nk}(u^{ex}) = \sum_{u \in \Delta u_{nk}} P(u|u^{ex}) \ , \tag{6.2}$$

where $P(u|u^{ex})$ is the error function in connection with fuzziness #1. The sum is taken over all values of u that lie in the subject's quantization interval Δu_{nk} for nk in an LB experiment. (For the equation that corresponds to (6.2) in the case of fuzziness #2 and 3, see Hisdal 1986b, sources of fuzziness #2a, 3a. Concerning nonexact MU experiments, see Hisdal 1986b, pp. 123, 134.)

$P(u^{ex}|nk)$ is the 'nk-qualified probability' that a randomly chosen object which has been assigned the label nk in the LB experiment has the exact attribute value u^{ex}.

When nk belongs to the first-order parent family of u^{ex}, then the numerical value of the 'is a' pointer from u^{ex} to nk (see def. 6.2) is defined to be equal to $P(nk|u^{ex}) = \mu_{nk}(u^{ex})$. In the case of an LDL membership structure (def. 6.3 and fig. 6.1), this quantity is also equal to the error function $P(u|u^{ex})$ in connection with fuzziness #1.

The value of the 'may be a' pointer from nk to u^{ex} is defined to be equal to $P(u^{ex}|nk)$.

end def. 6.4

It is easy to show that the value of $P(u^{ex}|nk)$ is equal to the normalized product of $P(nk|u^{ex}) = \mu_{nk}(u^{ex})$ and the prior distribution $P(u^{ex})$, considered as functions of u^{ex}. (Hisdal 1984a, def. 4.6; 1987a, part 1.4.).

In order to have a fixed terminology, we now set up a list of the items which define a membership structure.

Definition 6.5 of the items defining a membership structure. A membership structure is defined as being given by the following three items.

1. The originator tree.
2. The grades of membership of each of its nonexact nodes as functions of the exact, lowest nodes.

3. A statement as to whether a linguistic duplicate lowest node level, see def. 6.2, has been added to the originator tree.

end def. 6.5

Items 1 and 3 determine the potential 'is a' pointers according to defs. 6.1-6.3, although not their values. We shall see that item 2 determines the numerical values of the 'is a' pointers for SIM reference. For RR reference it determines these values as a function of u^{ex}.

Definition 6.6. The meaning of $P(nk|nj)$**, of** $\mu_{nk}(nj)$ **and of the values of the higher 'is a' pointers,** where $nk \in \Lambda k = \{nk_{lk}\}$ and $nj \in \Lambda j = \{nj_{lj}\}$ are two nonexact nodes of the membership structure, and Λk and Λj are generalized families of the originator tree. $P(nk|nj)$ is defined as the probability that an instance that has been labeled $nj \in \Lambda j = \{nj_{lj}\}$ in one nonexact LB experiment will be labeled $nk \in \Lambda k = \{nk_{lk}\}$ in another. When we refer to a subject's estimate of this probability, then we can also call it "S's subjective grade of membership of the nonexact class nj in the nonexact class nk",

$$\mu_{nk}(nj) = P^{est}(nk \mid nj) . \tag{6.3}$$

In Hisdal (1986a, p. 107) we identify this quantity with Zadeh's possibility measure for 'X is nk' given that 'X is nj' (Zadeh, 1978, eqn (2.16)).

When the $P(nk|nj)$ probabilities refer only to objects of a given value of u^{ex}, then we denote them by $P(nk|(nj, u^{ex}))$.

In subsections 6.3, 6.4 we show that both $P(nk|nj)$ and $P(nk|(nj, u^{ex}))$ are not uniquely defined, but depend on the SIM versus RR reference of the assignment of nk and nj. Furthermore, the expected $P(nk|nj)$ depends on the prior distribution $P(u^{ex})$ in the RR case.

When nj and nk belong to the i-th and $(i+1)$-st parent families of u^{ex} respectively, $i = 1, 2, \ldots$, then $P(nk|(nj, u^{ex}))$ is identified with the numerical value of the 'is a' pointer from nj to nk; and $P(nj|(nk, u^{ex}))$ is identified with the numerical value of the 'may be a' pointer from nk to nj. (See def. 6.2. Concerning the dependence of these pointers on u^{ex}, see eqns (6.7a), (6.7b) and subsections 6.3, 6.4.)

end def. 6.6

Requirement 4. Let $\{nk_{lk}\}$, $lk = 1, \ldots, Lk$ be a first or higher-order parent family of one of the exact nodes u^{ex}. From defs. 6.4, 6.6 it follows that the following equations (6.4)-(6.7a) must be satisfied for the 'is a' pointers of every membership structure:

$$\mu_{nk}(u^{ex}) = P(nk \mid u^{ex}) \in [0, 1] , \tag{6.4}$$

$$\sum_{lk=1}^{Lk} \mu_{nk_{lk}}(u^{ex}) = \sum_{lk=1}^{Lk} P(nk_{lk} \mid u^{ex}) = 1 , \tag{6.5}$$

$$\mu_{root}(u^{ex}) = P(\text{root} \mid u^{ex}) = 1 , \tag{6.6}$$

$\mu_{nk}(u^{ex})$ = sum over all possible upwards paths from u^{ex} to nk of the product of the $P(\,[n(j+1)] \mid (nj, u^{ex})\,)$ 'is a' pointers along each path. (6.7*a*)

Equations analogous to (6.4), (6.5) hold for the downwards pointers of def. 6.4.

In (6.7a), the nodes nj, $[n(j+1)]$ and nk belong to the j-th, $(j+1)$-st, and k-th order parent family of u^{ex} respectively, $k > j$. The first nj node of each path is $nj = u^{ex}$, and the last one is $nj = [n(k-1)]$. The first factor of each term on the right hand side of (6.7a) is thus always equal to $P(\ [n(j+1)] \mid (u^{ex}, u^{ex})\ ) = P(\ [n(j+1)] \mid (u^{ex})\ ) = \mu_{[n(j+1)]}(u^{ex})$ = membership value of the node $[n(j+1)]$ belonging to the lowest-but-one level.

The values of the other factors of (6.7a) (corresponding to the higher-up arcs) depend on whether they refer to a SIM or an RR situation (see subsect. 6.3, 6.4). We show in subsection (6.3) that their values are 1 for SIM reference when $[n(j+1)]$ is the tree parent of nj and 0 otherwise. Consequently the values of the SIM 'is a' pointers of a membership structure , def. 6.5, are unique, and $P(\ [n(j+1)] \mid (nj, u^{ex})\ )$ can be replaced by $P(\ [n(j+1)] \mid nj\ )$ in (6.7a). Figs. 6.1a and 4.2(b) illustrate the SIM case.

In the RR case (subsection 6.4) we have that $P(\ [n(j+1)] \mid (nj, u^{ex})\ )$ depends only on u^{ex}, not on nj. It can therefore be replaced by $P(\ [n(j+1)] \mid u^{ex}\ ) = \mu_{[n(j+1)]}(u^{ex})$. Eqn (6.7a) therefore reduces in the RR case to

$$\mu_{nk}(u^{ex}) = \text{sum over all possible upwards paths from } u^{ex} \text{ to } nk \text{ of the product of the membership values of the nodes along each path (including the top node } nk)\ . \qquad (6.7b)$$

Eqn (6.7b), taken by itself, has no connection with SIM or RR reference, and holds for every membership structure because of (6.5). The latter equation follows from the consideration that the subject who classifies the instance in an LB experiment assigns exactly one element of the nonredundant label set of the experiment; this label set being a generalized family of the tree. The RR case is illustrated in fig. 4.2(c) and 6.1b,c.

Finally we define the abbreviation of a tree and of a membership structure. This definition, together with eqns (6.7a), (6.7b), enables us to identify every $P(nk|u^{ex}) = \mu_{nk}(u^{ex})$ of def.. 6.4 with the value of the 'is a' pointer from u^{ex} to nk in the abreviated structure in which nk is a first-order parent of u^{ex}.

Definition 6.7 Abbreviations of a tree and a membership structure. Consider a tree T0, and another tree T1 obtained from TO by deleting the child family of any one of the nodes of T0 except those of the lowest-but-one level. In T1, the parent of the deleted family is connected downwards to its grandchildren in T0.

Let M0 and M1 be the membership structures constructed from T0 and T1 respectively according to def. 6.2. Then M1 is called a first order abbreviation of M0. An (i+1)-st order abbreviation of T0 and M0, $i = 1, 2, \ldots$ is obtained by abbreviating an i-th order abbreviation.

Fig. 4.2(a) illustrates an abbreviation of the membership structures 4.2(b), 4.2(c).

The following is a summary of those points of subsection 6.2 which have not been mentioned in subsection 6.1.

The lowest nodes of a membership structure refer always to exact LB (labeling) experiments, the other nodes refer to LB experiments performed in the presence of fuzziness.

Each node nk of a membership structure (except the lowest ones) has a unique membership function of the exact nodes u^{ex} attached to it. This function is a subjective one for a given subject S. It is interpreted as $\mu_{nk}(u^{ex})$ or $P(nk|u^{ex})$; the μ value being the subject's estimate of $P(nk|u^{ex})$ when the LB experiment concerning $\{nk_{lk}\}$ is performed under conditions of fuzziness.

Let the node nj of a membership structure belong to the j-th order parent family, $j = 0, 1 \ldots,$ of an exact node u^{ex}. Then the potential 'is a' pointers from nj point to all the elements nk of the $(j+1)$-st order parent family of u^{ex}. The numerical value of an 'is a' pointer from nj to nk is interpreted as $\mu_{nk}(nj)$ or $P(nk|nj)$, see defs. 6.4, 6.6. When nj is a nonexact node, then both $nj \in \{nj_{lj}\}$ and $nk \in \{nk_{lk}\}$ refer to two LB experiments performed in the presence of fuzziness.

The membership value $\mu_{nk}(u^{ex})$ of any nonexact node nk of a membership structure can be obtained by multiplying the values of the 'is a' pointers along an upwards path from the leaf u^{ex} to nk, and summing over the different possible upwards paths from u^{ex} to nk. (eqn (6.7a), (6.7b)). These paths must always go from an element of an i-th order parent family of u^{ex} to an element of its $(i+1)$-st order parent family (see figs. 4.2, 6.1a-c and caption to fig. 4.2). The 'summing-up to MU formula' does not hold for the *expected* RR 'is a' pointers of fig 6.1c and eqns (6.12), (6.13).

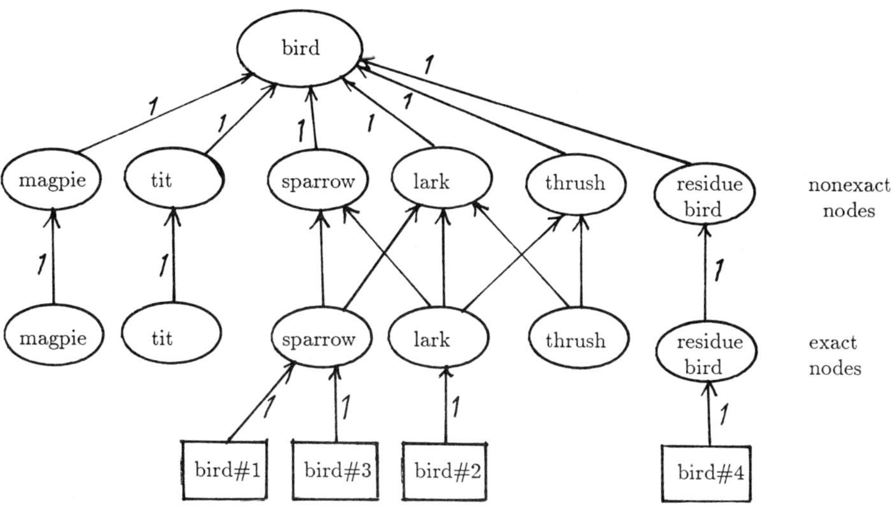

Fig. 6.2. A membership structure for the classification of birds. The lowest oval nodes are exact ones. They are assigned by the experimenter in an exact experiment in which the different attribute values which determine the species of the bird are measured exactly. The assignment of the label of a nonexact node is performed by the subject in a nonexact LB (labeling) experiment, or in a MU (grade of membership) experiment in which he specifies the grade of membership of a given bird (rectangular nodes) in the class whose label is that of the nonexact node (see caption to fig. 6.1). The pointers are valid both for SIM and for RR reference because we have only one level of nonexact nodes.

6.3 SIM Pointers

In subsection 6.2 we showed that the 'is a' pointer from an exact leaf u^{ex} of a memberhsip structure to a node nk belonging to the next higher level has a unique numerical value equal to $\mu_{nk}(u^{ex})$, the grade of membership of the node to which it points.

This uniqueness does not hold for the higher 'is a' pointers from a *nonexact* node nj to a nonexact node nk. For a given membership structure, as defined in def. 6.5, there exist two main categories of 'is a' and of 'may be a' pointers referring to SIM and RR situations respectively. The numerical values of the 'is a' pointers referring to these two situations are generally different.

To define the meaning of SIM and RR situations, we start by defining a composite experiment.

Definition 6.8. of a composite experiment for the assignment of the labels of the two nodes $nj \in \Lambda j$ and $nk \in \Lambda k$ to an object. Let $\Lambda j = \{nj_{l_j}\}$, $\Lambda k = \{nk_{l_k}\}$ be two generalized families (or complete, nonredundant label sets in the terminology of the TEE-model papers) of the originator tree of the membership structure. And let $\exp j$ be an LB experiment referring to the label set Λj and $\exp k$ an LB experiment referring to the label set Λk. Both experiments are performed on the same set of objects. The two experiments considered as a whole are called a composite LB experiment with components $\exp j$ and $\exp k$ respectively.

Definition 6.9 of a SIM composite experiment. A composite experiment, def. 6.8, is called a SIM one in connection with fuzziness #1 when the subject's estimate u of the attribute value of each of the objects is the same in $\exp j$ and in $\exp k$. The given object is then observed SIMultaneously in connection with both experiments, i.e. under the same point of the set of conditions of observation (see Hisdal 1986b, p. 118 concerning this set). In connection with fuzziness #3, the composite experiment is called a SIM one when $\exp j$ is performed on a given object by the same subject as $\exp k$.

We now wish to find the numerical values of the 'is a' and 'may be a' pointers of a membership structure referring to a given originator tree, see def. 6.5. We show below that the following theorem holds for these pointers.

Theorem 6.3.1 concerning the 'is a' and 'may be a'pointers of a membership structure for SIM reference. Let a SIM composite experiment (defs. 6.8, 6.9) with components $\exp j$, $\exp k$ be performed on a set of objects. $\exp j$, $\exp k$ are LB experiments referring to the generalized families $\Lambda j = \{nj_{l_j}\}$, $\Lambda k = \{nk_{l_k}\}$ of the originator tree. We require that the elements of these two families are nonexact nodes in the membership structure. The theorem refers to two cases:

Case (a). nk is the first or a higher-order tree parent of nj.

Case (b). nk is not the first or a higher-order tree parent of nj, and nj is not the first or a higher-order tree parent of nk.

The theorem says that the following eqns (6.8), (6.9), (6.10) hold. All of them can be applied to the special case that Λj, Λk are the i-th and $(i+1)$-st parent families of u^{ex} respectively, $i = 1, 2, \ldots$. Eqn (6.8) then applies to the 'is a' pointer from nj to nk, and eqns (6.9), (6.10) to the 'may be a' pointer from nk to nj.

The 'is a' equation:

$$P(nk \mid (nj, u^{ex})\ ) = P(nk \mid nj) = \begin{cases} 1, & \text{in case (a);} \\ 0, & \text{in case (b).} \end{cases} \tag{6.8}$$

The 'may be a' equation for objects of a given u^{ex} :

$$P(nj \mid (nk, u^{ex})\ ) = \begin{cases} P(nj \mid u^{ex})/P(nk \mid u^{ex}) = \mu_{nj}(u^{ex})/\mu_{nk}(u^{ex}), & \text{in case (a);} \\ 0, & \text{in case (b).} \end{cases} \tag{6.9}$$

The expected 'may be a' equation:

$$P(nj|\ nk) = \begin{cases} P(nj)/P(nk) = \sum_{u^{ex}} P(u^{ex})\mu_{nj}(u^{ex})/\sum_{u^{ex}} P(u^{ex})\mu_{nk}(u^{ex}) & \text{in case (a);} \\ 0, & \text{in case (b).} \end{cases} \tag{6.10}$$

The proof of these equations is as follows. In all cases (a) we have that the relation $\Delta u_{nj} \in \Delta u_{nk}$ holds for the quantization intervals of the two labels. Consequently, an object which is labeled nj in exp j , is also labeled nk in exp k for SIM reference, def. 6.9. This proves eqn (6.8), case (a).

Eqn (6.9a) follows from the consideration that the event in $\Lambda j \times \Lambda k$ [that an object of attribute value u^{ex} is assigned the label nj in exp j] is a subset of the event [that the object is assigned the label nk in exp k]. The same holds for (6.10a), except that we have removed the qualification that the object must have the attribute value u^{ex} .

The expression after the second equality sign in (6.9a) follows from def. 6.4; and that after the second equality sign in (6.10a) is obtained by replacing $P(nj)$ by $\sum_{u^{ex}} P(u^{ex})P(nj|u^{ex})$, and substituting $\mu_{nj}(u^{ex})$ for $P(nj|u^{ex})$. An analogous expression is substituted for $P(nk)$.

In all cases (b) the quantization intervals Δu_{nj} , Δu_{nk} are disjoint. An object can therefore never be labeled nj in exp j and nk in exp k for a SIM reference situation. We have therefore a '0' value for both the 'is a' and the 'may be a' pointers.

Finally we prove the expression after the first equality sign in (6.8). Since the conditional probability $P(nk \mid (nj, u^{ex})\ )$ is always 1 in case (a) and always 0 in case (b), it follows that it is independent of the value of u^{ex} and can therefore be written as $P(nk|nj)$.

The following is a summary of subsection 6.3. For pointers with SIM reference, there exists only a single upwards (is a) pointer from any of the nonexact nodes of a membership structure, namely that of the originator tree. Its value is 1. By reversing the direction of these pointers, we obtain the only potentially nonzero-valued downwards (may be a) pointers. The values of the downward pointers are the same as their values in the originator tree. They depend on the u^{ex} value of the object, see eqn (6.9). Their expectation over objects of any u^{ex} depends on $P(u^{ex})$.

The only exception to the pure tree structure occurs for the pointers between the exact leaves and the next higher level (see def. 6.4 and figs. 6.1a and 6.2). The exact nodes are, in general, fuzzily classified, i.e., they have more than one parent.

6.4 RR Pointers

We show in this subsection that the 'is a' and 'may be a' pointers of a membership structure are, in general, much more numerous for RR reference than for SIM reference because, in contrast to the SIM case, also the nonexact nodes may have more than one parent. A further difference between the two cases is that not only the numerical values of the 'may be a' pointers, but also those of the 'is a' pointers depend on the u^{ex} value of the object in the RR case. These values can be averaged over u^{ex}.

We start with the definition of an RR experiment.

Definition 6.10 of an RR experiment. A composite experiment, def. 6.8, is of the RR type in connection with fuzziness #1 when each object is ReRandomized with respect to the points of the set of conditions of observation between the performance of exp j and exp k. Consequently, the subject's estimate u of the object's attribute value is, in general, not the same in the two component experiments. In connection with fuzziness #3, the experiment is of the RR type if the component experiments are performed on a given object by two, generally different subjects chosen at random from the set of subjects. The two subjects will then generally have slightly different quantization intervals for the same label.

Theorem 6.4.1 concerning the 'is a' and 'may be a'pointers of a membership structure for RR reference. Let an RR composite experiment (defs. 6.8, 6.10) with components exp j, exp k be performed on a set of objects. exp j, exp k are LB experiments referring to the generalized families $\Lambda j = \{nj_{l_j}\}$, $\Lambda k = \{nk_{l_k}\}$ of the originator tree. We require that the elements of these two families are nonexact nodes in the membership structure.

The theorem says that the following eqns (6.11)-(6.14) hold. All of them can be applied to the special case that Λj, Λk are the i-th and $(i+1)$-st parent families of u^{ex} respectively, $i = 1, 2, \ldots$. Eqn (6.11) then represents the value of the 'is a' pointer from 'nj' to 'nk' for an object of exact attribute value u^{ex}, and eqns (6.12), (6.13) the expection over all objects of the value of the 'is a' pointer. The corresponding values of the 'may be a' pointer from nk to nj are obtained by exchanging nj and nk in eqns (6.11)-(6.14).

$$P(nk \mid (nj, u^{ex})\ ) = P(nk \mid u^{ex}) = \mu_{nk}(u^{ex})\ , \tag{6.11}$$

$$P(nk \mid nj) = \sum_{u^{ex}} P(u^{ex}|nj)\ P(nk|u^{ex})\ , \tag{6.12}$$

or equivalently

$$\begin{aligned} P(nk \mid nj) &= [1/P(nj)] \sum_{u^{ex}} P(u^{ex})\ P(nj|u^{ex})\ P(nk|u^{ex}) \\ &= [1/P(nj)] \sum_{u^{ex}} P(u^{ex})\ \mu_{nj}(u^{ex})\ \mu_{nk}(u^{ex})\ , \end{aligned} \tag{6.13}$$

where

$$P(nj) = \sum_{u^{ex}} P(u^{ex})\ P(nj|u^{ex}) = \sum_{u^{ex}} P(u^{ex})\ \mu_{nj}(u^{ex}) \tag{6.14}$$

is the 'probability of the fuzzy event nj in Zadeh's (1968) terminology.

The reason for the first equality sign in (6.11) is that when the u^{ex} value of the object is given, then the choice of label in one component experiment is not influenced by the choice of label in the other one due to the rerandomizing of the object with respect to conditions of observation.

Eqns (6.12), (6.13) are practically self-evident. They are discussed in more detail in Hisdal (1984, eqns (11.123), (11.124)).

In the following we present a qualitative discussion of pointers with RR reference, and of the implications of theorem 6.4.1.

Let us start with case (b) of theorem 6.3.1 in which the quantization intervals Δu_{nj}, Δu_{nk} are disjoint. Because of the rerandomizing of the object with respect to the condition of observation, it may happen in an RR situation that the estimate u of the attribute value of the object falls into Δu_{nj} in exp j and into Δu_{nk} in exp k in spite of the disjointness. (This can happen when the distance between the two nearest points of the two quantization intervals is smaller than the size of the u-interval for which $P(u|u^{ex}) > 0$ for a given u^{ex}.) Consequently, an object that is labeled nj in exp j may be labeled nk in exp k. In contrast to the SIM pointers, the RR 'is a' pointers can thus be nonzero in case (b).

In case (a), when $\Delta u_{nj} \subset \Delta u_{nk}$, it may happen that the estimated attribute value u of the object falls into Δu_{nj} in exp j, and that the new u value of exp k falls outside Δu_{nk} in an RR situation. Consequently the 'is a' pointer from nj to its tree parent may have a value that is smaller than 1. Furthermore it may happen that the i-th and $(i+1)$-st order parent families of an exact node u^{ex} contain one or more common elements (see fig. 4.2(c)). Consequently the nodes at the tail and at the head of an 'is a' pointer may sometimes be identical. However, $P(nk = nj \mid (nj, u^{ex})\ )$ is, in general, smaller than 1 for RR reference because the estimated u value of an object may fall into Δu_{nj} in exp j, and outside this interval in exp k.

The wider the $P(u|u^{ex})$ error curve (for a given u^{ex}), the bigger is the deviation of the numerical values of the RR 'is a' pointers from the 0 and 1 values of the SIM 'is a' pointers an cases (b) and (a) respectively.

It follows from eqn (6.11) that for a given u^{ex}, the values of both the RR 'is a', and the RR 'may be a' pointers are simply equal to the grade of membership of the node at the head of the pointer for that u^{ex}. Thus all the RR pointers that point towards a given node n have the same numerical value $\mu_n(u^{ex})$. Consequently the values of the RR 'is a' and 'may be a' pointers depend on the u^{ex} value of the object, and there exists, in principle, one set of values of the 'is a' pointers, and one set of values of the 'may be a' pointers for each exact node u^{ex}. However, there also exist many unique 'is a' pointers for a membership structure with RR reference. Those from the exact nodes are unique because they start at u^{ex}. Furthermore, the 'is a' pointers towards the root have all the value 1, independent of the value of u^{ex}. Fig. 6.2 shows an example in which the membership structure contains only these pointers which are thus valid for SIM reference, as well as for RR reference independent of the value of u^{ex}. In general, the upwards 'summing up to MU' eqn (6.7a) holds for the u^{ex}-dependent RR 'is a' pointers.

As an example of the dependence on u^{ex} of the existence or nonexistence of a given 'is a' pointer, consider the two exact leaves 'virus' and 'cottage' in fig. 2.1. For u^{ex} =virus, there exists an 'is a' pointer from 'non-organism' to 'organism' because these two nodes belong to the parent and the grandparent family of 'virus' respectively. No such 'is a' pointer exists for u^{ex} =cottage because there exist no i-th and $(i+1)$-st order parent families of 'cottage' which contain 'non-organism' and 'organism' respectively as elements.

Eqns (6.12)-(6.14) for the expected $P(nk|nj)$ values are generally valid for two nodes belonging to two, given, generalized partitions of the tree. If they are to be applied to an 'is a' pointer from nj to nk, or to a 'may be a' pointer from nk to nj, then nj and nk must belong to an i-th and $(i+1)$-st order parent family of u^{ex} respectively. As we have just seen, this belongingness may depend on u^{ex}. In connection with the 'is a' and 'may be a' pointers, the summation in eqns (6.12)-(6.14) must therefore be performed only over those values of u^{ex} for which nj and nk belong to an i-th and $(i+1)$-st order parent family of u^{ex} respectively. The 'summing up to MU' formula (6.7a) does not hold for the expected 'is a' pointers of (6.12), (6.13).

The expected RR 'is a' and 'may be a' pointers (for objects of any u^{ex}) depend on the prior distribution $P(u^{ex})$. This result for the 'is a' pointers is in contrast to the SIM case. The prior distribution $P(u^{ex})$ is discussed in the next subsection.

6.5 The Prior or Unqualified Distribution $P(u^{ex})$

The probability distribution $P(u^{ex})$ in the equations of subsections 6.2, 6.3 is called the 'prior' or 'unqualified' probability distribution over U^{ex} for the set of instances to which the semantic structure refers. It is 'unqualified' by a label such as nk; in contrast to the 'qualified' distribution over U^{ex} of def. 6.4 et seq. .

We have seen that the $\mu_n(u^{ex})$ membership values are independent of the prior distribution. So are the values of the 'is a' pointers of the membership structure for SIM reference.

However, the values of the expected 'is a' pointers for RR reference, the values of the expected 'may be a' pointers for both SIM and RR reference, and the values of the qualified probabilities $P(u^{ex}|nk)$ (these being the lowest 'may be a' pointers) all depend on $P(u^{ex})$.

The unqualified distribution $P(u^{ex})$ is a natural function for a structure such as that depicted in fig. 6.1. It signifies the probability distribution over height for the set of women to which the structure is applied.

As another example of a prior distribution, suppose that you are a bird-lover, and observe all the birds that sit on a particular tree of your garden at a particular time of the year (see fig. 6.2). Your semantic structure has the root 'bird'. The children of 'bird' are the fuzzy nodes of the structure whose labels are different types of bird, such as 'sparrow'. The lowest non-instance nodes are the exact bird classes as assigned by the experimenter in an exact experiment.

Also in this case $P(u^{ex})$ has a straightforward meaning. It is the probability that a randomly observed bird on your tree is a sparrow, magpie, etc. .

For the tree of fig. 4.3, the assumption of a prior distribution seems less natural. We would, e.g., have to specify the probability that a randomly chosen instance is a dog, versus the probability that it is a virus. But without assuming some prior distribution, we cannot specify the numerical values of the expected 'may be a' pointers, of the expected RR 'is a' pointers, and of the qualified probabilities of def. 6.4.

However, this way of presenting the problem is extremely hypothetical. In a real-world case we will almost always have additional information about our physical object; e.g., information concerning its size. This information will rule out 'dog' and many other leaves of the membership structure when we apply it to a physical object observed under a microscope. Consequently, the set of acceptable $P(u^{ex})$ distributions is drastically narrowed down. Only those distributions are left for which $P(u^{ex}) = 0$ for all leaves whose size is macroscopic. To find $P(\text{virus}|\text{physical object})$ (i.e. the probability that our microscopic object is a virus), assuming that the 'is a' pointers of fig. 4.3 are given, we would, however, still have to specify the prior distribution for 'organism' versus 'non-organism', and for the leaves of fig. 4.3 that have not been ruled out.

What we *can* do in this case is to specify the values of the 'may be a' pointers (and of the RR 'is a' pointers) under the condition that a specific prior distribution holds. Alternatively, we can specify or estimate a probability value for a given prior distribution; implying the same probability value for the resulting set of values for the 'may be a' pointers etc. . A third possibility is to assume a uniform prior distribution over the acceptable u^{ex} nodes, and to find the set of numerical values derived under this assumption.

The assumption of a uniform prior distribution for the purpose of computing probabilities such as those mentioned here is usually referred to as Bayes' postulate, and has been the subject of heated arguments (Fisher Box 1978, pp. 66-68, 70, 80). However, we do not really have to 'invoke Bayes' postulate'. We can simply say that our values for the 'may be a' pointers etc. are computed under the assumption of a uniform prior distribution over the pertinent interval in U^{ex} .

6.6. Conclusion

We conclude section 6 by noting that the $\mu_n(u^{ex})$ membership function of every nonexact node n of a membership structure is uniquely determined from the values of the 'is a' pointers through eqn (6.7a). The conditioning variable u^{ex} on the right hand side of this equation can be deleted for SIM reference. The converse statement does not hold. There exist two valid sets of 'is a' pointers when the $\mu_n(u^{ex})$ membership functions of all the nodes are given, one for SIM reference, and one for RR reference. The RR set may, in addition, depend on u^{ex} . Furthermore the expected RR 'is a' pointers, and the expected RR and SIM 'may be a' pointers, depend also on the prior distribution $P(u^{ex})$.

It is not necessary to store the numerical values of the 'is a' pointers in the knowledge base or lexcicon. We need only store the 'is a' connections of the originator tree. Def. 6.1 then allows us to construct an algorithm for finding the different order parent families of a leaf. These families determine the potential 'is a' pointers of the membership structure according to def. 6.2. In addition we must store the membership functions of each node in order to determine the numerical values of the RR 'is a' and 'may be a' pointers. For SIM reference the parent families algorithm is unnecessary because the 'is a' connections are the same as those of the originator tree except for the lowest pointers. The only membership functions that need to be stored are those of the nodes above the exact leaves. These data are sufficient to determine the membership functions of the remaining nodes and the values of the SIM is a pointers, also those of the abbreviated structure of def. 6.7.

7. The Storage Structure of the Lexicon Entries

7.1 Overview

In section 2.2 we already mentioned that the basic 'pretty print' structure of a lexicon entry is structurally equivalent to a tree. Each line of a 'pretty print' structure represents the name of one node of the tree. Its parent is the first line in the upwards direction whose margin is displaced one step to the left. The lexicon term itself is thus the root of the tree. In the sequel we will call a line of the 'pretty print' structure a 'node' of the entry tree or, as before, a subentry of the lexicon term.

In the next subsection we discuss shortly how we can represent this two dimensional structure succintly on a file in such a way that we can retrieve the structural information. E.g., when we ask 'What is a mammal?', we must be able to find the '(is a)' node of the < term > =mammal lexicon entry, and its child 'vertebrate' in the entry tree. This example illustrates also our remark in subsection 2.2 that the entry tree is completely different from the semantic one. In the above example, both 'is a', and 'mammal' and 'vertebrate' are nodes of the entry tree, and 'vertebrate' is a descendant of 'mammal'. In contrast, it is the parent of 'mammal' in the semantic tree of fig. 2.1.

There are many ways of representing a tree on a file. The main criterion for an aceptable representation is, of course, that we must be able to retrieve the structural information contained in the tree. In this connection we must distinguish between trees which may have nodes with identical names, and those which may not. If we allow two nodes with identical names, then we may still be able to find the correct node and its children, provided that the rootpath of the two identically named nodes are different, and predefined in the program. The device of predefining the rootpath for the purpose of finding a given node was chosen in the Alex2 system (Ohren, 1986).

However, a predefinition of the complete rootpath results in a system which is not flexible enough for our purposes. E.g., we may wish to be able to insert an optional node such as 'time' to indicate the time at which the information contained in some subtree was supplied. Furthermore, we may wish to store information supplied at a later time about the same items, without deleting the old information. Similarly we may wish to insert an extra node called 'informant' to store information supplied by different informants about the same items. The 'partition' and '(partion tree)' nodes of fig. 3.2 are another illustration of such optional nodes (compare with fig 2.2). All these examples serve to illustrate two difficulties. 1) The optional nodes modify the rootpath so that a rigidly predefined rootpath cannot be used in the procedures which specify how a particular node is to be located. 2) The tree has not only identically-named nodes, but their rootpaths consist in most cases of the same, or of identically-named elements. A specification of this path is therefore of no help for the purpose of locating a specific node.

In the present Alex3 system, we do allow nodes with identical names in some cases in which this ambiguity does not cause any difficulties. Such a case may occur for two nodes, both of which are the value of an attribute; e.g., '(195 cm)' or 'large' as a child of both 'length' and 'breadth'. This type of identically named nodes makes no difficulties in the retrieval of information because the program only looks for their parents 'length' and 'breadth'. The children of the latter can then be found irrespective of whether they have identical names or not.

In contrast we have the case of the entry structure of fig. 3.2 which has two '(is a)' nodes, one or both of which will be the object of a search when the system is asked 'What is a < term > ?'. Furthermore both of these nodes have the same rootpath. This difficulty is solved by the nested-tree structure of sect. 7.4. In this structure, the two '(is a)' nodes of fig. 3.2 do not belong to the outer-level tree. Instead they belong to two different inner-level trees, each of which is nested inside a leaf of the outer tree. An analogous nesting of subtrees holds for the upper two braces of fig. 3.2.

In summary, the Alex3 system solves the problem of finding a node by searching through the nodes of the entry tree. In those cases in which identical node names cause difficulties, the device of wrapping up alternative subtrees (each of which contains the ambiguous nodes once only) in different leaves of the outer tree is used. In addition, a requirement that the rootpath must contain one or more prespecified elements can be used for the purpose of locating an ambiguous node in those cases in which the rootpaths of the two identically-named nodes are different. Such a partial prespecification of the rootpath, possibly combined with the wrapping-up device, can also be used more generally for the purpose of improving the efficiency of the search.

In subsection 7.2 we discuss the storage and retrieval of information in pure entry trees. The storage and retrieval of the special information terms is discussed in subsection 7.3, and the nested-*tree* structure in 7.4.

7.2 The Structure of Pure Entry Trees

In the Alex3 system we have tried two storage structures for the entry trees, both of which work satisfactorily. We call the first of these the 'numeral structure, and the second the 'nested-*list*' structure. The latter has no connection with the 'nested-*tree*' structure of subsection 7.4.

The numeral structure is analogous to the so-called 'decimal system' often used for library catalogues. Since this structure has no connection with the number '10' (*decas*=ten in latin), we call it 'numeral structure' here. In this structure, every node n has a key k. k is a list of numbers (positive integers) separated by 'decimal points' or, in our system, by spaces. Thus the keys of the children of a node n with key k are $k.1$ $k.2$, etc. . And the children of, e.g., $k.2$ have keys $k.2.1$, $k.2.2$, ... The root of the tree (i.e., < term >) has the key 1. The numeral structure of the whole entry of < term > can be looked upon as a table of two columns, one column consisting of the keys, and the other of the corresponding nodenames. This table is stored as a list of pairs. E.g., a partial numeral tree representation of fig. 2.2 is,

(((1) vertebrate) ((1 1) (syntactic category)) ((1 2) IS)
((1 1 1) (count noun)) ((1 2 1) (may be a))) . (7.1)

The rootpath of a given node is easily found by successively deleting the last element of its key. Vice versa, the children of a node n with key k are found by searching for keys consisting of the key k with one additional element (number) inserted at the end of the key.

The keys of some of the nodes may be updated as information is added to a given lexicon entry. Consequently the keys must be used only for searches inside a given entry. There must be no references from one lexicon entry to the keys of another entry.

An alternative to the 'numeral' storage structure for a pure tree is the 'nested-list' one which is recursively defined. Each node of the tree is stored in the form (x) below, which is a two-element list (or a three-element list when the node has a special information term, see next subsection),

$$(x) = (<\text{node}> \quad (\text{nest})\ )\ . \tag{7.2}$$

The first element is the name of the node (which may be an atom, or a general LISP list). The second element, (nest), is a list of the children of < node >, where each child has again the complete structure (7.2), including the outer brackets. I.e., each child of < node > is represented by a two-element list, the first element being its name, and the second being a list of *its* children, where each child again contains its subtree in its nest. E.g., if < node > =1 has two children, 1.1 and 1.2 respectively (using nodenames which would be their keys in a numeral tree), then (nest) in eqn (7.2) is replaced by,

$$(\text{nest}) \Longrightarrow (\quad (1.1\ (\text{nest}1.1)\ )\ \ (1.2\ (\text{nest}1.2)\ )\quad), \tag{7.3}$$

where the two (nest)'s on the right hand side are replaced by new expressions in the next nesting step. When we reach a node which is a leaf, then its two-element list, (leaf ()), has the empty list as its second element, and the recursion stops.

This structure allows us to perform a depth-first or breadth-first search for a node of a given name. The parent and the child-family of a node are easily found from this structure also.

The single top-level element of a lexicon entry nested-list structure is always the name of the root, namely the term itself. The second element contains the whole subtree of the root. As an example, that part of the entry tree of fig. 2.2 whose numeral-tree structure is shown in eqn (7.1) has the following nested-tree structure,

$$\begin{array}{ll}(\text{ vertebrate } (& (\ (\text{syntactic category})\ (\quad (\ (\text{count noun})\ ()\ )\quad)\ ) \\ & (\ \text{IS}\ (\quad (\ (\text{may be a})\ (\ldots)\ )\quad)\ )\quad)\ )\ . \end{array} \tag{7.4}$$

The spaces between successive parenthesis have been put into (7.4) merely for visual clarity. In spite of this, the structure looks quite complicated. In the program it is, of course, created automatically via the information supply dialogue of the Alex system whose input is used by the algorithm which constructs the entry tree.

7.3 The Special Information Term

In sections 2 and 3 we had use for a 'special information term' for some of the subentries of a lexicon entry. The special information terms are enclosed in double square brackets in figs. 2.2 and 3.2. The brackets do not appear in this form in the storage structure, they are used only for purposes of visualization in the figures.

It is the stringent requirement of two list-elements for the representation of a node and its subtree (or nest) respectively (see eqn (7.2)) which allows us to add a special information term to any node. (7.2) is then replaced by

$$(x) = (< \text{node} > \quad (\text{nest}) \quad (\text{special info})\ )\ . \tag{7.5}$$

Whenever a nest has three outer-level elements in a nested-tree structure then the third element is identified as being a special information term. The meaning or signal which this term conveys is usually related to the node itself, i.e. to the first of the three elements. As an example, the nest associated with the node 'residue' in fig. 2.2 has the nested-list storage form,

$$(\quad \text{residue} \quad () \quad (\text{empty (yes)})\ \ ) \tag{7.6}$$

In a numeral-tree structure, special information terms are listed in a third column of the table; i.e. a pair of (7.1) is converted to a 3-element list when the node has a special information term.

7.4 Nested Entry Trees

Finally we come to the storage of entry structures with the groupings indicated by the braces in fig. 3.2. Two or more braces with a common parent (such as '(partition tree)' in fig. 3.2) represent alternative subtrees of that parent in the entry structure, just like the alternative subtrees of 'organism' in the multiply partitioned semantic structure of fig. 3.1. An entry structure with two or more neighboring braces is thus the structural analogue of a semantic structure with a multiply partitioned node.

That a multiply partitoned entry structure is used for the representation of information concerning the multiple partitoning of a node of a semantic structure (as in fig. 3.2) should, however, be considered largely as a coincidence. Multiply partitioned *entry* structures can also be used for the representation of a lexicon term with two or more meanings; or, as mentioned in subsection 7.1, for the representation of information supplied at different times or by different informants.

The storage principle for the groupings represented by the braces is simple. The parent of the groups, e.g. 'partition' for the first two braces in fig. 3.2, is supplied with a child family whose number of elements is equal to the number of groups or braces. Each of these children is a leaf in the outer tree. The name of this leaf is identical with the complete subtree or nest which 'IS' would have had if there had been only one group and a non-nested tree structure. Thus each leaf-child of 'partition' is a package in which a whole subtree (excluding its root 'partition') is wrapped up.

For example, if the second brace in fig. 3.2 had not been present, then the second child of $< \text{term} >$ would be

$$(\text{IS} \quad (\ (\text{child1}) \ \ (\text{child2})\ \ )\ \ )\ , \tag{7.7}$$

where

$$(\text{child1}) \Rightarrow (\quad (\text{may be a}) \ \ (\text{nest1}) \quad)\ , \tag{7.8}$$

$$(\text{child2}) \Rightarrow (\quad \text{attributes} \ (\text{nest2}) \quad) , \tag{7.9}$$

and the right hand sides of (7.8) and (7.9) refer to the top brace in fig. 3.2.

With the two top braces of fig. 3.2 we have instead of (7.7),

$$(\text{IS} \quad (\ (\text{partition} \ ((\text{newchild}^{1}) \ (\text{newchild}^{2})) \ (\text{true}) \ ) \ ) \quad) , \tag{7.10}$$

where 'true' is the special information term of 'partition' in fig. 3.2 which, together with the node-name 'partition', signals to the system that a multiply partitioned entry follows. (In addition it is a signal for the semantic analysis, see, e.g., subsection 3.2.)

The two 'newchild' nodes in (7.10) should be replaced by

$$(\text{newchild}^{1}) \Rightarrow (\quad (\ (\text{child1}^{1}) \ (\text{child2}^{1}) \ ) \ ) \ () \quad) \tag{7.11}$$

$$(\text{newchild}^{2}) \Rightarrow (\quad (\ (\text{child1}^{2}) \ (\text{child2}^{2}) \ ) \ ) \ () \quad) \tag{7.12}$$

where (7.11) and (7.12) refer to the first and second brace of fig. 3.2 respectively. E.g., (child1) and (child2) in (7.11) should be replaced by their expressions (7.8), (7.9).

The big difference between (7.7) and (7.11) or (7.12) is that in (7.7) the list ((child1) (child2)) occupies the place of a nest and thus signifies a sibling *family* of nodes, each of which starts its own subtree. In contrast, in (7.11) the whole list ((child1) (child2)) occupies the place of a node-name (see eqn (7.2)), signifying a single node. This node is a leaf in the outer tree as is indicated by the empty nest in (7.11). The subtree corresponding to the second brace is wrapped up in a different leaf as is shown by equations (7.10), (7.12).

The nested-*tree* structure is independent of whether the outer or inner pure trees are represented as a nested-*list* structure (as in the above examples) or as a numeral or some other tree structure. In the numeral representation of fig. 3.2, the leaf-children of 'partition' in the outer tree are now themselves subtrees represented by a two-column table of keys and node-names. I.e., 'partition' has two children, one for each brace in fig. 3.2. The two pairs of (7.2) whose keys belong to these children have a complete numeral structure (list of pairs) as their second element.

The reconstruction of one of the pure trees contained in a multiply partitioned entry structure (e.g., that of the first two braces of fig. 3.2) is performed by identifying the correct leaf of the outer tree (via the 'partiton' node) and replacing the nest of 'IS' in (7.10) by this subtree.

The following is a summary of the situation with respect to multiply partitoned entry trees (see the braces in fig. 3.2). A multiply partitioned tree structure can be represented as a nested tree in the sense that alternative subtrees are wrapped up in (used as node-names for) different leaves of the outer tree. These leaves are supplied with an extra common parent containing a special information term in order to signal to the program that a nested structure follows. Nested entry trees can be used to represent information concerning multiply partitioned nodes in a semantic tree (see fig. 3.2). They can also be used to represent lexicon terms with more than one meaning, and information supplied at different times, or by different informants, about the same items.

8. The Structures of the Alex System, and Some Observations Concerning Traditional Mathematical Logic versus AI Structures

8.1 Three Ways of Representing Trees

Seen from a formal, mathematical point of view, we have presented in this paper three possible ways of storing and retrieving a tree in a computer program. Namely the numeral and the nested-list structures of subsection 7.2, used as alternative representations for the entry tree; and the lexicon with its parent, '(is a)', and child, '(may be a)', pointers for each node of the semantic tree. The last structure requires more storage space because all non-leaf and non-root nodes are stored several times; once as the lexicon term itself, once as a '(may be a)' pointer of another term, and one or more times as an '(is a)' pointer of other terms. If the '(may be a)' pointers are not needed, then their storage can, of course, be avoided.

The drawback of the multiple storage in the lexicon is, however, compensated for by many other considerations. If it is a question of representing a single semantic tree only and no other information than that pertaining to its structure, then we can store the semantic parent and children of a lexicon entry as the first and the remaining elements of a list respectively, and save the storage of the '(is a)' and '(may be a)' rows of fig. 2.2. Compared with a hypothetical nested-list structure for the *semantic* tree, we save the many levels of parenthesis; and compared with the numeral structure, we save the storage of the keys. Furthermore, the retrieval of information is simpler and more efficient in the 'lexicon and pointer' structure, especially for semantic trees with many nodes.

The crucial arguments for the 'lexicon and pointer' structure for the semantic information in the Alex system are however quite different ones. In the first place we need a lexicon anyhow for a natural language system which also stores syntactic categories such as verbs, prepositions, etc. .

In the second place, a single lexicon entry can contain the parent and child pointers pertaining to many different semantic trees; namely 1) the different trees contained in a multiple partition structure; and 2) the pointers of '(IS IN)' and 'PART' trees (see fig. 2.2 and end of subsection 2.1). Thus the lexicon may be said to represent, among other things, a structure of intertwined trees; in the sense that a single node (namely the lexicon term) can belong to many different trees.

In contrast, in a nested-list or numeral structure for the *semantic* trees, we would have to store each semantic tree separately, thereby repeating the given node for each tree. And the information retrieval algorithm would have to search through the nodes of one or more of the appropriate semantic trees until the specified term is found.

Furthermore, there is nothing in the structure of the lexicon entries which prevents us from having cross references that are not arcs in any of the semantic trees. For example, a ((refers to attribute) ((height))) subentry under the lexicon term 'tall'; or a ((is measured in units of) ((kilogram))) subentry under 'weight'. These subentries do not destroy the structure of the semantic tree as long as the program does not identify them with parent or child pointers of the latter.

Thus the lexicon, which contains the semantic trees and much other information in its strict entry-tree structure, is well suited for the storage and retrieval of the pertinent information contained in a natural-language sentence.

8.2 The Semantic Structures of the Alex System

The network or graph of a nonfuzzy or pure semantic tree can be considered either as a 'downward network' with all the arcs pointing in the downward direction and having the meaning of '(may be a)' pointers (or their equivalent for 'IS IN' and 'PART' trees, see sect., 2.1); or it can be considered to be an 'upward' network, with all the arcs pointing in the upward direction, and having the meaning of '(is a)' pointers (or their equivalents).

The multiple partition and membership structures described in this paper are modifications of the pure tree structure. However, these modifications do not change the above strict description of a network which has either upward or downward arcs, but no sideways connections. It is the strict upward or downward structure which makes it possible to draw probabilistic inferences from the membership structure concerning the grade of membership of a subnode in a supernode or vice versa (eqns (6.7)-(6.14)). The possibility of the inference of eqns (6.7) depends on the definition of the first and higher-order parent families of a node in a pure tree (see sect. 1.3 and defs. 6.1, 6.3). In contrast to present-day fuzzy set theory, the numerical value of the upward pointer between two given nodes is, in general, quite different from the numerical value of the downward pointer between these nodes.

To perform numerical downward inferences, we must know the prior distribution (see sections 6.3-6.5). In many cases this distribution is a meaningful one and can be elicited in a statistical experiment; resulting in definite values for the '(may be a)' pointers. When the prior distribution is not known, or when it is not clear how it should be ascertained, then we can usually narrow down the set of possible prior distributions by considerations of attribute values. This results in the ruling-out of certain sets of values of the '(may be a)' pointers, such that we are left with a subset of the collection of all apriori possible such sets. Each '(may be a)' pointer will thus, in general, be interval-valued when the prior ditribution is not known precisely. This vagueness of the values of the '(may be a)' pointers is not due to a deficiency of our method. It is the best that we can do with the knowledge that we have at hand. The interval-values are a more correct representation of the possib/probab(subnode | supernode) than the assumption that this quantity is numerically equal to possib/probab(supernode | subnode).

On the semantic side, the main difference between the Alex system and semantic networks such as those described by Charniak and McDermott (1985, p. 26) is that while '(is a)' pointers are structurally equivalent to attribute pointers such as 'color' in the latter network, '(is a)' and '(may be a)' pointers are sharply distinguished in the Alex system from the attributes of the nodes. This difference between '(is a)' and '(may be a)' pointers on the one hand versus attributes and their values on the other is reflected in the structure of the lexicon entries in which all attributes and their values are collected into a single group which does not contain the '(is a)' and '(may be a)' pointers. Furthermore, the '(is a)', '(may be a)' and '(attribute)' entries are collected together under a single superentry entitled 'IS'.

This grouping is very important in connection with the multiple partition semantic structure described in section 3 which combines several pure substructure trees in a single structure. In such a structure it may happen that the applicability of some attributes to a given node depends on the particular partition of that node (see the 'profession' example in subsection 3.2).

The Alex system thus allows on the theoretical-semantic side a stepwise building-up of a logical structure, starting from nonfuzzy trees and partitions (referring to LB or YN experiments), via a multiple partiton structure to the membership structure described in sections 4-6 which allows probabilistic inferences not only in the downward, but also in the upward direction.

8.3 AI Structures Versus those of Predicate Calculus

Returning to the other extreme of the most unsophisticated semantic structure, namely a pure semantic tree without attributes of the nodes, we saw in section 2.3 that the automatic building up of this structure in the Alex system can replace the quantification and implication structures of mathematical logic, and the inferences to be drawn from them. We believe that the simple mechanism for the drawing of transitive or '(is a)' inferences from such a tree, as well as the mechanisms for negative inferences, and for probabilistic inferences in the downward direction, are simpler and more straightforward than the much-discussed implication of mathematical logic (Hisdal, 1981, section 2).

The tree and array structures for the representation of quantification and implication and of AND and OR connectives respectively (see Hisdal, 1982, 1984b for the latter) exploit the human ability to think in terms of two-dimensional geometrical structures. The human brain has thus at its disposal a high-level language which simplifies enormously the task of building up logical structures suited for purpose of information retrieval and inference.

To represent such a structure in a computer, we must convert it to a one-dimensional representation such as that described in section 7. The partial examples of eqns. (7.4) and (7.10)-(7.12) show that this representation, which is built up automatically by the computer algorithm, looks very complicated. If we had tried to build it up directly, without the intermediate step of our human high-level geometric language, we would probably have been hopelessly lost.

In contrast to Sowa's conceptual graphs (Sowa, 1984, 1987) and to the Alex system, traditional mathematical logic does not make use of the intermediate step of a high-level geometric language. As John F. Sowa remarked in a recent talk in Oslo, the methods of mathematical logic correspond to programming in assembly language.

The difference between the Alex system and Sowa's conceptual graphs is, on the one hand, that the latter cover an enormously wider range of the conceptual structures of natural language. On the other hand, the Alex system keeps constantly in mind the essential 'compiler' step of translating the high-level geometric language to a storage representation in the computer. The practical philosophy of the Alex system is thus to take very small steps at a time, but to make sure that each of them is unambiguously representable in the computer for purposes of information retrieval and inference.

It is my belief that the intermediate step of the higher-level languages mentioned here results in an intrinsically simpler logic, nearer to the human processing of logical structures, than the traditional one. This intrinsic simplicity will, in the last resort, allow us to perform more compicated logical operations than those which can be performed in the traditional 'assembly language' of mathematical logic. The nested entry-tree structure described in sect. 7.4 for the representation of a multiple-partition semantic structure, as well as the semantic membership structure may be examples of such a case. These structures pay extra dividends in the form of an infinite-valued or fuzzy logic due to the possibility of numerical probabilistic inferences in the upward as well as in the downward direction.

Acknowledgments

This work is based on several years of experience with the Alex system. I would like to thank my former and present students,

Sverre Andreassen, Geir Haug, Per-Willy Johansen,
Oddrun Ohren and Egil Utne

for their work with this system and for many stimulating discussions.

References

Andreassen, S. (1988). Alex-3. Hovedoppgave, Institute of Informatics, University of Oslo, Box 1080 Blindern 0316 Oslo 3, Norway.

Charniak, E. & McDermott, D. (1985). Introduction to Artificial Intelligence. Addison-Wesley.

Curtis, H. (1968). Biology. Worth Publishers, New York 10016.

Fisher-Box, J. (1978). "R.A. Fisher: The Life of a Scientist." John Wiley.

Harris, M.D. (1985). Introduction to Natural Language Processing. Reston Publishing Company, Reston, Virginia.

Haug, G. & Johansen, P.W. (1983). Alex, Prossessering av informasjon i naturlig språk. Hovedoppgave, Institute of Informatics, University of Oslo, Box 1080 Blindern 0316 Oslo 3, Norway.

Hisdal, E. (1981). The IF THEN ELSE Statement and Interval-Valued Fuzzy Sets of Higher Type. Int. J. of Man-Machine Studies, 15, 385-455.

Hisdal, E. (1982). Sets and Natural Language. In Ballester, A., Cardus, D. and Trillas, E., Eds, Second World Conference on Mathematics at the Service of Man, pp. 346-350, Canary Islands: Universidad Politecnica de Las Palmas.

Hisdal, E. (1984a). A Theory of Logic Based on Probability. (ISBN 82- 90230-60-5), Res. Rep. No. 64, Institute of Informatics, University of Oslo, Box 1080, Blindern, 0316 Oslo 3, Norway.

Hisdal, E. (1984b). Decisions Based on Statements in Natural Language. In Gaines, B.R., Zimmermann, H.J., Zadeh. L.A., Eds, TIMS 20: Fuzzy Sets and Decision Analysis, pp. 357-381. North Holland.

Hisdal, E. (1986a). Infinite-Valued Logic Based on Two-Valued Logic and Probability, Part 1.1. Difficulties with Present-Day Fuzzy Set Theory and their Resolution in the TEE Model. Int. J. of Man-Machine Studies, 25, 89-111.

Hisdal, E. (1986b). Infinite-Valued Logic Based on Two-Valued Logic and Probability, Part 1.2.Different Sources of Fuzziness. Int. J. of Man-Machine Studies, 25, 113-138.

Hisdal, E. (1987a). Infinite-Valued Logic Based on Two-Valued Logic and Probability, parts 1.3-1.6. To appear in the Int. J. of Man-Machine Studies.

Hisdal, E. (1987b). Are Grades of Membership Probabilities? To appear in a Special Issue of Fuzzy Sets and Systems on 'Interpretations of Grades of Membership'.

Kleene, S.C. (1968). Mathematical Logic. John Wiley.

Koestler, A. (1968). The Sleepwalkers. A History of Man's Changing Vision of the Universe. Pelican Books.

Ohren, O. (1986). Alex-2, Et system for interaktiv oppbygging av en representasjon for leksikalsk struktur i et fragment av engelsk språk. Hovedoppgave, Institute of Informatics, University of Oslo, Box 1080 Blindern 0316 Oslo 3, Norway.

Quirk, R. & Greenbaum, S. (1979). A University Grammar of English. London: Longman. Ninth edition.

Sowa, J.F., (1984). Conceptual Structures, Information Processing in Mind and Machine. Addison-Wesley.

Sowa, J.F. & Foo, N.Y. (1987). Conceptual Graphs for Knowledge Systems. Preliminary draft.

Utne, E. (1988). Alex-3. Hovedoppgave, Institute of Informatics, University of Oslo, Box 1080 Blindern 0316 Oslo 3, Norway.

Zadeh, L.A., (1968). Probability Measures of Fuzzy Events. J. Math. An. Appl., 23, 421-427.

Zadeh, L.A. (1978). Fuzzy Sets as a Basis for a Theory of Possibility. Fuzzy Sets and Systems, 1, 3-28.

Fuzzy Logic in Knowledge-Based Systems,
Decision and Control
M.M. Gupta and T. Yamakawa (Editors)

SINUSOIDAL-BASED COGNITIVE MAPPING FUNCTIONS

M.M. Gupta, G.K. Knopf and P.N. Nikiforuk

Intelligent Systems Research Laboratory, College of Engineering, University of Saskatchewan, Saskatoon, Saskatchewan, Canada S7N OWO.

ABSTRACT

This paper presents a new class of cognitive mapping functions which are generated using generalized sinusoids. These mapping functions are easy to generate and are found to be very useful in problems dealing with vision perception, cognitive information processing and cognitive (fuzzy logic) controllers.

KEYWORDS

Cognitive mapping functions, Fuzzy mapping functions, Graded membership function, S-function, π - function, Cognitive information, Cognitive uncertainty.

1. INTRODUCTION

This paper is concerned with the development of a new class of cognitive mapping functions which are generated using sinusoidal functions.

The processing of 'cognitive-like' information is receiving increasing attention from researchers working in many fields of engineering, the medical sciences and other scientific disciplines which deal with the processing of *soft data*. One of the problems encountered in the processing of this type of information is the mapping (conversion) of physical information (measurements) into *relative grades* over the real interval [0, 1].

When the information from a physical phenomenon is acquired by our natural sensors (eyes, nose, ears etc.) the brain views this new information in some appropriate context, recalls the previous experience, and then converts the physical attributes of this phenomenon into relative grades. The *natural cognitive processor* processes this information based upon these *relative grades* and yields a perception of that physical phenomenon. The perception, thus, appears in the form of relative grades. For example, a person from a tropical country visiting Saskatoon in January when the temperature is around $-5^o C$, will grade this experience as *'very-very cold,'* while a person from Ottawa may grade it as *'fairly warm.'* Similarly, the temperature of a human body in the range 36.5^o to $37.5^o C$ may be considered by a physician as normal with membership 1, while a temperature of $35^o C$ may again be considered normal but with the graded membership of 0.8.

The human brain, which is the natural cognitive processor, possesses a very robust attribute which converts (maps) a physical phenomenon into *relative grades* over the real interval [0, 1], and then carries out the information processing based upon this mapped (converted) information. If we want to emulate some of the attributes of the natural cognitive processes of the brain on a computer we must be able to map the absolute physical measurements of a physical phenomenon into relative grades. However, the value of relative grades is context dependent.

In the past, several types of such mapping functions, usually called fuzzy mapping functions, have been introduced [1-5]. In this paper, we call them *cognitive mapping functions.* These mapping functions are generated using sinusoidal functions. As the name implies, cognitive mapping functions transform some attributes of a physical phenomenon over the real interval [x_{min} to x_{max}] into the cognitive domain over the interval [0, 1]. Thus, the light intensity in the neighborhood of x_s may be considered as *bright* with a value of graded membership equal to 0.7.

In Section 2 of this paper we define ambiguity in the context of fuzzy set theory and in Section 3, we derive several S - and π - cognitive mapping functions. We derive also multi-region cognitive mapping functions useful for vision-perception and cognitive control problems. In Section 4, we give various fuzzy mathematical operations such as concentration, dilation and contrast intensification. Finally, in Section 5 we give various measures of fuzzy information.

2. AMBIGUITY AND FUZZY SET THEORY

Consider as an example the gray-levels from the intensity profiles of an image. The problem of edge detection by a natural vision system deals essentially with the *'perception'* of these gray-level intensities and the act of perceiving *'significant'* changes at the macroscopic level. The natural vision system perceives the image at an 'aggregate' or 'granule' level rather than on an individual pixel-by-pixel basis. This *'granularity'* or *'aggregation'* is one of the important attributes found in human perception and thinking and, therefore, in natural cognitive information processing.

It is possible to emulate the functions of natural cognition in a machine [5, 6] and to design an artificial perceptor in a machine vision system which transforms the absolute gray intensity levels of an image found in the x-y plane over the real interval [0, 1]. This transformation into perceived intensity profiles can be achieved using fuzzy set theory. In fuzzy set theory the sets are not defined crisply as in the case of ordinary set theory *(it belongs to,* membership grade = 1; *it does not belong to,* membership grade = 0). Instead, fuzzy set theory deals with the vague cognitive phenomena whose grade of membership is defined over the interval [0, 1].

There are many types of uncertainties that we encounter in our scientific work. In this paper we classify these uncertainties into two broad categories: (i) random type uncertainty and (ii) cognitive uncertainty.

The first type of uncertainty arises from the random behaviour of *physical systems,* for example, random behaviour of gas molecules in a thermal field. This type of random uncertainty has been studied for centuries, and we have a very rich statistical theory to characterize such a random phenomenon. The measure of such a random phenomenon can be described using the notion of *entropy* which is a well-known measure in the fields of thermodynamics and statistical communication theory.

The second type of uncertainty, unlike the first one, deals with the uncertain phenomenon arising from human thinking, reasoning, cognition and perception processes. We will call such an uncertainty as *cognitive uncertainty* and its cognate as *cognitive information,* and this is the subject of these studies. The uncertainty associated with cognitive information can be characterized using fuzzy logic tools and associated entropy measures.

Graded Membership Function, $\mu_A(x)$

A fuzzy set is a class of objects or phenomena that accepts the notion of partial membership over the interval [0, 1]. Let $\mathbf{X} = \{x\}$ be a set of objects, then a fuzzy set $A \in \mathbf{X}$ is a set of ordered pairs

$$A = \{x, \mu_A(x)\}, \quad x \in \mathbf{X}, \tag{1}$$

where $\mu_A(x)$ is called the 'characteristic function,' or 'graded membership' of x in A. The membership function $\mu_A(x)$ maps the fuzzy set A onto the interval [0, 1]

$$\mu_A: \ A \rightarrow [0, 1].$$

In ordinary (nonfuzzy) set theory the membership function of A in a universe $\mathbf{X}$ is defined as

$$\mu_A(x) = \begin{cases} 1 & \textit{iff } x \in A \\ 0 & \textit{otherwise}, \textit{ie } x \notin A. \end{cases} \tag{2}$$

Thus, in ordinary set theory the valuation set is $\{0, 1\}$; that is, μ is either 0 or 1, whereas in fuzzy set theory it is the real interval over [0, 1], where the grades 1 and 0 respectively represent the *'full membership'* and *'non-membership'* of a fuzzy set $A \in \mathbf{X}$ and the grades in the interval $0 < \mu_A < 1$ represent its partial membership. It is interesting to note that the grade of membership $\mu_A(x)$ of a phenomenon in A can be interpreted as the 'degree of compatibility' of the phenomenon in $\mathbf{X}$, or the degree of possibility that a phenomenon x is restricted to A. In natural vision, for example, $\mu_A(x)$ may represent the level of *perception* in *seeing* certain attributes of a scene.

In general, fuzzy set theory is based on the grade of membership, $\mu_A(x)$, and deals with an *'inexact phenomenon.'* This *'inexactness'* in cognitive information may arise due to the following three situations or combination thereof:

(i) Generality: This concept refers to a variety of possible situations of a phenomenon; that is, the defined universe is not just a point.

(ii) Ambiguity: This concept describes more than one distinguishable sub-phenomena such that the membership will have several local maxima.

(iii) Vagueness: This concept reflects a set of phenomena with non-precise or non-crisp boundaries.

In other words, the membership function $\mu_A(x)$ describes the strength of our *'perception'*, *'confidence'* or *'belief'* that $x \in A$. If $\mu_A(x) = 1$, it is *certain* that x is in A, and if $\mu_A(x) = 0$, then it is *certain that* x is *not* in A. For x over $0 < \mu_A(x) < 1$, there is an uncertainty associated with x; that is, x belongs to A with the possibility $\mu_A(x)$. However, it should be noted that the assignment of a membership function to a fuzzy set is a subjective one and, in general, reflects the context in which the problem is viewed.

3. COGNITIVE MAPPING FUNCTIONS

Consider a fuzzy phenomenon **X**. This phenomenon may be a real continuous phenomenon or may represent a discrete countable event. If **X** is a continuous phenomenon, then the set A can be represented as

$$A = \int_x \mu_A(x) \mid x \quad = \quad \int_x \mu_A(x), \tag{3}$$

and if it is one with countable discrete events, then

$$A = \underset{i}{U} \mu_A(x_i) \mid x_i \quad = \quad \underset{i}{U} \mu_A(x_i) \tag{4}$$

where $\mid x$, or $\mid x_i$ represents the support of the set.

Let the fuzzy phenomenon **X** be defined over a real interval $[x_m, x_M]$, where x_m and x_M correspond respectively to the lower and upper bounds of the set **X**. In the gray-level image processing problems, for example, x_m and x_M correspond respectively to the minimum and maximum intensity levels of the image. In order to transform the image from the spatial-intensity domain over $[x_m, x_M]$ to the spatial-perception domain over [0, 1], a variety of fuzzy mapping functions is required.

There are many mapping functions described in the literature [1, 2, 4, 5] which map a subjective and ambiguous (fuzzy) phenomenon **X** on a real line into a membership domain μ: $X \rightarrow [0, 1]$. Several such special mapping functions, generated by a sinusoidal function, will be described now. The development of the **S** - and π - mapping functions given here follows a sequential pattern and is accompanied with a discussion of some of the important attributes of each function. Table 1 gives a summary of the mapping functions which are developed in this section.

(i) The S_1 – Mapping Function*

Consider again a fuzzy phenomenon **X** described over the interval $[x_m, x_M]$. Define a mapping function S_1 as

$$S_1 = \sin\theta, \quad 0 \le \theta \le \frac{\pi}{2}, \tag{5a}$$

where θ is given by

$$\theta = \frac{\pi}{2}\left(\frac{x - x_m}{x_M - x_m}\right), \quad x_m \le x \le x_M. \tag{5b}$$

Using this S_1 function, the membership mapping function can be defined as

$$\mu_{S_1}(x) = S_1(x, x_m, x_M) = \sin\frac{\pi}{2}\left(\frac{x - x_m}{x_M - x_m}\right), \quad x_m \le x \le x_M. \tag{5c}$$

The function S_1 in (5c), thus, maps x, $x_m \le x \le x_M$, into an equivalent membership function $\mu_{S_1}(x)$, $0 \le \mu_{S_1}(x) \le 1$, of a fuzzy phenomenon **X** as illustrated in Figure 1.

$$\mu_{S_1}: \ \mathbf{X} \rightarrow [0, 1].$$

Define now a crossover point $x = x_c$ for this mapping function as a point for which

$$\mu_{S_1}(x)\,|_{x = x_c} = \mu_{S_1}(x_c) = 0.5. \tag{5d}$$

For the mapping function S_1 defined in (5c), the crossover point x_c is given by

$$x_c = \frac{1}{3}(x_M + 2\,x_m). \tag{5e}$$

As shown in Figure 1, the mapping function S_1 is not symmetrical about the crossover point x_c. For a symmetrical mapping function about x_c, another S function, S_2, must be defined.

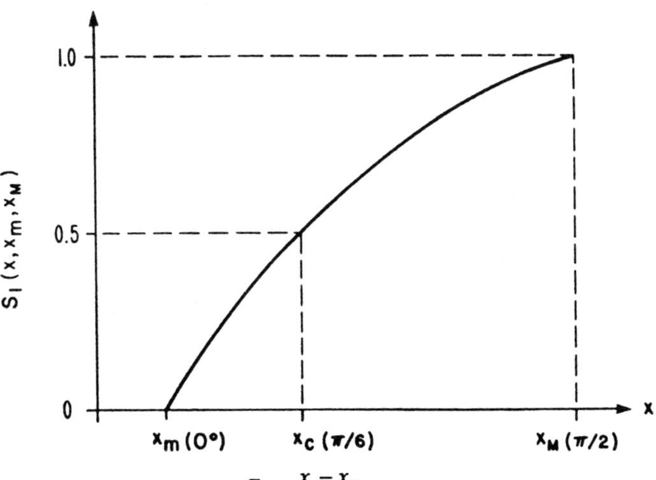

Fig. 1. $S_1 = \sin\frac{\pi}{2}\left(\frac{x - x_m}{x_M - x_m}\right)$, Eqn. (5c).

*The shape of the letter S approximately represents the shape of the mapping function.

(ii) The S_2- Mapping Function

For the fuzzy set **X** over the real interval $[x_m, x_M]$, define a symmetrical crossover point x_c as

$$x_c = \frac{1}{2}(x_m + x_M), \tag{6a}$$

with a corresponding mapping function S_2 as

$$\mu_{S_2}(x) = S_2(x, x_c, x_M) = \frac{1}{2}\left[1 + sgn(x - x_c) \mid \sin\frac{\pi}{2}\left(\frac{x - x_c}{x_M - x_c}\right)\mid^g\right],$$

$$x_m \le x \le x_M, \tag{6b}$$

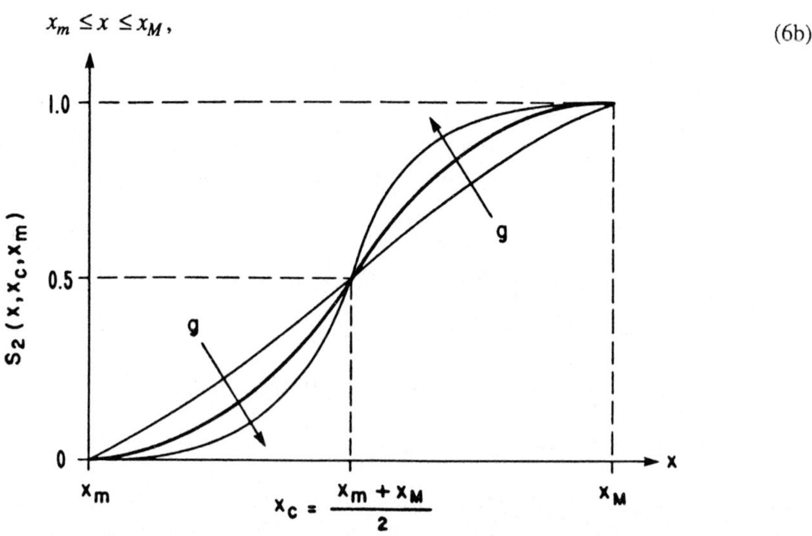

Fig. 2. Symmetrical S - Mapping Function, $S_2(x, x_c, x_M)$, Eqn. (6b).

where the power index g is a positive real constant, $g > 0$. The mapping function S_2 assigns low membership values, between $0 \le \mu \le 0.5$, to x for $x \le x_c$ and higher values, between $0.5 \le \mu \le 1$, to x for $x \ge x_c$. The assignment of membership values can be enhanced by the power index $g > 1$ as shown in Figure 2.

(iii) The S_3- Mapping Function

The $S_2(x, x_c, x_M)$ mapping function given in (6) is symmetric about the crossover point x_c. Define now another $S-$ mapping function which is also symmetrical about the crossover point x_c. Define $S_3(x, x_m, x_M)$ as

$$\mu_{S_3}(x) = S_3(x, x_m, x_M)$$

$$= \sin^2\frac{\pi}{2}\left(\frac{x - x_m}{x_M - x_m}\right), \quad x_m \le x \le x_M. \tag{7a}$$

The corresponding crossover point is the mid-point of x_m and x_M; that is,

$$x_c = \frac{1}{2}[x_m + x_M] \tag{7b}$$

and

$$\mu_{S_3}(x = x_c) = 0.5. \tag{7c}$$

This S_3- mapping function, shown in Figure 3, can be considered as a special case of S_2 in (6b) for a power index of $g = 2$.

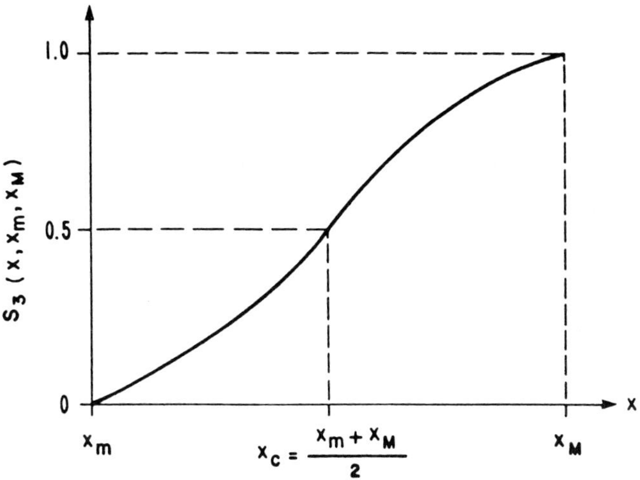

Fig. 3. Symmetrical S - Mapping Function, $S_3(x, x_m, x_M)$, Eqn. (7a).

(iv) The π_1- Mapping Function*

Define now, as illustrated in Figure 4, a symmetrical π_1- mapping function for a fuzzy phenomenon **X** on the real line over the interval $[x_m, x_M]$. For this mapping function we define a symmetrical point x_s as the mid-point of x_m and x_M,

$$x_S = \frac{1}{2}[x_m + x_M], \tag{8a}$$

and the lower and upper crossover points as

$$x_{c_1} = \frac{1}{2}(x_m + x_s) = \frac{1}{4}(3x_m + x_M), \tag{8b}$$

and

$$x_{c_2} = x_s + (x_{c_1} - x_m) = \frac{1}{4}(x_m + 3x_M). \tag{8c}$$

Using the definitions of the points x_s, x_{c_1} and x_{c_2}, a symmetrical π_1 function can now be defined as

$$\mu_{\pi_1}(x) = \pi_1(x, x_m, x_M)$$

$$= \pi_{1A}(x, x_{c_1}, x_s)\ U\ \pi_{1B}(x, x_{c_2}, x_M), \tag{8d}$$

*The shape of the letter π approximately represents the shape of the mapping function.

which, as is shown in Figure 4, $\pi_{1A}(x, x_{c_1}, x_s)$ is defined over $(x_m \le x < x_s)$ and is given by

$$\pi_{1A}(x, x_{c_1}, x_s) = S_2(x, x_{c_1}, x_s)$$

$$= \frac{1}{2}[1 + sgn(x - x_{c_1}) \mid \sin\frac{\pi}{2}(\frac{x - x_{c_1}}{x_s - x_{c_1}}) \mid^g]; \quad g > 1, \tag{8e}$$

and $\pi_{1B}(x, x_{c_2}, x_M)$ is defined over $(x_s \le x \le x_M)$ and is given by

$$\pi_{1B}(x, x_{c_2}, x_M) = 1 - S_2(x, x_{c_2}, x_M)$$

$$= \frac{1}{2}[1 - sgn(x - x_{c_2}) \mid \sin\frac{\pi}{2}(\frac{x - x_{c_2}}{x_M - x_{c_2}}) \mid^g]; \quad g > 1. \tag{8f}$$

Figure 4 shows that $\mu_{\pi_1}(x)$ is a convex mapping function which increases monotonically from 0 to 1 over the interval $[x_m, x_s]$ and decreases monotonically from 1 to 0 over $[x_s, x_M]$. At the crossover points x_{c_1} and x_{c_2}

$$\mu_{\pi_1}(x_{c_1}) = \mu_{\pi_1}(x_{c_2}) = 0.5.$$

Thus, $\mu_{\pi_1}(x)$ maps a function $\mathbf{X} = [x_m, x_M]$ into [0, 1]

$$\mu_{\pi_1}: \ \mathbf{X} \to [0, 1]. \tag{8g}$$

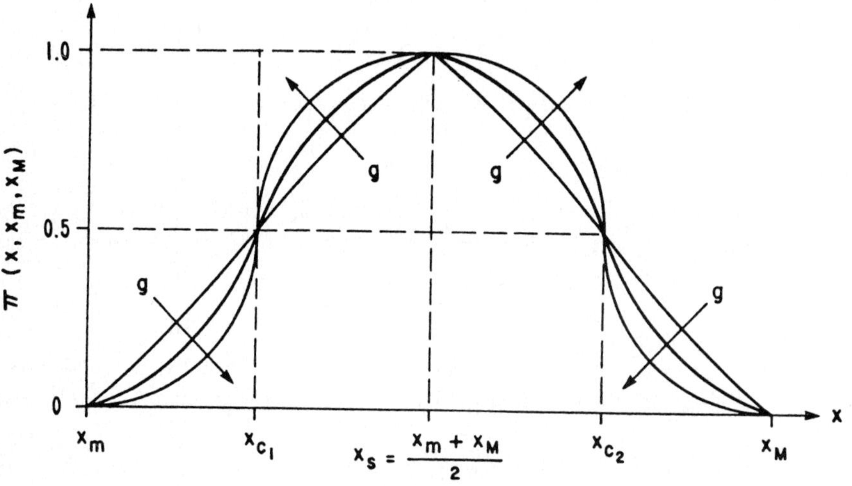

Fig. 4. π - Mapping Function, $\pi_1(x, x_m, x_M)$, Eqn. (8d).

(v) The π_2 – Mapping Function

Let a symmetrical π – mapping function now be defined, as shown in Figure 5, over the interval $(x_m \le x \le x_M)$

$$\mu_{\pi_2}(x) = \pi_2(x, x_m, x_M)$$

$$= \sin^g \pi(\frac{x - x_m}{x_M - x_m}). \tag{9a}$$

This π – mapping function is symmetrical around

$$x_s = \frac{1}{2}\,[x_m + x_M]. \tag{9b}$$

For $g = 2$, the crossover points are also symmetrically located, that is

$$\pi_2(x\,, x_m\,, x_M) = \sin^2 \pi(\frac{x - x_m}{x_M - x_m}) \tag{9c}$$

with crossover points,

$$x_{c_1} = \frac{1}{4}[3x_m + x_M], \quad \text{and} \quad x_{c_2} = \frac{1}{4}[x_m + 3x_M]. \tag{9d}$$

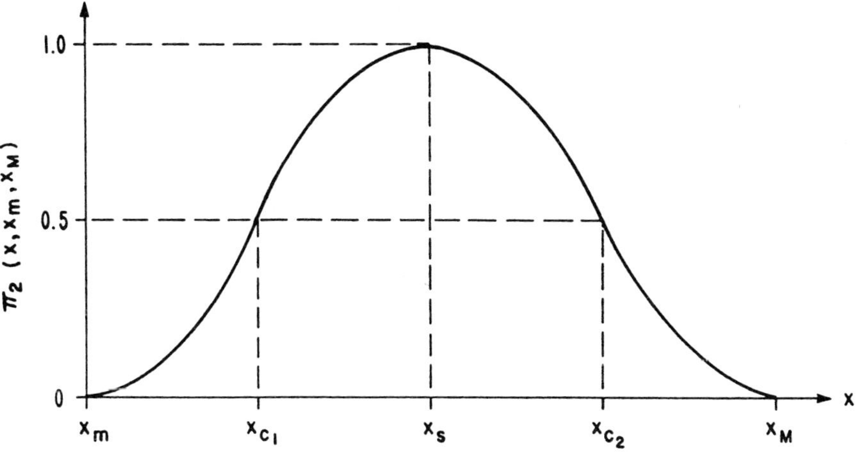

Fig. 5. $\pi_2(x\,, x_m\,, x_M)$ - Mapping Function, Eqn. (9c).

(vi) The Half - S - Mapping Functions

In certain image processing problems it is useful to define a half-mapping function only over the membership values [0, 0.5], or [0.5, 1] as shown in Figure 6. Such functions are defined as follows:

(a) Lower - half S - mapping function, $S_{LH}(x,\ x_1, x_2)$, Figure 6(a),

$$\mu_{S_{LH}}(x) = S_{LH}(x\,, x_1, x_2)$$

$$= [1 - \sin^2 \frac{\pi}{4}(\frac{x_2 + x - 2x_1}{x_2 - x_1})], \quad \text{for} \quad x_1 \le x \le x_2. \tag{10a}$$

(b) Upper - half S - mapping function, $S_{UH}(x,\ x_1, x_2)$, Figure 6(b),

$$\mu_{S_{UH}}(x) = S_{UH}(x\,, x_1, x_2)$$

$$= \sin^2 \frac{\pi}{4}\,(\frac{x_2 + x - 2x_1}{x_2 - x_1}), \quad \text{for} \quad x_1 \le x \le x_2. \tag{10b}$$

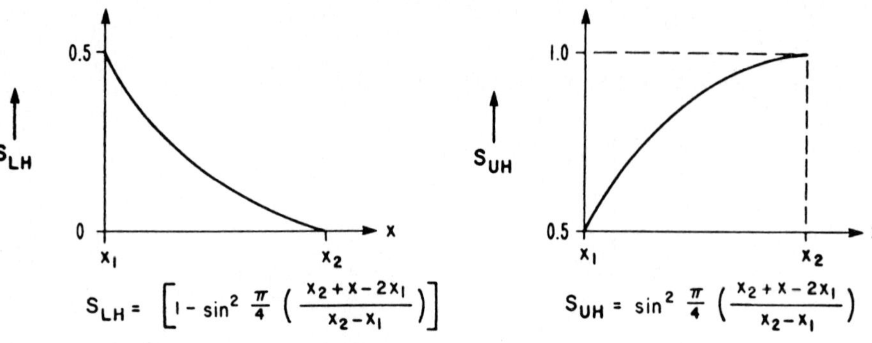

a) Lower - Half S - Mapping Function b) Upper - Half S - Mapping Function

Fig. 6. Half S - Mapping Function.

(vii) The Half π - Mapping Functions

Figure 7 illustrates the upper-half and lower-half symmetrical π – mapping functions which are defined mathematically as follows:

(a) Upper-half π – mapping function, $\pi_{UH}(x, x_1, x_2)$, Figure 7(a),

$$\mu_{\pi_{UH}}(x) = \pi_{UH}\ (x, x_1, x_2)$$

$$= \frac{1}{2}[1 + \sin^2 \pi\, (\frac{x - x_1}{x_2 - x_1})], \tag{11a}$$

with the mid-point as

$$x_s = \frac{1}{2}[x_1 + x_2],$$

with

$$\mu_{\pi_{UH}}(x = x_s) = 1, \quad \text{and} \quad \mu_{\pi_{UH}}(x = x_1, x_2) = 0.5.$$

(b) Lower-half π – mapping function, $\pi_{LH}(x, x_1, x_2)$, Figure 7(b),

$$\mu_{\pi_{LH}}(x, x_1, x_2) = \pi_{LH}(x, x_1, x_2)$$

$$= \frac{1}{2}\ [1 - \sin^2 \pi\, (\frac{x - x_1}{x_2 - x_1})], \tag{11b}$$

with the mid-point at

$$x_s = \frac{1}{2}\ (x_1 + x_2),$$

with

$$\mu_{\pi_{LH}}(x = x_s) = 0, \quad \text{and} \quad \mu_{\pi_{LH}}(x = x_1, x_2) = 0.5.$$

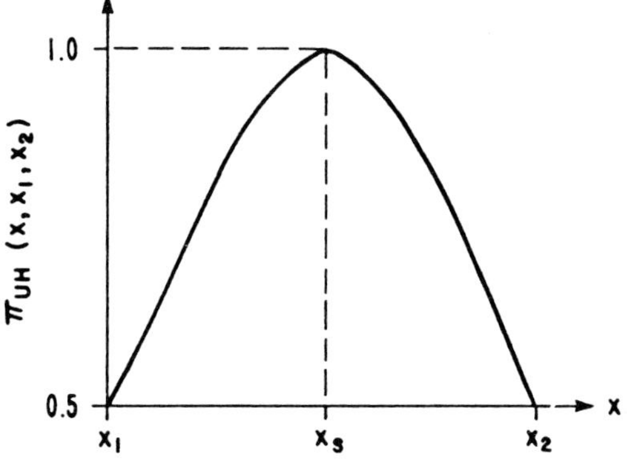

a) Upper - Half π - Mapping Function

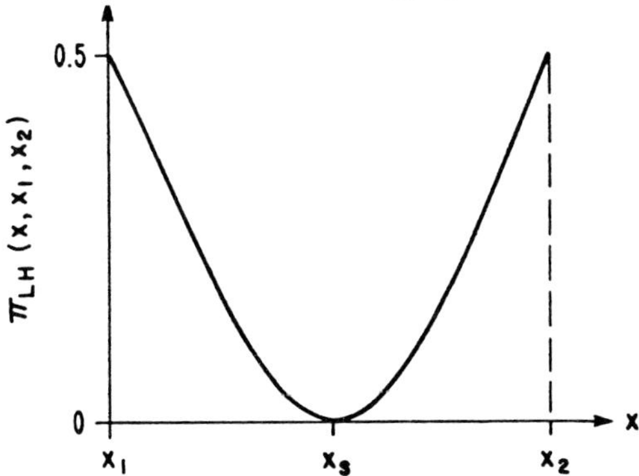

b) Lower - Half π - Mapping Function

Fig. 7. Half - π Mapping Functions.

Remarks

The S - and π - cognitive mapping functions defined above map the real values of $x \varepsilon X$ defined over the interval $[x_m, x_M]$ into the *cognitive-domain* over the real interval [0, 1]. Thus, for example, the S-mapping functions, $S_1(x, x_m, x_M)$, $S_2(x, x_c, x_M)$ and $S_3(x, x_m, x_M)$, map the interval $[x_m, x_M]$ onto the interval [0, 1].

$$S: \quad [x_m, x_M] \rightarrow [0, 1] \tag{12}$$

The S-mapping function is a monotonically increasing function and essentially gives a perception of a physical phenomenon such as *'the image is bright'* for intensity profiles over the intensity levels $[x_m, x_M]$. The π– mapping function, on the other hand, monotonically increases over the interval $[x_m, x_s]$ and then decreases over $[x_s, x_M]$. This function, thus, essentially represents a perception of a physical phenomenon such as the *'brightness of the image is x_s'* around the intensity profile centered at x_s over the intensity levels $[x_m, x_M]$. The S-mapping function thus covers the entire region from x_m to x_M, whereas the π– mapping function decimates this region into two sections, x_m to x_s and x_s to x_M. This is important in vision perception problems since situations may arise where the entire intensity range, from x_m to x_M, has to be decimated into several smaller regions in order to isolate one intensity level from the adjacent one. For example, in edge perception problems, two adjacent regions which are characterized by a 'significant' change in intensity levels have to be decimated. A multiple-region-mapping function is defined, therefore, which helps to decimate a range of intensity profiles into a number of distinct regions within the *'cognitive-domain.'* To explain this phenomenon, we will give an example using a vision perception problem.

Decimation into Multi-Region Intensity Levels,

Consider as an example a gray-level digital image with varying intensities over the range $[x_m, x_M]$ and with a possible k-modal intensity profile in the histogram as shown in Figure 8. From this histogram one can distinctly identify k intensity levels, each corresponding to a peak in the histogram. This histogram profile can be viewed, therefore, as a composition of several intensity levels, $x_1, x_2, \cdots, x_q \cdots x_k$, and this intensity profile can also be described *linguistically (fuzzily)* as follows:

'Over the intensity range $[x_m, x_M]$, *the gray-level digital image is composed of a set of intensity levels* $(x_o, x_1, x_2 \cdots x_q \cdots x_k)$' $= \bigcup_{q=0}^{k} x_q$.

In the above, the intensity levels $x_q, q = 0, 1 \cdots, k$, represent a set of fuzzy numbers. These intensity levels are distinctly identified (with membership grade = 1) at their peaks $x_q, q = 1 \cdots k$. A maximum amount of uncertainty exists, however, at the valleys $x_{c_1}, x_{c_2} \cdots$ in the sense that it is not clear whether the intensity level at a valley belongs to one or another of the adjacent intensity levels. For example, the intensity level at the valley x_{c_2} may belong equally to the intensity x_1 or the intensity x_2 with an equal possibility of $\mu = 0.5$.

Use shall now be made of the fact that a maximum amount of certainty (no ambiguity, therefore, $\mu = 1$) exists at the peaks and a maximum amount of uncertainty (therefore, $\mu = 0.5$) at the valleys of the histogram. These valleys should be considered, therefore, as the crossover points with $\mu = 0.5$, and the entire intensity range can then be decimated into several regions, $R_o, R_1 \cdots R_k$, with each region being separated by *points of maximum uncertainty*.

A histogram given over an intensity range $[x_m, x_M]$ with k valleys, $x_{cq}, q = 1, 2 \cdots k$, can be decimated, therefore, into $(k + 1)$ regions, $R_q, q = 0, 1 \cdots k$, where each region is bounded by the adjacent crossover points. Thus, the width of the *qth* region R_q is given by

$$w_q = (x_{cq+1} - x_{cq}) \tag{13}$$

where

$$\mu(x_{cq+1}) = \mu(x_{cq}) = 0.5.$$

The mid-point between the crossover points can be approximated as a point of maximum certainty, that is,

$$x_q = \frac{1}{2}(x_{cq} + x_{cq+1}). \tag{14}$$

Note that the actual peaks of the histogram are at $x = x_q$, which may or may not be the mid-point $x = x_q$ of the two adjacent crossover points defined in (14). Thus, there is a great need for multiple-region cognitive mapping functions as illustrated above in vision-perception problems and, similarly, in the design of cognitive controllers. In the following section, we will describe such a cognitive mapping function.

The Multi-Region Cognitive Mapping Function, $\Phi[x]$

As illustrated above, the main objective in a gray digital image edge perception problem is to decimate any two adjacent intensity levels. Decimation of two adjacent intensity regions can be done by letting the intensity level of one region go to 0 (dark) and the other to 1 (maximum brightness). This process of decimating two adjacent regions into two distinct intensity levels can be achieved by, alternatively, mapping the regions $R_0, R_1 \cdots R_k$ in the histograms into '*low* and *high*' or '*high* and *low*' intensity profiles as shown in Figure 8.

Thus, using the alternate '*low* and *high*' intensity profiles, a multi-region mapping function is defined, therefore, as follows:

$$\Phi[x] = \bigcup_{q=0}^{k} M_q(x) \tag{15a}$$

where $M_q(x)$, $q = 0, 1 \cdots k$ are defined as follows:

(i) For the extreme left region, R_o,

$$M_o(x) = S_{LH}(x, x_m, x_{c_1})$$
$$= \sin^2 \frac{\pi}{4}\left(\frac{x - x_m}{x_{c_1} - x_m}\right), \quad x_m \le x \le x_{c_1}. \tag{15b}$$

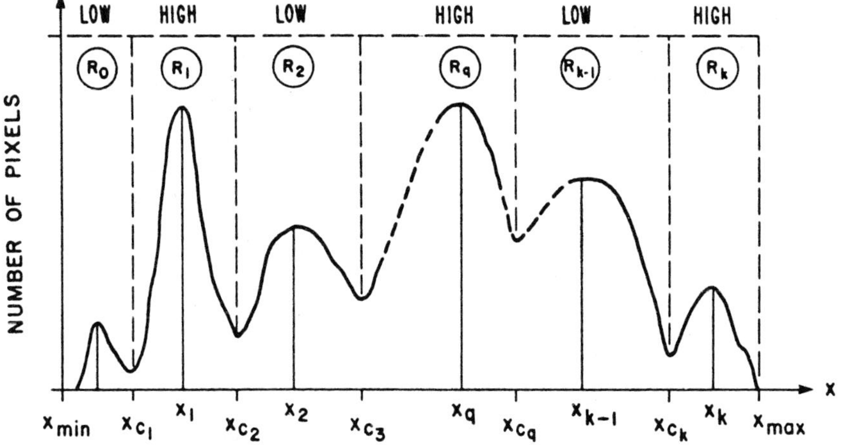

Fig. 8. K - modal Histogram of Intensity Profiles of a Gray Level Digital Image.

(ii) For regions R_1 to R_{k-1},

$$M_q(x) = \pi_{UH}\ (x, x_{cq}, x_{cq+1})$$
$$= \frac{1}{2}\ [1 + \sin^2 \pi\ (\frac{x - x_{cq}}{x_{cq+1} - x_{cq}})], \tag{15c}$$

for $q = 1, 3, 5 \cdots$ odd numbered regions, over $x_{cq} \le x \le x_{cq+1}$; and

$$M_q(x) = \pi_{LH}(x, x_{cq}, x_{cq+1})$$
$$= \frac{1}{2}\ [1 - \sin^2 \pi(\frac{x - x_{cq}}{x_{cq+1} - x_{cq}})], \tag{15d}$$

for $q = 2, 4, 6 \cdots$ even numbered regions.

(iii) For the extreme right region, R_k,

$$M_k(x) = \begin{cases} [1 - \sin^2 \frac{\pi}{4}\ (\frac{x_M + x - 2x_k}{x_M - x_k})], & \text{for } x_k \le x \le x_M, \text{ and } k = even \\ [\sin^2 \frac{\pi}{4}(\frac{x_M + x - 2x_k}{x_M - x_k})], & \text{for } x_k \le x \le x_M \text{ and } k = odd \end{cases} \tag{15e}$$

Figure 9 shows such a multi-region mapping function $\Phi[x]$.

Thus,

$$\Phi[x]: \quad [x_m, x_M] \quad \rightarrow \quad [0, 1]. \tag{16}$$

In the following section some fuzzy operations are defined.

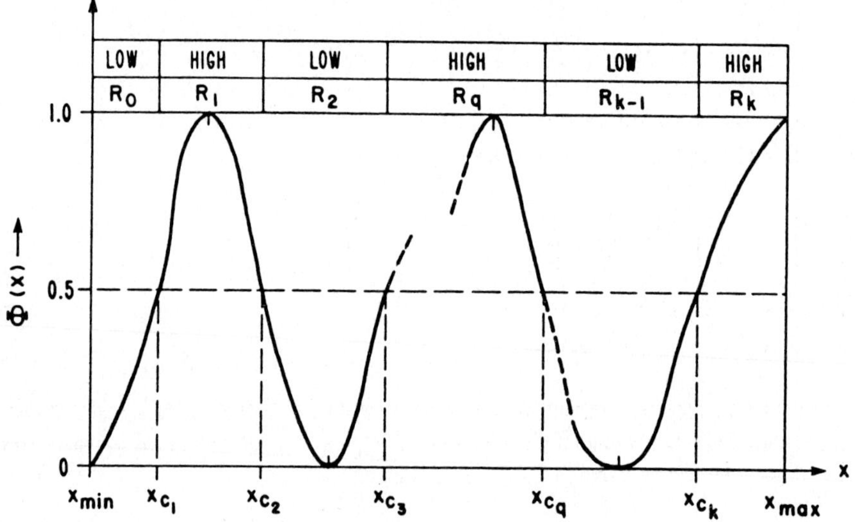

Fig. 9. Multi - Region Mapping Function, Φ (x).

4. SOME FUZZY MATHEMATICAL OPERATIONS

The notion of graded membership that was introduced in Section 3 plays an important role in many cognitive mathematical operations that are not found in ordinary set theory. Several such operations are defined in this section. For this purpose consider an example of a fuzzy set A as a *'representational image'*. This notion allows one to proceed with some fuzzy operations in the *cognitive domain* using the membership function $\mu_A(x)$. In this approach a membership is assigned to 'gray-levels' between the non-membership ($\mu_A(x) = 0$) and the full membership ($\mu_A(x) = 1$). For example, a black and white television picture is a fuzzy subset of the white screen (or the black screen). Thus, one can think of *focusing*, of *concentrating*, of *dilating*, of *contrast intensification*, and of *blurring* an image in the *cognitive domain*.

In this section, some important fuzzy operations are defined which are very useful in the field of cognitive information processing.

Let A be a fuzzy set of **X** with membership $\mu_A(x)$, $x \in \mathbf{X}$.

$\mathbf{M_1}$: The Power Set, $[A^g]$

$$A^g = \int_{x \in X} \{\mu_A^g(x) \mid x\}, \quad g > 0. \tag{17}$$

$\mathbf{M_2}$: Concentration, CON(A)

$$\mathrm{CON}(A) = \int_{x \in X} \{\mu_A^g(x) \mid x\}, \quad g > 1. \tag{18}$$

$\mathbf{M_3}$: Dilation, DIL(A)

$$\mathrm{DIL}(A) = \int_{x \in X} \{\mu_A^g(x) \mid x\}, \quad 0 < g < 1. \tag{19}$$

Usually, the values $g = 2$ for **CON**(A), and $g = 0.5$ for **DIL**(A) are taken.

Other fuzzy operations can be constructed using the primitive operations given in $\mathbf{M_1}$, $\mathbf{M_2}$, and $\mathbf{M_3}$, some of which will be discussed now.

$\mathbf{M_4}$: Contrast Intensification, INT(A)

$$\mathrm{INT}_1(A) = \begin{cases} \mathrm{CON}(A), & \text{for } \textit{all } x \textit{ such that } 0 \le \mu_A(x) < 0.5, \\ \mathrm{DIL(A)}, & \text{for } \textit{all } x \textit{ such that } 0.5 \le \mu_A(x) \le 1. \end{cases} \tag{20}$$

The **INT**$_1$(A) operation has the effect of increasing the membership if it is greater than 0.5, and decreasing it if it is less than 0.5 as illustrated in Figure 10. An alternate expression for the

INT(A) operation is

$$\mathbf{INT}_2(A) = \begin{cases} \int_{x \in X} 2\{\mu_A^2(x) \mid x\}, & \text{for } all\ x\ such\ that\ 0 \le \mu_A(x) < 0.5, \\ \int_{x \in X} \{1 - 2(1 - \mu_A(x))^2 \mid x\}, & \text{for } all\ x\ such\ that\ 0.5 \le \mu_A(x) \le 1. \end{cases} \tag{21}$$

Contrast intensification reduces the amount of ambiguity associated with a fuzzy set and, hence, the entropy associated with it. This operation is found to be very useful, therefore, in image processing problems such as image enhancement.

M_5: Blurring, BLR(A)

As the name implies, this function increases the ambiguity of a set A by increasing the membership if it is less than 0.5, and decreasing if it is greater than 0.5. Thus, **BLR**(A) is just the opposite of **INT**(A), as is illustrated in Figure 11,

$$\mathbf{BLR(A)} = \begin{cases} \mathbf{DIL(A)}, & \text{for } all\ x\ such\ that\ 0 \le \mu_A(x) < 0.5, \\ \mathbf{CON}(A), & \text{for } all\ x\ such\ that\ 0.5 \le \mu_A(x) \le 1. \end{cases} \tag{22}$$

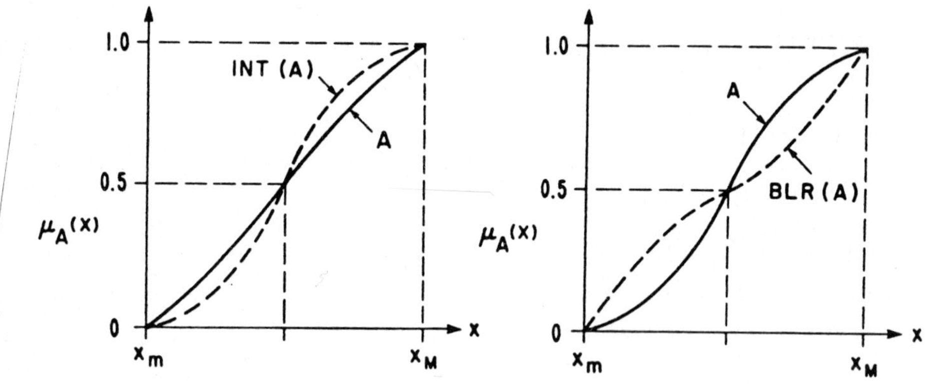

Fig. 10. INT (A) Operation.

Fig. 11. BLR (A) Operation.

5. MEASURE OF FUZZY INFORMATION

The notion of the measure of information is a well-established concept in communication theory and is based on the probabilistic approach. This concept has been extended in fuzzy set theory by de Luca and Termini [7] and others [1, 2, 4].

Consider a fuzzy set A defined over a real line $\mathbf{X}$ with membership function

$$\mu_A(x), \quad x \in \mathbf{X}.$$

Here, the membership function maps $\mathbf{X}$ into [0, 1],

$$\mu_A: \quad \mathbf{X} \rightarrow [0, 1].$$

Two metrics that are used extensively for measuring the information contained in the fuzzy set A are:

(i) the energy metric $e(A)$; and

(ii) the entropy metric $h(A)$.

These measures can be defined for both finite and infinite support on x, as indicated by Eqns. (3) and (4) using the notion of the membership function $\mu_A(x)$. For finite support, a general definition is given by

$$e(A) = \sum_{x \in X} E[x, \mu_A(x)], \tag{23}$$

and

$$h(A) = \sum_{x \in X} H[x, \mu_A(x)], \tag{24}$$

where E and H are respectively the energy and entropy mapping functions which map $\mathbf{X} \times [0, 1]$ into R_+ such that for any $x \in \mathbf{X}$,

(i) $E[x, \mu_A(x)] = 0$, *iff* $\mu_A(x) = 0$, and $E[\cdot]$ is a monotonically increasing function over $\mu_A \in [0, 1]$

(ii) $H[x, \mu_A(x),] = 0$, *iff* $\mu_A(x) = 0$, and $H[\cdot]$ is a monotonically increasing function over $\mu_A(x) \in [0, 0.5)$ and decreasing over $\mu_A(x) \in [0.5, 1]$.

Special cases for $E[\cdot]$ and $H[\cdot]$ are

$$E[x, \mu_A(x)] = w(x) E_1[\mu_A(x)], \tag{25}$$

and

$$H[x, \mu_A(x)] = w(x) H_1[\mu_A(\tilde{x})], \tag{26}$$

where $w: \ X \rightarrow R_+$ is a positive weighting function, and $E_1[\mu_A(x)]$ and $H_1[\mu_A(x)]$ are mapping functions from [0,1] to R_+. For a constant weighting, $w(x) = 1$, (23) and (24) can be written as

$$e(A) = \sum_{x \in X} E[\mu_A(x)], \tag{27}$$

and

$$h(A) = \sum_{x \in X} H[\mu_A(x)]. \tag{28}$$

It is to be noted that in (23) and (27), $e(A)$ is a measure of the total energy of the fuzzy set A, while in (24) and (28), $h(A)$ is a measure of total entropy (information content) of A.

The following three important properties of the entropy function $h(A)$ can now be postulated:

P_1: $h(A) \geq 0$;

P_2: $\max[h(A)] = 1$;

P_3: $h(A) = 0$ if A *is an empty set.*

It should be noted from postulates P_1, P_2 and P_3 that $h(A)$ is a non-negative and bounded function; $0 \leq h(A) \leq 1$. Postulate P_3 reflects that fact that an empty set A provides no information.

Some specific functions of the energy and entropy metrics which are useful for measuring the ambiguity in cognitive information will now be given.

Energy Measure

For the fuzzy set $A \varepsilon \mathbf{X}$ with graded membership function $\mu_A(x) \varepsilon [0, 1]$, the generalized energy function $e(A)$ is given by

$$e(A) = \sum_{x \varepsilon X} E\,[\mu_A(x)], \tag{29a}$$

where $E: \ \mu_A[0, 1] \rightarrow [0, 1]$.

One such function is,

$$e_q(A) = \sum_{x \varepsilon X} \mu_A^q(x); \tag{29b}$$

where, for $q = 1$

$$e_1(A) = \sum_{x \varepsilon X} \mu_A(x) \tag{29c}$$

gives a linear measure of energy, and for $q = 2$

$$e_2(A) = \sum_{x \varepsilon X} \mu_A^2(x) \tag{29d}$$

gives a quadratic measure of the energy function. The quadratic measure provides 'very small' weights to 'small' values. One can generate other types of energy functions which satisfy the properties given earlier.

Entropy Measure

For the fuzzy set $A \varepsilon \mathbf{X}$ with graded membership function $\mu_A(x) \varepsilon [0, 1]$, a generalized entropy function $h(A)$ is given by

$$h(A) = \sum_{x \varepsilon X} H[\mu_A(x)]. \tag{30}$$

This function must satisfy the postulates given earlier.

Several entropy functions which are based upon the generalized Euclidean metric and Shannon's function will now be described.

The generalized Euclidean metric between $\mu_A(x)$ and its nearest* ordinary set $\mu_{A_0}(x)$ is defined as

$$h_q[\mu_A(x)] = \alpha[\sum_{x \in X} \mu^q_{A \cap \bar{A}}(X)]^{\frac{1}{q}}. \tag{31a}$$

where $\bar{A}$ is the complement of A, q a positive integer, α a normalization constant ($2N^{\frac{-1}{q}}$) such that $h_q \rightarrow [\mu_A(x)]: \mu_A(x) \rightarrow [0, 1]$, and $A \cap \bar{A}$ is the intersection of A with its complement $\bar{A}$. Thus,

$$\mu_{A \cap \bar{A}}(x) = \mu_A(x) \wedge \mu_{\bar{A}}(x) = \min[\mu_A(x), 1 - \mu_A(x)].$$

Then, for $q = 1$,

$$h_1[\mu_A(x)] = \frac{2}{N} \sum_{i=1}^{N} \mu_{A \cap \bar{A}}(x_i), \tag{31b}$$

and for $q = 2$

$$h_2[\mu_A(x)] = \frac{2}{\sqrt{N}}[\sum \mu^2_{A \cap \bar{A}}(x_i)]^{\frac{1}{2}}. \tag{31c}$$

The sum of linear distances is given in (31b) while the distances in (31c) represent the standard Euclidean metric.

Another measure of the entropy function is based upon Shannon's function, Sh, and is given by

$$h_s(A) = \frac{1}{N \ ln \ 2} \sum_{i=1}^{N} Sh\ [\mu_A(x_i)], \tag{32}$$

where,

$$Sh[\mu_A(x_i)] = ln\{(\mu_A)^{-\mu_A} \cdot (1 - \mu_A)^{-(1-\mu_A)}\},$$

$$= -\mu_A(x_i)\ ln\ [\mu_A(x_i)] - [1 - \mu_A(x_i)] \ln [1 - \mu_A(x_i)],$$

ln is the natural logarithm, and $[\frac{1}{(N\ ln\ 2)}]$ is the normalization factor.

A typical mapping from $x \in X$ to $\mu_A(x)$, and from $\mu_A(x)$ to $h(A)$ is shown in Figure 12 from which the following important mapping properties of $h(A)$ can be observed:

(i) $h(A) = 0$, for $\mu_A = 0$ or 1 (minimum value);

(ii) $h(A) = 1$, for $\mu_A = 0.5$ (maximum value at crossover point);

(iii) $h(A)$ increases monotonically for $\mu_A(x) \in [0, 0.5)$, and decreases monotonically for $\mu_A(x) \in [0.5, 1]$.

---

*Nearest ordinary set A_0 is defined as the set for which

$\mu_{A_0} = 0$, for $0 \leq \mu_A(x) < 0.5$, and

$\mu_{A_0}(x) = 1$, for $0.5 \leq \mu_A(x) \leq 1$.

In this notion of entropy, $\mu_A(x)=0$ and $\mu_A(x)=1$ represent respectively the certainty of non-membership ($\mu_A=0$) and full membership ($\mu_A=1$) and, hence, there is no fuzziness (ambiguity) in the information. This implies that $h(A, \mu_A = 0 \; or \; 1) = 0$. When $\mu_A = 0.5$, there is a maximum amount of uncertainty associated with the function, thus making $h(A, \mu_A = 0.5) = 1$, the maximum value.

It should be noted that the operators **INT**(A) and **BLR**(A) defined in Section 4 essentially and respectively decrease and increase the entropy of A, thus decreasing and increasing the ambiguity that is associated with A.

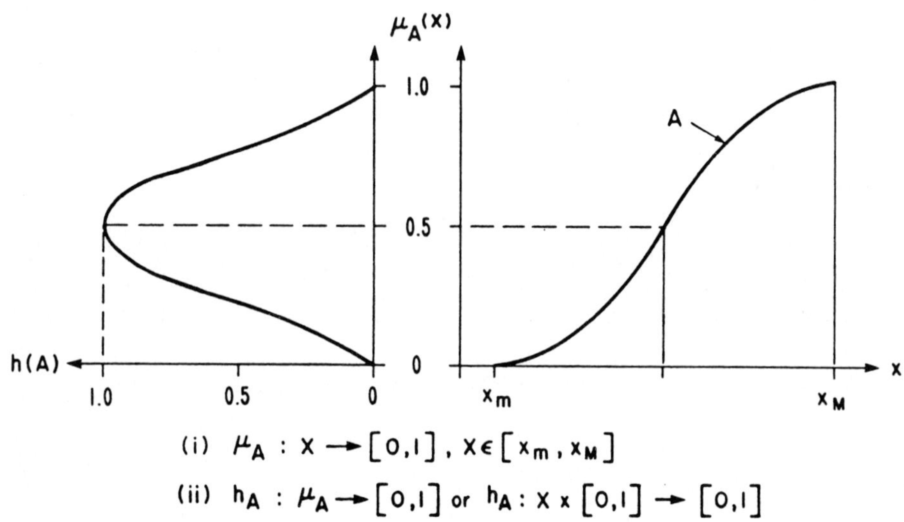

Fig. 12. Membership Function $\mu_A(x)$ and its Entropy Function h(A).

6. CONCLUSIONS

In this paper, we have given an important class of cognitive mapping fuctions that are derived using sinusoids. These mapping functions, along with other various fuzzy mathematical operations, give rise to many interesting properties and applications, which, hopefully, the readers will find useful in their studies in the field of *cognitive uncertainty* and *cognitive information*.

References

[1] R.K. Ragade and M.M. Gupta, "Fuzzy Set Theory: Introduction", *Fuzzy Automata and Decision Processes,* ed. M.M. Gupta, G.N. Saridis and B.R. Gaines, *North-Holland,* New York, pp. 105-131, 1977.

[2] L.A. Zadeh, "Outline of a New Approach to the Analysis of Complex Systems and Decision Process", *IEEE Trans. Syst. Man, Cybern.,* Vol. SMC-3, No. 1, pp. 28-41, January 1973.

[3] S.K. Pal, "Decision-Making Through Fuzzy Measure", *Approximate Reasoning in Expert Systems,* ed. M.M. Gupta, A. Kandel, W. Bandler, J.B. Kiszka, *North-Holland,* New York, pp. 179-194, 1985.

[4] A. Kaufman and M.M. Gupta, *Introduction to Fuzzy Arithmetic: Theory and Applications, Van Nostrand,* New York, 1985.

[5] M.M. Gupta, G. Knopf and P.N. Nikiforuk, "Computer Vision with Fuzzy Edge Perception", *IEEE International Symposium on Intelligent Control,* Jan. 18-20, Philadelphia, 1987.

[6] M.D. Levine, *Vision in Man and Machine, McGraw-Hill Inc.,* New York, 1985.

[7] A. DeLuca and S. Termini, "A Definition of a Non-Probabilistic Entropy in a Setting of Fuzzy Set Theory", *Information Control,* Vol. 20, pp. 301-312, 1972.

TABLE 1

SUMMARY OF COGNITIVE MAPPING FUNCTIONS

MAPPING FUNCTION	DESCRIPTION
(1) S_1 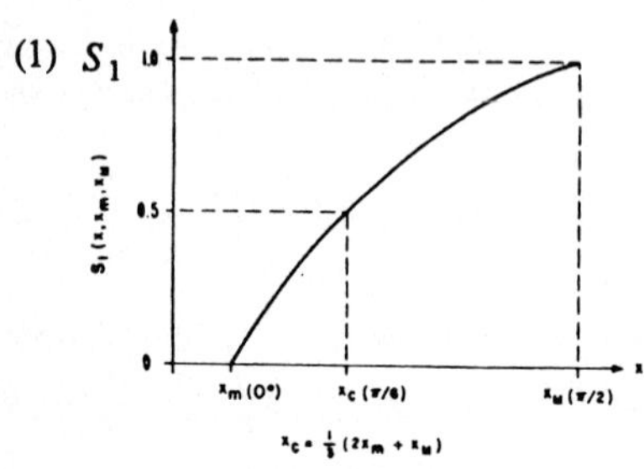 Fig. 1, Eqn. (5)	$\mu_{S_1}(x) = S_1(x, x_m, x_M) = \sin \frac{\pi}{2} (\frac{x - x_m}{x_M - x_m})$, $x_m \leq x \leq x_M$, $x_c = \frac{1}{3} (x_M + 2x_m)$.
(2) S_2 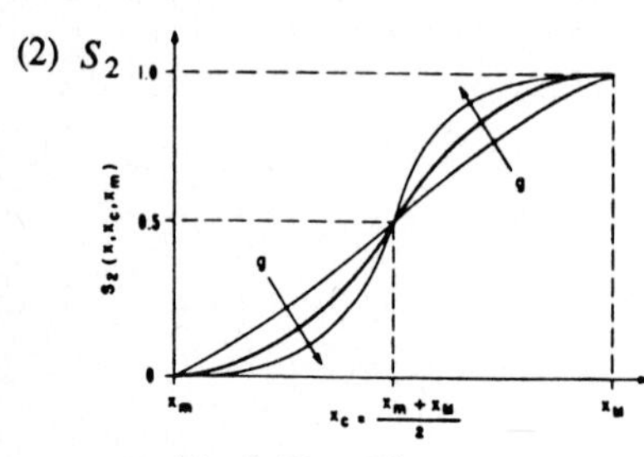 Fig. 2, Eqn. (6)	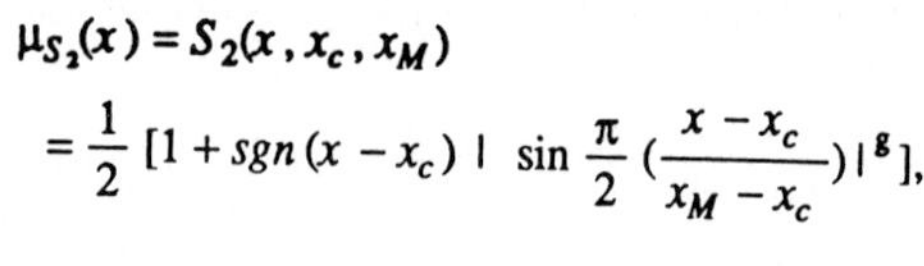 $\mu_{S_2}(x) = S_2(x, x_c, x_M)$ $= \frac{1}{2} [1 + sgn(x - x_c) \mid \sin \frac{\pi}{2} (\frac{x - x_c}{x_M - x_c}) \mid^g]$, $x_m \leq x \leq x_M,\ g > 1$, $x_c = \frac{1}{2} (x_m + x_M)$.
(3) S_3 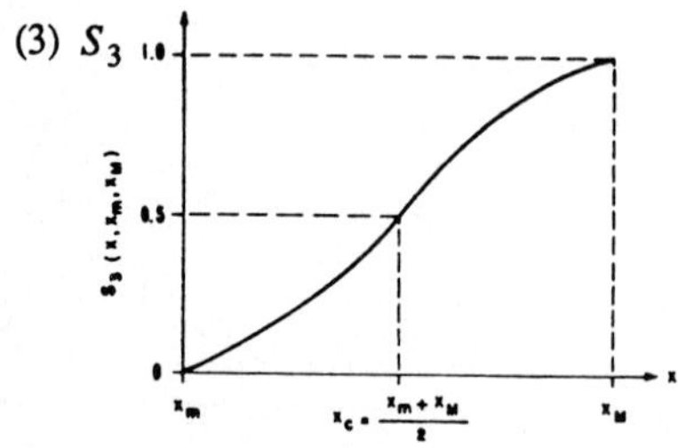 Fig. 3, Eqn. (7)	$\mu_{S_3}(x) = S_3(x, x_m, x_M) = \sin^2 \frac{\pi}{2} (\frac{x - x_m}{x_M - x_m})$, $x_m \leq x \leq x_M$, $x_c = \frac{1}{2} [x_m + x_M]$.

(4) π_1

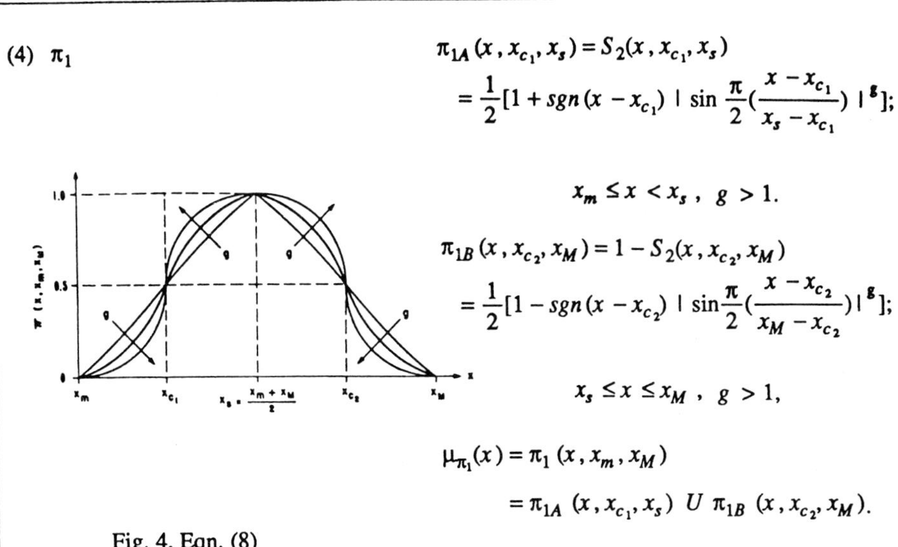

$$\pi_{1A}(x, x_{c_1}, x_s) = S_2(x, x_{c_1}, x_s)$$
$$= \frac{1}{2}[1 + sgn(x - x_{c_1}) \mid \sin \frac{\pi}{2}(\frac{x - x_{c_1}}{x_s - x_{c_1}}) \mid^g];$$

$$x_m \leq x < x_s\,, \quad g > 1.$$

$$\pi_{1B}(x, x_{c_2}, x_M) = 1 - S_2(x, x_{c_2}, x_M)$$
$$= \frac{1}{2}[1 - sgn(x - x_{c_2}) \mid \sin \frac{\pi}{2}(\frac{x - x_{c_2}}{x_M - x_{c_2}}) \mid^g];$$

$$x_s \leq x \leq x_M\,, \quad g > 1,$$

$$\mu_{\pi_1}(x) = \pi_1(x, x_m, x_M)$$
$$= \pi_{1A}(x, x_{c_1}, x_s) \; U \; \pi_{1B}(x, x_{c_2}, x_M).$$

Fig. 4, Eqn. (8)

(5) π_2

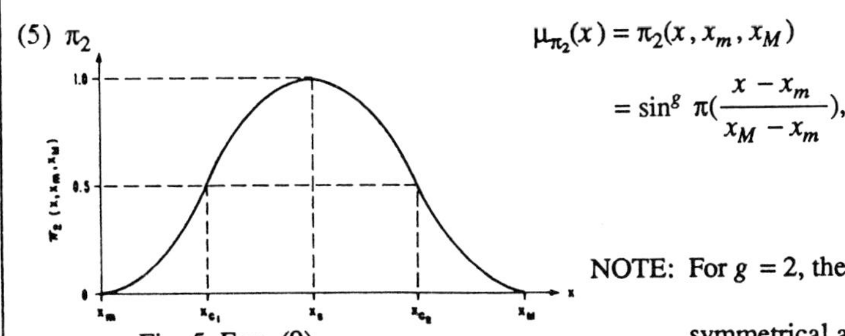

$$\mu_{\pi_2}(x) = \pi_2(x, x_m, x_M)$$
$$= \sin^g \pi(\frac{x - x_m}{x_M - x_m}), \qquad x_m \leq x \leq x_M.$$

Fig. 5, Eqn. (9)

NOTE: For $g = 2$, the crossover points are symmetrical about x_{c_1} and x_{x_2}.

(6a) S_{LH}

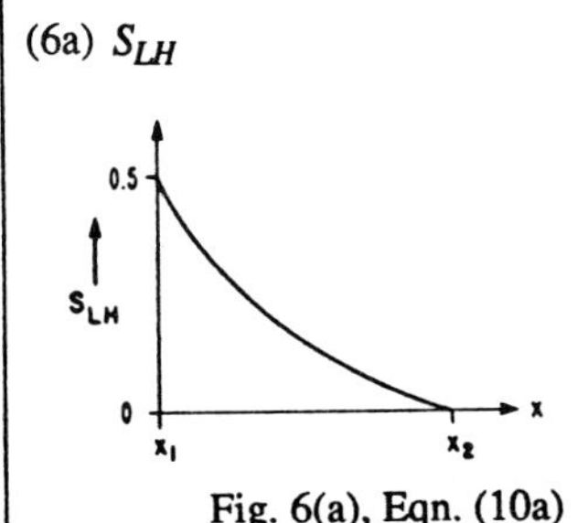

$$\mu_{S_{LH}}(x) = S_{LH}(x, x_1, x_2)$$
$$= [1 - \sin^2 \frac{\pi}{4}(\frac{x_2 + x - 2x_1}{x_2 - x_1})],$$

$$x_1 \leq x \leq x_2.$$

Fig. 6(a), Eqn. (10a)

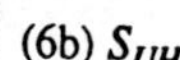

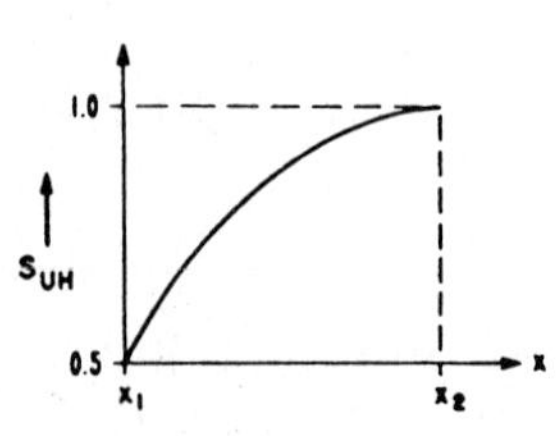

Fig. 6(b), Eqn. (10b)

$$\mu_{S_{UH}}(x) = S_{UH}(x, x_1, x_2)$$
$$= \sin^2 \frac{\pi}{4} \left(\frac{x_2 + x - 2x_1}{x_2 - x_1}\right),$$
$$x_1 \leq x \leq x_2.$$

(7a) π_{UH}

Fig. 7(a), Eqn. (11a)

$$\mu_{\pi_{UH}}(x) = \pi_{UH}\ (x, x_1, x_2)$$
$$= \frac{1}{2}[1 + \sin^2 \pi \left(\frac{x - x_1}{x_2 - x_1}\right)].$$

(7b) π_{LH}

Fig. 7(b), Eqn. (11b)

$$\mu_{\pi_{LH}}(x, x_1, x_2) = \pi_{LH}(x, x_1, x_2)$$
$$= \frac{1}{2}\ [1 - \sin^2 \pi \left(\frac{x - x_1}{x_2 - x_1}\right)].$$

NOTE:

x_c : crossoverpoint, $\mu = 0.5$

x_s : midpoint

Fuzzy Logic in Knowledge-Based Systems,
Decision and Control
M.M. Gupta and T. Yamakawa (Editors)
Elsevier Science Publishers B.V. (North-Holland), 1988

ENTROPIC MODELS : A GENERAL FRAMEWORK FOR MEASURES OF UNCERTAINTY AND INFORMATION

Bernadette BOUCHON

CNRS, LAFORIA, Université Paris VI, Tour 45
4 place Jussieu, 75252 Paris Cédex 05, France

Entropic models provide a general framework to deal with various quantities measuring the uncertainty that may occur in an experiment or a process. Such quantities have been introduced in a probabilistic or a possibilistic way. They can be called quantities of information, entropies, measures of uncertainty or fuzziness, fuzzy entropies, for instance. Their properties are various and different types of entropic models are defined to classify all these quantities and point out their differences of structure.
Keywords : Measures of uncertainty, measures of fuzziness, vagueness, information, entropy.

1. INTRODUCTION

Many quantities have been introduced in the past years to cope with problems of uncertainty or inaccuracy contained in the data involved in an observed phenomenon, a transmitting channel, a decision process, an expert system, for instance. The vagueness or the incertitude they take into account has various aspects. It can be probabilistic, and the tools to deal with them are classical, proposed by probability and information theory. It may also be non probabilistic because of the use of subjective criteria, of approximate variables, of observations difficult to describe, for example. The methods to manage this second type of uncertainties are related to fuzzy set or possibility theory.

All these quantities have been extensively introduced and studied in the few last years. Their nature, their properties are various and they do not always seem to belong to the same class of tools. A formal analogy between a measure of uncertainty and the Shannon information is clearly not sufficient to confer an interest or a vindication to this measure, to its utilization and to the handling of facts through this quantity. It does not mean that this measure has no intrinsic interest, but its good properties often stem from the introduction of the logarithmic function in its expression, they do not have any informational interpretation, as often claimed.

For this reason, we have defined entropic models [6], to provide a general framework allowing us to classify them, identify their variable nature and point out the different purposes of their utilization. Some operations using them, such as the additivity or the optimization, can be exhibited as completely different, even though they seem formally similar.

Entropic models of the first type can be considered to measure the information contained in the results of an experiment, in spite of the uncertainty occurring during the observation. Entropic models of the second type measure this uncertainty itself.
This work provides a unification of concepts which have been used in different ways with regard to some measures of fuzziness, entropy or information and it also exhibits the structural differences between quantities introduced in analogous ways but treating completely different aspects of uncertainty.

2. DEFINITIONS AND PROPERTIES OF ENTROPIC MODELS

2.1. Definitions

Let U be a given finite universe and C a family of characterized subsets of U, i.e. elements of **P**(U) associated with either a real coefficient or a mapping taking values in [0, 1]. We consider, for any integer n, families $P_n = \{(p_1, \ldots, p_n) / p_i \geq 0\ \forall i, \sum_i p_i = 1\}$ of probability distributions on (U, **P**(U)) by means of a probability law p defined on (U, **P**(U)) and families $F_n = \{(x_1, \ldots, x_n) / x_i \in C, \forall i\}$ of n-tuples of C. More particularly, we consider subfamilies of C, such as $C_1 = \mathbf{P}(U)$, $C_2 = \{(x,u) / x \in \mathbf{P}(U), u \in \mathbf{R}\}$ and $C_3 = \{(U,\mu) / \mu : U \to [0,1]\}$. We denote by $F_n{}^j$ the restriction of F_n to sequences of C_j, $1 \leq j \leq 3$. .
An entropic model is a sequence of mappings defined on $F_m \times P_m \times F_n \times P_n$ and ranging in **R** (for any integers m and n), satisfying several basic requirements, and denoted by

$$J_{mn}(X, Q, Y, P) = J_{mn} \left\| \begin{matrix} x_1, \ldots, x_m \\ q_1, \ldots, q_m \end{matrix} \right| \left| \begin{matrix} y_1, \ldots, y_n \\ p_1, \ldots, p_n \end{matrix} \right\| .$$

An interpretation of the part taken by the four elements of the 4-tuple (X, Q, Y, P) can be given in several ways. The pair (X, Q) can be regarded as the actual characteristics of the studied phenomenon, as opposed to the pair (Y, Q) representing the description of this phenomenon available through the observations. The pair (Q, P) corresponds to the probabilistic aspects of the data, when the pair (X, Y) collects the non probabilistic elements involved in the phenomenon. Another possible view of these parameters of an entropic model deals with the construction of a fictive pair (X, Q) from a given observation (Y, Q), which could help a deep study of the phenomenon.
We suppose that Q and P respectively depend on X and Y by means of a mapping R : $\cup_n F_n \to \cup_n P_n$, with Q = R(X) and P = R(Y). In this paper, we restrict ourselves, to the following cases : either the entropic model depends on only one of the pairs (X, Q) and (Y,P) (case 1) , or these two pairs are related together (case 2). We denote by X = T(Y) the result of the transformation T expressing this relationship (case 2) or indicating that X is not involved in the entropic model as soon as Y is taken into account (case 1). We can sum up the relations between the elements of the 4-tuple (X, Q, Y, P) as follows :

$$\begin{array}{ccc} & (T) & \\ X & \leftarrow & Y \\ (R)\downarrow & & \downarrow(R) \\ Q & & P \end{array}$$

2.2. Basic requirements

The basic properties required from an entropic model are described with regard to the pair (Y, P) since it can be considered either as the only pair involved in the value of the entropic model (case 1), or as the reference pair from which the other one is obtained (case 2). We denote by J_{*n} the entropic model, where * indicates the number of components in T(y), immediately deduced from the number of components in any y of $U_n F_n$ and from the transformation T. The list of basic requirements is the following [7]:

- *(S) symmetry*:
The replacement of $Y = (y_1, ..., y_n)$ by $Y' = (y_{r(1)}, ..., y_{r(n)})$, for any permutation r of {1, ..., n}, and the associated P = R(Y) by P' = R(Y'), does not change the value of the entropic model :

$$J_{*n}(T(Y'), R(T(Y')), Y', P') = J_{*n}(T(Y), R(T(Y)), Y, P)$$

- *(E) expansibility*:
The replacement of $Y = (y_1, ..., y_n)$ by $Y' = (y_1, ..., y_n, V)$, for an "empty" element V of C, does not change the value of the entropic model :

$$J_{*(n+1)}(T(Y'), R(T(Y')), Y', P') = J_{*n}(T(Y), R(T(Y)), Y, P)$$

This may mean that $V = \emptyset$ if $Y \in F_n^1$, that V = (w, 0), for $w \in C_1$, if $Y \in F_n^2$ and that $V = (U, \mu)$, with $\mu(w) = 0\ \forall w \in U$, if $Y \in F_n^3$. It may also mean that $R(Y) = (p_1, ..., p_n)$ and $R(Y') = (p_1, ..., p_n, 0)$.

- *(C) continuity*:
The value of the entropic model is continuous with respect to the weighting elements of the components of Y, i.e. the coefficients f_i, with $y_i = (w_i, f_i)$, if $Y \in F_n^2$ or the values of $\mu(w)$, $\forall w \in U$ if $Y = (U, \mu) \in F_n^3$. It is also continuous with respect to the probabilities p_i, $1 \le i \le n$.

These first properties can be considered as elementary and they are useful for an easy exhibiting of functions to define entropic models. The next requirements are more sophisticated and they deal with the refinement of the informations available about the studied phenomenon.

- *(R) recursivity*:
The replacement of $Y = (y_1, ..., y_n)$ par $Y' = (y_1 \cup y_2, y_3, ..., y_n)$, and $P = (p_1, ..., p_n)$ by $P' = (p_1+p_2, p_3 ..., p_n)$ reduces the value of the entropic model proportionally to the value of the entropic model corresponding to the 4-tuple (T(Y"), R(T(Y")), Y",

P")), with $Y" = (y_1, y_2)$ and $P" = (p_1/(p_1+p_2), p_2/(p_1+p_2))$. The operator U is the classical union if $Y \in F_n^1$ and, more generally satisfies $P' = R(Y')$ in any case.

Therefore, the following inequality holds :

$$J_{*(n-1)}(T(Y'), R(T(Y')), Y', P') \leq J_{*n}(T(Y), R(T(Y)), Y, P).$$

This property takes into account the fact that grouping two components of Y is equivalent to observing the studied phenomenon with less details in a second phase of the experiment than in the first one, since y_1 and y_2, separately involved in Y, become indistinguishable in Y'. It expresses the fact that the entropic model has a higher value when the description of the facts is more detailed.

If we interpret the recursivity property in the converse way, we remark that it allows to evaluate the value of the entropic model corresponding to Y in F_n from its value corresponding to Y' in F_{n-1} and, by a recursive process, from its value corresponding to elements in F_2.

More clearly, we can calculate J_{*n} by means of some mapping $\psi : F_2 \times P_2 \rightarrow [0,1]$, for instance :

$$J_{*n}(T(Y), R(T(Y)), Y, R(Y)) = J_{*(n-1)}(T(Y'), R(T(Y')), Y', R(Y')) + \psi(Y", R(Y"))\ J_{*2}(T(Y"), R(T(Y")), Y", R(Y")),$$

and :

$$J_{*(n-1)}(T(Y'), R(T(Y')), Y', R(Y')) = J_{*(n-2)}(T(Y'_1), R(T(Y'_1)), Y'_1, R(Y'_1)) + \psi(Y"_1, R(Y"_1))\ J_{*2}(T(Y"_1), R(T(Y"_1)), Y"_1, R(Y"_1)),$$

with $Y'_1 = (y_1 \cup y_2 \cup y_3, y_4 ..., y_n)$, $Y"_1 = (y_1 \cup y_2, y_3)$. It is easy to see that Y" and $Y"_1$ are in F_2 and that this method can be repeated until the only quantities appearing in the calculus of J_{*n} concern elements of F_2 and their correspondents by R and T. It must also be remarked that this construction is not unique and that we could consider, for example, $Y'_1 = (y_1 \cup y_2, y_3 \cup y_4, y_5, ..., y_n)$, $Y"_1 = (y_3, y_4)$ instead of the previous solution.

- *(A) additivity* :

If we consider elements $Y = (y_1, ..., y_n)$ associated with $P = (p_1, ..., p_n)$ and $Y' = (y'_1, ..., y'_{n'})$ associated with $P' = (p'_1, ..., p'_{n'})$ which are non-interactive, in some sense, with regard to an operation ∩ such that $Y*Y' = (y_1 \cap y'_1, ..., y_i \cap y'_j, ...)$ is associated by R with $P*P' = (p_1 p'_1, ..., p_i p'_j, ...)$, then this combined element Y*Y' provides a value of the entropic model equal to the sum of the two original values provided by Y and Y' :

$$J_{*nn'}(T(Y*Y'), R(T(Y*Y')), Y*Y', P*P') = J_{*n}(T(Y), R(T(Y)), Y, P) + J_{*n'}(T(Y'), R(T(Y')), Y', P')$$

This property presents an interest similar to the previous one. It says that the description of the studied phenomenon contains more details in the last experiment, corresponding to Y*Y', than in each of the two preliminary ones, respectively corresponding to Y and to Y'. As a matter of facts, the components y_i ($1 \leq i \leq n$) of the description, given in a first experiment are split into several indications $y_i \cap y'_1, ..., y_i \cap y'_{n'}$ about the observed phenomenon, in the last experiment. Consequently, the entropic model has a higher value in the final experiment than in the original ones. Further, the additivity property indicates that the value of the entropic model corresponding to an experiment crossing the informations provided by two non interactive experiments is exactly equal to the sum of the values that would be obtained by means of these two distinct experiments.

In both properties of recursivity and additivity, the elements of description of the studied phenomenon provided by the last experiment are made of combinations of those obtained during the preliminary experiments. The aspects of the phenomenon described through the observations are not the same during the different stages of the study.
The last basic property we propose to require from entropic models concerns a possible change in the indications obtained during two different experiments, the associated probability distributions remaining identical. It can be interpreted as an improvement in the description of the components y_i $(1 \le i \le n)$ observed about the studied phenomenon, the probability induced by these components not being altered. The observation is clearer, in some sense, in the second experiment than in the first one, but concerns the same aspects of the phenomenon. This property is the following :

- *(M) monotonicity :*
If some partial order $\tilde{<}$ is defined on F_n, for any n, and two elements $Y = (y_1, \ldots, y_n)$ and $Y' = (y'_1, \ldots, y'_n)$ are such that $Y \tilde{<} Y'$ and $P = R(Y) = R(Y')$, then the value of the entropic model corresponding to Y is not greater than the one corresponding to Y' :

$$J_{*n}(T(Y), R(T(Y)), Y, P) \le J_{*n}(T(Y'), R(T(Y')), Y', P)$$

2.3. Types of entropic models

The terminology is similar to the most classical names of conditions used in information theory [2], and different ways of introducing some of the above-mentioned requirements could be given. Nevertheless, they seem sufficient to classify the most important measures of uncertainty.
The basic requirements allow the following specification of entropic models.

An *entropic model of type 1* satisfies at least two of the properties of symmetry, expansibility, continuity and one of the properties of recursivity and additivity.
An *entropic model of type 2* satisfies at least two of the properties of symmetry, expansibility, continuity and also the property of monotonicity.

In other words, an entropic model of type 1 takes into account the choice of the characteristics used to describe an observation, their sharpness, their possible splitting or grouping. An entropic model of type 2 measures the quality, the accurracy of the characteristics provided by the observation, the reliability or credibility of the experimenter with regard to the observations. We can roughly remark that entropic models of type 1 describe an informational point of view, when entropic models of type 2 express a study of vagueness.
The three first properties (S), (E) and (C) are common to the two types of models. A difference stems from the fact that the properties of recursivity (R) or additivity

(A) compare the value of entropic models for elements of $U_n F_n$ with different numbers of components, Y and Y' for (R), with respectively n and n-1 components, Y and Y*Y' for (A), with respectively n and nn' components. Further, (R) needs that the components of any $Y \in F_n$ are choosen in a subset F' of $U_n F_n$ closed for a given operation U, in such a way that, if Y = (..., y', ..., y", ...) belongs to $F' \cap F_n$, then Y' = (..., y' U y", ...) belongs to $F' \cap F_{n-1}$. Similarly, (A) needs that the components of any $Y \in F_n$ are choosen in a subset F" of $U_n F_n$ closed for a given operation $\cap$, in such a way that, if Y = (..., y, ...) and Y' = (..., y', ...) belong to $F'' \cap F_n$, then Y*Y' = (..., y$\cap$y', ...) belongs to $F'' \cap F_{nn'}$.

These conditions are not always available, for instance because U is not possible for entropic models defined for elements $Y \in F_n$ such that $y_i = (\{w_i\}, f_i)$, for U = $\{w_1, ..., w_n\}$ and $f_i \in [0, 1]$, $\forall i \in \{1, ..., n\}$. Some examples of such models will be given in the last section. Consequently, they cannot be considered as entropies, in the classical meaning, even though they satisfy formulae formally similar to properties of the Shannon's entropy.

Conversely, the monotonicity property (M) requires the comparison of values of an entropic model for two elements Y and Y' of $U_n F_n$, with the same number of components, associated with the same probability distribution P = R(Y) = R(Y'). This is not possible, for instance, if $Y = (y_1, ..., y_n) \in F_m{}^1$ are such that $y_i \cap y_j = \emptyset$, $\forall i \neq j$, p_i is the probability of y_i, for any $i \in \{1, ..., n\}$ and the entropic model only depends on Y by means of P. Once more, examples will be given in the last section.

2.4. Example of an additional property, interpretable for entropic models of both types

Many other properties are satisfied by entropic models ; they could be added to the previous list, but they seem either less fundamental then these six ones, or equivalent to several of them. Some of these properties present the interest of being interpretable for entropic models of types 1 and 2 and then to establish a relationship between these two kinds of measures of uncertainty.

For instance, we consider the branching property, classical in information theory, which takes into account the combination of two processes, the first one providing partial results on the observed phenomenon, the second one complementing the informations on some particular aspects of the data. A conditional entropic model is introduced in the following way : let $Z = (z_1, ..., z_n)$ be the description of the first process yielding informations on the phenomenon, associated with the probability distribution $R(Z) = (r_1, ..., r_n)$, and $Y = (y_1, ..., y_k)$ be the description of the second process, corresponding to $R(Y) = (p_1, ..., p_k)$, with $k \geq 1$.

If we suppose that Y occurs in the only case where one of the components z of Z is realized, which means that further informations are required on the studied phenomenon when z is first obtained, then we represent by Y/z a k-tuple of elements of C corresponding to the knowledge deduced from Y, knowing that z holds.

The succession of the two processes will be described by $Z \boxed{z}\, Y = (z_1, \ldots, y_1 \cap z, \ldots, y_k \cap z, \ldots, z_n)$, where $\cap$ denotes an operation satisfying the condition indicated for the additivity property (A). The corresponding probability distribution is $R(Z \boxed{z}\, Y) = (r_1, \ldots, r\,p_1, \ldots, r\,p_k, \ldots, r_n)$, if r is the component of $R(Z)$ with the same index as z in Z. An entropic model may satisfy the following particular condition :

- *(B) branching property :*

The difference between the value of the entropic model deduced from the occurrence of the sequence of processes $Z \boxed{z}\, Y$ (where z is a component of Z and Y appears after z has been obtained during the first experiment) and the value of the entropic model corresponding to Z, is proportional to the value of the entropic model corresponding to Y, knowing that z is realized, whatever Y and Z may be.
Further, we can write the branching property as follows :

$$J_{*(n+k-1)}\,(T(Z \boxed{z}\, Y), R(T(Z \boxed{z}\, Y)), Z \boxed{z}\, Y, R(Z \boxed{z}\, Y)) = J_{*n}\,(T(Z), R(T(Z)), Z, R(Z)) + \Phi(z, r)\, J_{*k}\,(T(Y/z), R(T(Y/z)), Y/z, R(Y/z)),$$

for a mapping $\Phi : C \times [0, 1] \to \mathbf{R}$.

In the case where Y and Z are non interactive with regard to $\cap$, the conditional process Y/z is equivalent to Y itself, and we obtain :

$$J_{*(n+k-1)}\,(T(Z \boxed{z}\, Y), R(T(Z \boxed{z}\, Y)), Z \boxed{z}\, Y, R(Z \boxed{z}\, Y)) = J_{*n}\,(T(Z), R(T(Z)), Z, R(Z)) + \Phi(z, r)\, J_{*k}\,(T(Y), R(T(Y)), Y, R(Y)).$$

If the second experiment, corresponding to Y, provides informations which are necessary after any occurrence of a component of Z, i.e. the observations obtained during the first experiment are not enough detailed and must be complemented, then the repetition of a branching process after any z_i provides the combined experiment $(\,(Z \boxed{z_1} Y) \boxed{z_2} Y) \ldots) \boxed{z_n} Y) = Z{*}Y$. The value of the entropic model corresponding to the series of experiments Z and Y equals, from (B):

$$J_{*(n+k-1)}\,(T(Y{*}Z), R(T(Y{*}Z)), Y{*}Z, R(Y{*}Z)) = J_{*n}\,(T(Z), R(T(Z)), Z, R(Z)) + J_{*k}\,(T(Y), R(T(Y)), Y, R(Y))\,[\textstyle\sum_{1 \le i \le n} \Phi(z_i)].$$

It is clear that this equality is equivalent to the expression of the additivity property (A) in the particular case where the following normalization condition (N) :

$$\textstyle\sum_{1 \le i \le n} \Phi(z_i, r_i) = 1,$$

for any k-tuple $z_1, \ldots, z_n$ in F_n and $r_1, \ldots, r_n$ in P_n.

Then, the branching property (B) is equivalent to the additivity property (A) and determines entropic models of type I, together with two of the properties (S), (E), (C), if there exists a mapping Φ satisfying the normalization condition (N).

Another point of view concerns the case where $k=1$, which means that the utilization of a second process Y, after some result z of the first experiment Z, modifies the knowledge about z but does not split the informations obtained through the first process Z.
This entails the following :

$$J_{*n}(T(Z \boxed{z} Y), R(T(Z \boxed{z} Y)), Z \boxed{z} Y, R(Z \boxed{z} Y)) = J_{*n}(T(Z), R(T(Z)), Z, R(Z)) + \Phi(z, r)\, J_{*1}(T(Y/z), R(T(Y/z)), Y/z, R(Y/z)).$$

We have $p_1 = 1$ and then $R(Z \boxed{z} Y) = R(Z)$. Thus, let us consider the partial order $<$ on F_n for any integer n, such that $Z \tilde{<} Z \boxed{z} Y$. The entropic model satisfies the monotonicity property (M) if there exists a mapping Φ satisfying the following positiveness condition (P) :

$$\Phi(z, r)\, J_{*1}(T(y), R(T(y)), y, R(y)) \geq 0,$$

for any y and z in C and any r in [0, 1].

Consequently, the branching property (B) implies the monotonicity property (M), and defines entropic models of type 2, together with two of the properties (S), (E), (C), if there exists a mapping Φ satisfying the positiviness condition (P).

The branching property can thus be regarded from two points of view, according to the characterization of the involved mapping Φ. If an entropic model satisfies two of the properties (S), (E), (C) and the branching property (B) for a mapping Φ, then it is of type 1 if Φ satisfies (N) and of type 2 if Φ satisfies (P).

3. EXAMPLES OF ENTROPIC MODELS

We will denote by $J_n : F_n \times P_n \to \mathbf{R}$, for any integer n, entropic models only depending on (Y, P) and by J_{mn} (X, Q, Y, P) those for which there exists a transformation T yielding X from Y. We give hereunder the main measures of uncertainty studied the the literature under various names.

◇ Conditions : $Y = (y_1, \ldots, y_n) \in F_n^1$ are such that $y_i \cap y_j = \emptyset$ for any $i \neq j$.
Name : inset entropy defined by Aczel and Daroczy [1].
Expression :

$$J_n(Y, P) = g(\cup_{i=1,\ldots,n} y_i) - \Sigma_{i=1,\ldots,n}\, p_i\, g(y_i) - A\, \Sigma_{i=1,\ldots,n}\, p_i \log p_i,$$

for a mapping g and a real parameter A.
Properties : (S), (C), (R).
Type : 1.
Particular case :

◇◇ Name : Shannon entropy [20].
Expression :

$$J_n(Y, P) = - \Sigma_{i=1,\ldots,n}\, p_i \log p_i.$$

Properties : (S), (E), (C), (R), (A).
Type : 1 (independent of Y).

◇ Conditions : $Y = (y_1, \ldots, y_n) \in F_n^1$ are such that $y_i \cap y_j = \emptyset$ for any $i \neq j$.
Name : Renyi entropy of order α, for a real parameter α ($\alpha \geq 0$, $\alpha \neq 1$) [19].
Expression :

$$J_n(Y, P) = (1/(1-\alpha)) \log (\Sigma_{i=1,\ldots,n}\, p_i^{\alpha}).$$

Properties : (S), (E), (C), (A).
Type : 1 (independent of Y).

◇ Conditions : $Y = (y_1, ..., y_n) \in F_n^2$ with $y_i = (v_i, u_i)$, for $v_i \in \mathbf{P}(U)$ such that $v_i \cap v_j = \emptyset$ for any $i \neq j$, and $u_i \in \mathbf{R}^+$, $\forall i \in \{1, ..., n\}$
Name : "useful" entropy.
Expression :

$$J_n(Y, P) = J_m{}^u \begin{vmatrix} u_1, ..., u_n \\ p_1, ..., p_n \end{vmatrix}, \text{ only depending on } y_i \text{ through } u_i, \forall i \in \{1, ..., n\}$$

Properties : (S), (E), (C), other properties depending on the combination law defined for the coefficients u_i.
Particular case :

◇◇ Name : qualitative and quantitative entropy, introduced by Belis and Guiasu [3].
Expression :
$J_n(Y, P) = - \sum_{i=1,...,n} u_i\, p_i \log p_i$.
Properties : (S), (E), (C), (R).
Type : 1.

◇ Conditions : $U = \{w_1, ..., w_n\}$ and the elements $Y = (y_1, ..., y_n)$ of F_n^2 are such that $y_i = (\{w_i\}, f_i)$, for $w_i \in U$ and $f_i \in [0, 1]$, $\forall i \in \{1, ..., n\}$.
Name : "fuzzy" entropy.
Expression :

$$J_n(Y, P) = J_n{}^f \begin{vmatrix} f_1, ..., f_m \\ p_1, ..., p_m \end{vmatrix}, \text{ only depending on } y_i \text{ through } f_i, \forall i \in \{1, ..., n\}.$$

Properties : (S), (E), (C), other properties depending on the combination law defined for the coefficients f_i.
Particular cases :

◇◇ Name : entropy of a fuzzy event, introduced by L.A. Zadeh [24].
Expression :
$J_n(Y, P) = - \sum_{i=1,...,n} f_i\, p_i \log p_i$.
Properties : (S), (E), (M) with respect to the fuzzy set inclusion $\tilde{<}_{\subseteq}$.
Type : 2.

◇◇ Name : entropy measure, introduced by De Luca and Termini [9].
Expression :
$J_n(Y, P) = - \sum_{i=1,...,n} p_i \log p_i + \sum_{i=1,...,n} p_i\, \Pi(f_i)$,
for a mapping $\Pi : [0, 1] \rightarrow \mathbf{R}$, non decreasing in $[0, 1/2]$, non increasing in $[1/2, 1]$.
Properties : (S), (E), (C), (M) with respect to the order defined as $y_i \tilde{<}_{1/2} y'_i$ if and only if $f_i \leq f'_i$ when $f'_i \leq 1/2$, and $f_i \geq f'_i$ when $f'_i \geq 1/2$.
Type : 2.

◇◇ Name : entropy of the fuzzy set defined by $f_1, \dots, f_n$, introduced by De Luca and Termini [8].
Expression :
$J_n(Y, P) = -\sum_{i=1,\dots,n} (f_i \log f_i + (1-f_i) \log (1-f_i))$.
Properties : (S), (E), (C), (M) with respect to the order $\tilde{<}_{1/2}$ defined as above.
Type : 2, independent of P.

◇◇ Name : max-min entropy, introduced by Trillas and Riera [21].
Expression :
$J_n(Y, P) = \max_{i=1,\dots,n} \min (W(w_i), \Pi(f_i))$,
for a weight $W(w_i)$ connected to any w_i and a mapping Π defined as above.
Properties : (S), (E), (C), (M) with respect to the order $\tilde{<}_{1/2}$.
Type : 2, independent of P.

◇ Conditions : $Y = (y_1, \dots, y_n) \in F_n{}^1$ are such that $y_i \cap y_j = \emptyset$ for any $i \neq j$, and p_i is the prior probability of y_i, when q_i is its posterior probability.
X = Y.
Name : inset-deviation, introduced by Kannappan [15].
Expression :

$$J_{nn}(X, Q, X, P) = g(\cup_{i=1,\dots,n} x_i) - \sum_{i=1,\dots,n} p_i\, g(x_i) - A \sum_{i=1,\dots,n} p_i \log p_i + D \sum_{i=1,\dots,n} p_i \log q_i,$$

for a mapping $g : \mathbf{P}(U) \to \mathbf{R}$, and two real parameters A and D.
Properties : (S), (C), (R).
Type : 1.
Particular case :

◇◇ Name : directed divergence, introduced by Kullback [17].
Expression :
$J_{nn}(X, Q, X, P) = -\sum_{i=1,\dots,n} q_i \log (q_i/p_i)$.
Properties : (S), (C), (R), (A).
Type : 1, independent of Y.

◇ Conditions : $Y = (y_1, \dots, y_n) \in F_n{}^1$ are such that $y_i \cap y_j = \emptyset$ for any $i \neq j$, and p_i is the prior probability of y_i, when q_i is its posterior probability.
X = Y.
Name : gain of information, introduced by Renyi [19].
Expression :

$$J_{mm}(X, Q, X, P) = (1/(\alpha-1)) \sum_{i=1,\dots,n} q_i{}^{\alpha} / p_i{}^{\alpha-1},$$

for a real parameter $\alpha \geq 0$, $\alpha \neq 1$.
Properties : (S), (C), (A).
Type : 1, independent of Y.

◇ Conditions : $U = \{w_1, \dots, w_n\}$ and $Y = (y_1, \dots, y_n) \in F_n{}^2$, with $y_i = (\{w_i\}, f_i)$, and $f_i \in [0, 1]$, $\forall i \in \{1, \dots, n\}$,

$X = \{x_1, x_2, x_3\}$, with $x_1 = \{ w_i \in U / f_i > 1/2 \}$, $x_2 = \{ w_i \in U / f_i = 1/2 \}$, $x_3 = \{ w_i \in U / f_i < 1/2 \}$ and $q_i = p(x_i)$, for $1 \le i \le 3$.

Name : total uncertainty, introduced by Xie and Bedrossian [22].

Expression :

$$J_{3n}(X, Q, Y, P) = -(q_1+0.5q_2) \log(q_2+0.5q_2) -(q_3+0.5q_2) \log(q_3+0.5q_2) - 1/n \sum_{i=1,\dots,n} (f_i \log f_i + (1-f_i)\log(1-f_i)).$$

Properties : (S), (C), (M) with respect to the fuzzy set inclusion $\tilde{<}'_{1/2}$ defined by $y_i \tilde{<}'_{1/2} y'_i$ if and only if $f_i \le f'_i$ when $f'_i < 1/2$, $f_i \ge f'_i$ when $f'_i > 1/2$ and $f_i = f'_i$ when $f'_i = 1/2$

Type : 2.

◇ Conditions : $U = \{w_1, \dots, w_n\}$ and $Y = (y_1, \dots, y_n) \in F_n^2$, with $y_i = (\{w_i\}, f_i)$, and $f_i \in [0, 1]$, $\forall i \in \{1, \dots, n\}$,
$X = \{x_1, \dots, x_m\}$, with $x_j = \{w_i \in U / f_i \ge L_j\}$ for any $j \in \{1, \dots, m\}$, if $L = \{L_1, \dots, L_m\}$ is the set of possible values of coefficients f_i, $1 \le i \le n$.

Name : possibilistic uncertainty, introduced by Higashi and Klir [13].

Expression :

$$J_{mn}(X, Q, Y, P) = (1/\max_{i=1,\dots,n} f_i) \sum_{k=1,\dots,m-1} (L_{k-1}-L_k) \log_2 |x_{k+1}|.$$

Properties : (S), (E), (C), (A), (M) with respect to the fuzzy set inclusion $\tilde{<}_c$.

Types : 1 and 2.

◇ Conditions : $Y = (y_1, \dots, y_n) \in F_n^2$, with $y_i = (U, \mu_i)$, $\forall i \in \{1, \dots, n\}$, for membership functions $\mu_i : U \to [0, 1]$,
$X = (x_1, \dots, x_m)$, with $x_i \cap x_j$ for any $i \ne j$, defined in such a way that any element w of U belongs to one x_j for which $\mu_j(w) \ge s$ ($j \in \{1, \dots, m\}$), for a given threshold s in $[0, 1]$,
$q_j = p(x_j)$, $\forall j \in \{1, \dots, m\}$.

Name : fuzzy gain of information, introduced by Bouchon [4].

Expression :

$$J_{mn}(X, Q, Y, .) = \sum_{i=1,\dots,m} q_i \log (q_i/\sum_{w \in x_i} f_i(w)\ p(w)).$$

Properties : (S), (E), (C), (R), (M) with respect to the order $\tilde{<}_s$ obtained by replacing the value 1/2 by s in the definition of $\tilde{<}_{1/2}$.

Types : 1 and 2.

We can sum up the presented quantities in the following list :

Type 1 :
Shannon entropy
qualitative and quantitative entropy
directed divergence
inset entropy
Renyi entropy
inset-deviation
gain of information

Type 2 :
max-min entropy
entropy of a fuzzy event
total uncertainty

Types 1 and 2:
possibilistic uncertainty
fuzzy gain of information

This list of examples is not exhaustive, but it represents the main directions of research presenting entropies, informations or measures of uncertainty which may be regarded as entropic models of type 1 or 2. There exist other quantities of the same kind introduced in various directions [1], [9], [12], [21] for instance.
Several different quantities have been presented, and they are neither of type 1 nor of type 2, see [14], [23] for instance. They take into account measures of uncertainty, specificity, precision or fuzziness and they belong to other types of entropic models, which should be introduced according their most important common properties.

REFERENCES

[1] J. ACZEL, Z. DAROCZY : A mixed theory of information. I : symmetric, recursive and measurable entropies of randomized systems of events, RAIRO Theoretical Computer Science, 12, 2, 1978, pp. 149-155.

[2] J. ACZEL, Z. DAROCZY : On measures of information and their characterization ; Academic Press, 1975

[3] M. BELIS S. GUIASU : A quantitative-qualitative measure of information in cybernetic systems ; IEEE Trans. Inform. Th. 14,1968, pp. 593-594.

[4] B. BOUCHON : Comparison of experiments and improvement of models ; 2nd World Conf. on Math. at the Service of Man, Las Palmas, 1982.

[5] B. BOUCHON : On measures of fuzziness and information ; 5th Intern. Conf. on Mathematical Modelling, Berkeley, 1985.

[6] B. BOUCHON : Entropic models ; Cybernetics and Systems, 17, 1986.

[7] B. BOUCHON : Mesures d'information relatives à des observations imprécises, 11ème Congrès International de Cybernétique, Namur, 1986.

[8] A. DE LUCA, S. TERMINI : A definition of a nonprobabilistic entropy in the setting of fuzzy sets theory ; Information and Control, 20, 1972, pp. 301-312.

[9] A. DE LUCA, S. TERMINI : Entropy and energy measures of a fuzzy set ; in Advances in Fuzzy Set Theory and Applic., M.M. Gupta et al. (eds), North Holland,1979, pp. 321-338.

[10] Z. DAROCZY : Generalized information functions ; Information and Control 16, 1970, pp.36-51.

[11] B. FORTE, C. SEMPI : Pruning and measures of uncertainty ; RAIRO Theoretical Computer Science 12, 2, 1978, pp. 157-168.

[12] R.V. HARTLEY : Transmission of information ; Bell System Techn. Journ. 7, 1928, pp. 535-563.

[13] M. HIGASHI, G. KLIR : Measures of uncertainty and information based on possibility distributions ; Int. J. General Systems 9,1982, pp. 43-58.

[14] P. HOHLE : Entropy with respect to plausibility measures ; Proc.12th IEEE Int. Symp. Multiple-Valued Logic, Paris, 1982, pp. 167-169.

[15] P. KANNAPPAN : A mixed theory of information. IV - Inset inaccuracy and directed divergence ; Metrika 27, 1980, pp. 91-98.
[16] G. KLIR : Where do we stand on measures on uncertainty, ambiguity, fuzziness, and the like ; Fuzzy Sets and Systems, to appear.
[17] S. KULLBACK : Information theory and statistics ; Wiley, 1959.
[18] H. T. NGUYEN : On entropy if random sets and possibility distributions ; in The analysis of fuzzy information (J. Bezdek, ed.), CRC Press, 1985.
[19] A. RENYI : Probability theory ; North Holland, 1970 (Appendix : Introduction to information theory).
[20] C.E. SHANNON : A mathematical theory of communication, Bell System Tech. J. 27, 1948, pp. 379-423, pp. 623-656.
[21] E. TRILLAS, T. RIERA : Entropies in finite fuzzy sets, Information Sciences 15, 1978, pp. 159-168.
[22] W.X. XIE, S.D. BEDROSIAN : An information measure for fuzzy sets ; IEEE Trans. Systems, Man and Cybern. 14, 1, 1984, pp. 151-156.
[23] R.R. YAGER : Entropy and specificity in a mathematical theory of evidence ; Int. J. General Systems 1, 1, 1983, pp. 249-260.
[24] L.A. ZADEH : Probability measures of fuzzy events ; J. Math. Anal. Applic., 23, 1968, pp. 421-427.

Fuzzy Logic in Knowledge-Based Systems,
Decision and Control
M.M. Gupta and T. Yamakawa (Editors)

PARALLELISM, INTEGRATION, AUTOCOORDINATION AND AMBIGUITY IN HUMAN SUPPORT SYSTEMS

Milan ZELENY

The Joseph A. Martino Graduate School of Business Administration, Fordham University at Lincoln Center, New York, NY 10023, U.S.A.

It is important to realize how the advances in technological hardware are at the same time causing and are themselves being caused by the advances in technological software. Parallel processing, autonomous coordinability and distributed intelligence in the hardware are being complemented and further stimulated by the emerging integrated knowledge-based software. Even further: socio-economic macro-trends themselves reflect similar shifts towards self-management, self-organization, autonomy and knowledge (as well as labor) reintegration. Consequently, hardware (the means), software (the know-how) and "brainware" (the know-what and -why) developments, see [17], are not proceeding independently and separately, but in step with the rapidly unfolding socio-economic dynamics.
What is the role of fuzziness and ambiguity of human language in all this? It is argued that fuzziness and ambiguity are not just "necessary evils" to be overcome and controlled by mathematical transformations, but indispensable strategic human "inventions" which make the spontaneity, autonomy, self-coordination and self-management finally possible.

KEYWORDS: Fuzziness; autopoiesis; knowledge; artificial intelligence; high technology.

> The phenomenon connoted by the world perception cannot be one of grasping features of an independent object world, but consists of bringing forth a world of actions.
>
> *Humberto R. Maturana*

1. INTRODUCTION

Parallel processing is not just a simple <u>hardware</u> advancement but a reflection of rather specific and urgent needs brought about by rapidly changing theory and praxis of knowledge, self-organization and self-management, distributed intelligence, autonomous coordinability, multiple criteria decision making (MCDM) as well as accompanying theories of autopoiesis (self-production), ambiguity and approximate reasoning.

<u>Remark</u>. Referring to "ambiguity and approximate reasoning" we wish to draw a distinction from the traditional theories of "fuzzy sets." Ambiguity and approximate reasoning are powerful human <u>systems</u> concepts, in fact they are strategic tools for achieving coherence and coordination of human action. In this sense they are indispensable if self-organization and autocoordination are to take place in human systems: they have to be reinforced and amplified in their function, not limited or perhaps even removed through pre-fixed and rigid symbolic transformations.

In addition to recognizing the positive role of ambiguity and approximate reasoning in systems autonomous coordination, we also progress from data processing through information systems to knowledge-acquisition engineering and support. While data and information are piecemeal components, partial and atomized by their very nature, knowledge and wisdom are "holistic" concepts, related to and expressed through systemic network patterns and thus integrative by definition.

While data and information (being components) can be generated *per se* and without explicit human interpretation, knowledge and wisdom (being relations) cannot. Knowledge and wisdom are human- and context-dependent and cannot be even contemplated without explicit involvement of human (not machine-like) comparisons, decision making and judgment.

In this paper we stress the coordinative and integrative nature of knowledge and the need for developing the "integration-friendly" and knowledge supportive systems. We further argue that ambiguity and fuzziness are necessary for self-coordination, self-management and autopoiesis of social orders to take place.

2. KNOWLEDGE

What is knowledge?

Maturana [5] sees perception as an operation of distinction, bringing forth a system of operational coherences, the world of coordinated action. Such coordination of actions takes place and is distinguished in language. Language then is a consensual (conversational) coordination of consensual (individual) coordinations of action.

Knowledge refers to observer's distinction of "objects" (wholes or unities) through which he brings forth from the background of experience a coherent and self-consistent set of coordinated actions.

Through the operation (or process) of distinction, individual pieces of data and information (components, concepts) become connected with one another (i.e. organized) in a network of relations. Knowledge is contained in the overall organizational pattern of the network and not in any of the separate components (nor in their simple aggregates or collections).

Thus emerging network of relations in turn determines which and how additional components are further brought forth ("produced", interpreted) from the background. Network organization is necessarily circular (organizationally closed), self-maintaining and self-producing. Its components are subject to external (and internal) structural perturbations and they "excite" or "inhibit" each other in a simultaneous (or parallel) fashion [7,12].

Knowledge is therefore an autopoietic (self-producing) system. See reviews in [5,13,14] for the necessary foundations of autopoiesis.

Knowledge cannot refer to a "given and fixed" set of objects "out there", which are to be simply "captured", represented or modeled. Such old idea of symbolic computation (associated with the names of H. Simon, A. Newell, N. Chomsky or M. Minsky) – the idea that knowledge and mental processes involve structures of symbols and transformations of symbolic expressions – is not in harmony with the emerging societal transformations. They might be quite suitable for artificial intelligence, but they are quite unrelated to human intelligence.

For example, human vision does not work as a camera, merely recording an image, as film does, and passing it to brain for analysis. Yet, machine vision is generally accomplished by recording and image with a camera. Why are cognitive sciences, autopoiesis and principles of neuronal organization "bypassed" by artificial intelligentsia?

Similarly, we have to abandon such familiar programming concepts as data, procedures and subroutines, and build programs that function as a community of near-autonomous individuals – the objects of knowledge. Each object is actually a piece of software that behaves like an expert in its own particular domain: a database is an object that knows how to take statistical information about itself and display it. Newly emerging *object-oriented programming* is now starting to approach the new ideal of autopoietic networks.

People do not "compute their way through" a given objective world: knowledge refers to the process of active network configuration and reconfiguration of a particularly human world of objects and their relations.

Because the objects of knowledge are not given and fixed a priori, but are repeatedly brought forth via the operation of human distiction, their definitional boundaries cannot be crisp or sharp. Consequently, the language which we use to coordinate our knowledge-derived actions must appear to be "imprecise and fuzzy" from any the information-processing perspective. In reality, fuzzy language is not only adequate, effective and "precise" for the purposes of action coordination – it is necessary.

In Figure 1 the crucial differences are outlined. In Fig. 1a, the observer-computor [C] perceives objects and concepts given "out there" in a crisp, well-defined fashion. He computes his way through the independent world of precisely measurable "scientific" concepts. In Fig. 1b we have a similar but fuzzified situation: "observer-computor" now ploughs through a world of fuzzy concepts "out there", computing crisply through their hypothesized fuzziness.

The capturing (crisp or fuzzy) of the crisp or fuzzy world "out there" allows treating knowledge through symbolic rules manipulation of disconnected components in essentially context-free fashion..

In Fig. 1c we finally acknowledge that the objects and concepts are brought forth from the background by the observer-coordinator [C]. He does not compute the world "out there", whether crisp or fuzzy", but creates his own world of coordinated action by bringing forth the components and interconnecting them into autopoietic

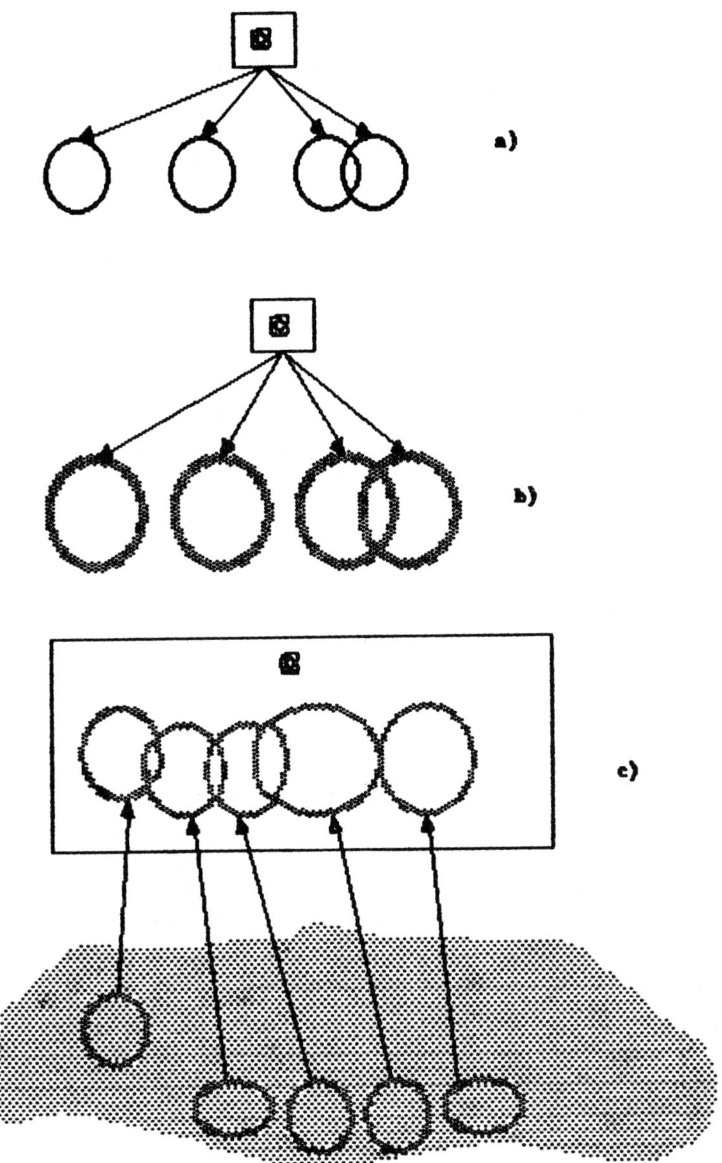

Figure 1.

networks of meaning, knowledge and coordinated action. Immediate, intertemporal, individual and local contexts become crucial [see also 5] to what is being "brought forth" at any moment. There is no a priori and per se meaning to be attached to any component "out there": only through the operation of distinction (bringing forth) and embedding in the system network of coordinated action can any component attain its meaning in the observer-coordinator domain. The set of coherent actions brought forth through the operation of distinction is coordinated by and in language. Language must be an integral part of any knowledge system based model, not as "processor of information" but as a "coordinator of action". The study of language as a means for effective coordination of human action has so far been neglected.

Because the objects of knowledge are not given and fixed a priori, but are repeatedly brought forth via the operation of human distiction, their definitional boundaries cannot be crisp or sharp. Consequently, the language which we use to coordinate our knowledge-derived actions must be "imprecise and fuzzy" from any the information-processing perspective. In reality, fuzzy language is not only adequate, effective and "precise" for the purposes of action coordination - it is necessary.

3. KNOWLEDGE AND AMBIGUITY

Human perception, knowledge and understanding maintain their effectiveness in spite of incomplete, ambiguous and fuzzy pieces of information. This is the consequence of the fact that knowledge is an autopoietic relational pattern and not a particular component(s).

A network of distributed but interconnected components (that excite and inhibit one another in a parallel fashion) is capable of dealing with the ambiguity and fuzziness of components (both internal-constitutive and external-perturbational). Network structural flexibility and openness to multiple sources of perturbations (information), while maintaining the organization of relations, is an effective way of dealing with the "incoming" ambiguity [7].

The requisite factor-flexibility and source-multiplicity have been difficult to simulate in the Simonian symbolic computation. By concentrating on the ambiguity and fuzziness of network components we do lose sight of the whole and thus, by definition, miss the phenomenon of knowledge. Traditional efforts to transform the (information) source ambiguity into "crisp" mathematical symbolism does not help. Abstracting from the source ambiguity entirely and simply re-transforming the traditional "crisp" symbolism into the "fuzzy" one ("fuzzification" of the fuzzy sets "theory") is quick and expedient, but does remain components-oriented and therefore a poor model of knowledge [15]: humans do not deal with information ambiguity through its symbolic computation and re-computation. Humans deal with ambiguity through evolving an autopoietic network of relations interconnecting the available pieces of information.

Because knowledge is a network and because this network is self-producing and self-maintaining, no "central information processing" is necessary. From this vantage point, the fuzziness and ambiguity of concepts (building blocks of knowledge) are not only factual but necessary.

Any complex system can be self-organizing and self-coordinating only if there is a conceptual (not just logical) overlap between its components. This "overlap", this meshing of conceptual "fields of vision", allows for any autonomous communication and propagation of "in-formation" (in the proper sense of indentation or perturbance) to take place over the network. Only through such overlaps of meaning can any "parallel processing" take place. Destroying, removing or narrowing down such fields of overlap would reduce system's ability for self-coordination. Crisp and non-overlapping concepts could then be coordinated only through a central processing agent, as in machines for sequential computing.

Vagueness, blurring, ambiguity, indefinitness, fuzziness, etc., are not therefore simplistic deficiencies or imprecisions to be tinkered with by expeditiously self-appointed manipulators of symbols; rather, they are the grand strategic innovations of human mind allowing for knowledge to emerge, persist and expand in an autonomous and autopoietic fashion.

Jan Christiaan Smuts, the father of holism, was also the first to recognize the importance of "concepts and their fields of ambiguity" [8, pp.16-19]:

"We have to return to the fluidity and plasticity of nature and experience in order to find concepts of reality. When we do this we find that round every luminous point in experience there is gradual shading off into haziness and obscurity. A 'concept' is not merely its clear luminous centre, but embraces a surrounding sphere of meaning or influence of smaller or larger dimensions, in which the luminosity tails off and grows fainter until it disappears. Similarly a "thing" is not merely that which presents itself as such in clearest definite outline, but this central area is surrounded by a zone of intuitions and influences which shades off into the region of indefinite. The hard abrupt contours of our ordinary conceptional system do not apply to reality and make reality inexplicable."

Smuts did not miss the communicative-propagative role of ambiguity, especially in his study of causation. The cause-effect connection, for example, has to be treated as a whole situation where the fields of "cause" intersect with those of "effect", one passing into another gradually and often imperceptibly. Reducing both participating aspects into crisp, disconnected and opposing forces makes it ultimately impossible to understand how one passes into another in actual causation.

Cause and effect are interlocked and mutually influencing each other throught the interpenetration of their two fields. Without this, no simultaneous, parallel or self-coordinated action would be possible. The "crisp" and isolated concrete things (or concepts) could never come into real contact or into active (or creative) relations with any other things (or concepts), but for their fields of ambiguity. There would be no knowledge, the real world of matter and life would become quite unintelligible and inexplicable. The world to us would appear to be a mere collection of *membra disjecta* : scattered data and barren pieces of information.

It is only in man-made artificial contrivances (like sequential machines), the products of abstract, computing and symbol manipulating intelligence, where no ambiguity of meaning is allowed and precision, exactitude and crispness are necessary.

4. OBJECTS OF KNOWLEDGE

Objects of knowledge are observer distinguishable entities (unities, systems) brought forth from their background either as wholes or as composites. The wholes are characterized only by properties imparted through the operation of distiction which brought them forth. No components are distinguished. Composites are brought forth through additional applications of (nested) operations of distiction: their components and their relationships are distinguished in relation to the whole they integrate.

Thus the wholes and their composites are related and differentiated only by the number of levels of distinction applied. One can see that it is observer's discretion whether to bring forth an object of knowledge as a whole or as a composite. Any component can itself be brought forth as a whole. The reality of objects of knowledge is thus being continually constructed, degraded and re-constructed by human observers.

To speak of knowledge, the distinction itself (whole or

composite) would be insufficient: it must be accompanied be some "empirical test" – it must bring forth a coherent set of actions.

Composite objects of knowledge can be distinguished at two separate levels: in terms of their organization and in terms of their structure [see for example [5,12,13,14].

Organization refers to general relations between the components that define a composite in terms of its class identity. Changes in organization result in changes of object class identity.

Structure refers to actual or specific components and relations that characterize a composite as a particular unity (system) of its kind. Changes in structure do not alter object class identity as long as the organization is maintained.

Note that the same organization can manifest (reveal) itself through a large variety of structural embodiments. A composite exists only through its structural embodiment: through components and their relationships. As long as object organization is preserved it can undergo a history of structural changes without the loss of class identity. It is also conceivable that different organizations could manifest themselves in relatively indistinguishable structures.

To know an object of knowledge (in the sense of bringing it forth from its background) can occur at two levels: (1) one can know the particular and specific by bringing forth object structure; (2) one can know the general, the identity class itself, by bringing forth object organization.

At the first level only a "surface knowledge" of the particulars is involved; at the second level also the "deep knowledge" (see for example [1, 2, 6,10]) is engaged. Observe that in order to to deal with the unexpected, novel and unprecedented, one has to know the organization, the identity class of a system in question; knowing its structure alone would be insufficient.

Any "expert" system of the production type, where a unit of "knowledge" is represented by an IF-THEN rule, operates on a structure level only and can work only for the specific problem solving in a very narrow domain.

5. MODELING KNOWLEDGE

One possibility of representing the relationship of both organization and structure of objects of knowledge is to combine ISA and PARTOF relations in constructing the appropriate knowledge base [10].

For example, statement "an automobile is a (ISA) physical object" connects the knowledge object with a particular identity class. An ISA arrow is directed from AUTOMOBILE to PHYSICAL OBJECT. Next, "engine is a part of (PARTOF) an automobile" represents components and their relation to the whole. An PARTOF arrow goes from ENGINE to AUTOMOBILE. One can also write: "an automobile has parts (HASPARTS): engine, steering, body, etc." with HASPARTS arows directed from AUTOMOBILE to individual parts (ENGINE, STEERING, BODY, etc.).

All components connected by HASPARTS arrows must be included in a single "frame" of slots (components), all identified by unique frame and slots names. The contents of identity class frame "Automobile" can then be "inherited" by for example frame "Porsche". Concrete and specific (structural) components are the values of "frames": LENGTH, WIDTH, WEIGHT, CAPACITY, etc.

In the table below we present (adapted and simplified from Ueno [10]) a typical frame which combines both organization and structure of a given system:

<u>Frame: AUTOMOBILE</u>

ISA	FRAME	PHYSICAL OBJECT
HASPARTS	FRAME	BODY, ENGINE, STEERING, ELECTRO
RELATIONS	PR	(INCLUDES BODY, ENGINE), (TOGETHER ENGINE, ELECTRO,
....		
LENGTH	NV	(200-500)
WIDTH	NV	(130-200)
WEIGHT	NV	(500-5000)
CAPACITY	NV	(2-50), (D4)

....

<u>Frame: BODY</u>

ISA	FRAME	PHYSICAL OBJECT
PARTOF	FRAME	AUTOMOBILE
HASPARTS	FRAME	FLOOR, ENGINE SPACE, TRUNK, ...
RELATIONS	PR	(FRONTOF ENGINE SPACE, FLOOR),

....

<u>Frame: PORSCHE</u>

ISA	FRAME	AUTOMOBILE
LENGTH	NV	165
WIDTH	NV	169
WEIGHT	NV	1140
CAPACITY	NV	4

From such relational and multi-level knowledge base the user can "bring forth", according to his subjective characteristics and particular circumstances, a set of "objects of knowledge" which are most effective in coordinating a set of actions. Each user, under different sets of circumstances, would use such a knowledge (or expert) support system differently.

Allowing "experts" to become experts requires broadening and expanding of their expertise, not narrowing it down. When an expert becomes too specialized (knows a lot about too little) he ceases to be an expert: he becomes unable to deal with any sort of novelty.

So called "knowledge explosion" (also "information explosion" or "information society") are now obviously misnomers limited to the structural ("surface") knowledge only. Organizational or "deep" knowledge is <u>not</u> as explosive as it may appear, if it is "explosive" at all.

6. DIVISION OF LABOR

Any task can be progressively broken down into a number of subtasks and operations. As long as all these subtasks are performed by a single worker or automated machine, there is no need to speak about the division of labor. It is sufficient to refer simply to the <u>division of task</u>.

This task disaggregation makes possible <u>parallel</u> processing of subtasks and therefore translates directly into increased productivity. In order to realize parallel (rather than serial) processing of subtasks, they have to be carried out by <u>different</u> workers or machines: now labor too has become divided and we speak of the <u>division of labor</u>.

The division of task and labor is accompanied by disaggregation, division and distribution of knowledge necessary to carry out the overall task. The knowledge itself becomes divided and distributed among different production agents (humans and machines): it is appropriate to speak of the <u>division of knowledge</u>.

Traditional economist's concept of the "division of labor" thus effectively hides at least three important, separate and relatively distinct dimensions: division of task, division of labor and division of knowledge.

When we divide a system into its components, whether mentally or physically, we have to uncover, postulate or design an overall controlling or coordinating organizational principle.

We divide the task into a thousand of subtasks. It is then appropriate to ask: How difficult and how costly is it to assure that the coherence of their proper sequencing, scheduling and interaction is maintained? We divide the labor among a thousand of "partial" workers (or machines): How difficult and how costly is it to maintain their coordination, motivation and functioning? We divide the knowledge into a thousand of tiny bits: How difficult and how expensive is it to maintain its necessary integration, storage and update as a whole?

The answer to all three questions is: progressively more difficult and progressively more expensive.

When the complexity, difficulty as well as costs implied by further divisions become too large, we have to turn, by necessity, in the direction of reaggregation of labor and reintegration of knowledge. Observe that the underlying division of task itself is not as crucial; it is the way the subtasks are carried out (in parallel, in series, by machines, by men) which is of importance.

Originally, one person performs a whole task: e.g., making clothes from hunting for the animal to sewing. As the process of division of labor sets in, more people are becoming involved and their tasks progressively more specialized. Coordinative agents and leaders (precursors of today's management) soon emerge. As markets grow, the division of labor and specialization grow correspondingly. It almost appears (as it did to Adam Smith) that barring the size of the market there are no limits to the process of division of labor.

But there are limits: not only labor but also human knowledge is becoming divided and distributed in the process. Coordination becomes more difficult, more costly and more complex. Markets continue to grow and press for ever higher productivity, but further division of labor simply "cannot deliver": the process slows down and ultimately must reverse itself [18].

7. DIVISION OF KNOWLEDGE

Any process of the division of labor is accompanied by the parallel process of the division of knowledge. Usable knowledge now rarely exists in its concentrated or integrated form: it has become dispersed into bits of incomplete and often contradictory pieces possessed by separate individuals.

How the spontaneous interaction of a number of people, each possessing only bits of knowledge, brings about a state of affairs which could be brought about by deliberate direction only by somebody who possessed the combined knowledge of all those individuals [16].

No single individual possesses the requisite knowledge of building a space shuttle. Yet the space shuttles get built. How? Where is the complete and requisite knowledge "stored" and how? In order to deal with the problem of increasingly atomized and widely distributed knowledge, humans have evolved, quite spontaneously, complex coordinative hierarchies of management and command. It is these hierarchies which in their totality represent the requisite "social memory", the way of "coding and storing" human knowledge about wholes.

Problem of coordinative hierarchies CANNOT be solved by first gathering all relevant data and information, then communicating all this "knowledge" to a central agency which, after properly integrating all and everything, issues its orders and sets to communicate them back to the points of their origin. This is how centrally planned societies and enterprises are organized. The problem must be solved by some form of decentralization, like for example the very effective distributed autonomy [3, 4, 9], to ensure that the crucial knowledge of the particular circumstances of time and place will be properly and promptly used by the system.

But the individual "on the spot" (anywhere in the hierarchy and least of all at the top) cannot act solely on the basis of his intimate but still limited and atomized knowledge of the facts of his immediate surroundings. There is still the problem of communicating to him the additional information which allows him to fit his decisions into the whole pattern of changes of the larger economic system.

8. REINTEGRATION OF KNOWLEDGE

If one person starts performing subtasks previously carried out by two persons, or if one person starts controlling two instead of one machine, or even if one machine starts performing functions previously done by two machines – we can be sure that no further division of either labor or knowledge is taking place. The process of reintegration has started.

These two apparently and fundamentally different responses (division and integration) to the same stimulus (growth of market demand) cannot be characterized as a "cycle" or "wave", nor as a "revolution" or "transformation", not even as a "metamorphosis" or "growth". The most fitting seems to be Giambattista Vico's conception of *corsi e ricorsi* in the evolution of social systems [11].

Any real origin in human affairs – and the process of division of labor is of a real origin – must meet with a real end. After each *corso* there follows a different and yet organically related *ricorso*.

There is always a course and recourse, outswing and rebound, disaggregation and reaggregation. The dual process of *corso e ricorso* should not be taken apart: it is a whole, a self-organizing and self-renewing whole. The success spawns failure and growth is also a decline. Every *corso* in human affairs is self-binding and self-limiting, transforming itself, gradually and at first imperceptibly, into its inevitable *ricorso*.

The *corso* of the division of labor is an old process, an *ancien régime* initiated concurrently with the emergence of human history. It has been expanding, exploding and spreading itself until today. Only now, at the very end of the twentieth century, the first manifestations of its *ricorso* can be observed. It is the longest of all long economic "cycles".

As the division of labor and knowledge progressed, the specialized efficiency of "partial" manual and knowledge workers was continually enhanced by dedicated and single-purpose machines. These are possibly complex machines designed to perform a small or narrowly defined task as quickly and as efficiently as possible. A worker becomes an operator and thus a necessary attachment to the machine. Worker supports and enhances the machine, not the other way around. Details are to be found in [21].

9. SUMMARY

In order to achieve autonomous coordinability of systems, their components (concepts, processors, objects, workers) have to be interconnected in a circular autopoietic network of relations. Their boundaries have to be sufficiently ambiguous (fuzzy) so that there exist interconnected fields of overlap among them. Only under such conditions of built-in ambiguity can possibilities of self-management emerge and the traditional centralized controlling component disposed of at all levels of system design and management, ranging from hardware all the way up to societies of humans.

"Integration" is now rapidly becoming a byword of the newly emerging, high-technology & knowledge-based management systems. Machines are becoming universal or multi-purpose, workers multifunctional, and managers (at least some) aspiring "renaissance men". Job and functional rotation are increasingly common at all levels of the hierarchy, quality circles and teamwork are mushrooming and education replaces training. Hierarchies are becoming flatter and workers are becoming managers. Autonomy, participation and self-management are replacing the traditional "managements" by objectives, motivation, control and reward.

Knowledge is being reintegrated: people are learning broader, not narrower tasks; workers are being rewarded according to the number of tasks they can carry out; machines are designed for multifunctionality; products are redesigned with sharply decreased number of parts.

These and related phenomena indicate that a broad transition toward parallelism, integration, self-management and autonomous coordination has already started at all levels of systems analysis and design.

REFERENCES

[1] Chandrasekaran, B. and S. Mittal, Deep versus compiled knowledge approaches to diagnostic problem-solving, in M. J. Coombs (ed.), Developments in Expert Systems (Academic Press, London, 1984) 23-34.

[2] Hart, P. E., Direction for AI in the eighties, SIGART Newsletter 79 (January 1982).

[3] Kobayashi, S., Ono, O., A Distributed Problem Solving Approach to Control Asynchronous and Concurrent Processes, Preprints of the VII-th International Conference on Multiple Criteria Decision Making, Kyoto, August 18-22 (1986) 319-328.

[4] Koizumi, M., Mori, K., Autonomous Coordinability of Decentralized System Considering Subsystems Failures, Preprints of the VII-th International Conference on Multiple Criteria Decision Making, Kyoto, August 18-22 (1986) 895-904.

[5] Maturana, H. R., What is it to see?, Arch. Biol. Med. Exp. 16 (1983) 255-269; also Maturana, H. R. and F. J. Varela, The Tree of Knowledge (Shambhala, Boston, 1987)

[6] Michie, D., High-road and low-road programs, AI Magazine 3 (1982) 21-22.

[7] Rumelhart, D. E., McClelland, J. L., et al., Parallel Distributed Processing (MIT Press, Cambridge, Mass., 1986)

[8] Smuts, J. C., Holism and Evolution (The Macmillan Company, New York, 1926)

[9] Tamura, S. et al., Intellectual Distributed Processing System Development, Autonomous Coordinability of Decentralized System Considering Subsystems Failures, Preprints of the VII-th International Conference on Multiple Criteria Decision Making, Kyoto, August 18-22 (1986) 772-781.

[10] Ueno, H., Object Model for a Deep Knowledge System, Preprints of the VII-th International Conference on Multiple Criteria

Decision Making, Kyoto, August 18-22 (1986) 792-800.
[11] Vico, G., Scienza nuova (1725); Scienza nuova seconda (1730/1744); in M. H. Fisch, T. G. Bergin (trans.), The Autobiography of Giambattista Vico, (Cornell Universisty Press, Ithaca,1944; Cornell Paperbacks ed.,1975).
[12] Zeleny, M., Self-Organization of Living Systems: A Formal Model of Autopoiesis, General Systems 4 (1977)13-28.
[13] Zeleny, M. (ed.), Autopoiesis, Dissipative Structures, and Spontaneous Social Orders (Westview Press, Boulder, CO, 1980)
[14] Zeleny, M. (ed.), Autopoiesis: A Theory of Living Organization (Elsevier North Holland, New York,1981)
[15] Zeleny, M., On the (Ir)Relevancy of Fuzzy Sets Theories, Human Systems management 4(1984) 301-306.
[16] Zeleny, M., Spontaneous Social Orders, General Systems 11(1985)117-131.
[17] Zeleny, M., High Technology Management, Human Systems Management 6(1986)109-120.
[18] Zeleny, M., Management Support Systems: Towards Integrated Knowledge Management, Human Systems Management 7(1987)1, pp.59-70.
[19] Zeleny, M., Fuzziness and the Theory of Knowledge, Proceedings of International Symposium on Fuzzy Systems and Knowledge Engineering,Guangzhou-Guiyang, July 10-16,1987, Guangdong Higher Education Publishing House, pp. 36-45.
[20] Zeleny, M., Fuzziness as a Referential Distance, Preprints of Second IFSA Congress, July 20-25,1987, Tokyo, pp. 614-617.
[21] Zeleny, M., The Grand Reversal: On the *corso* and *ricorso* of human way of life," in: Physis/To Inhabit the Earth, edited by Mauro Ceruti and Ervin Laszlo (Feltrinelli, Milano, 1987).

Fuzzy Logic in Knowledge-Based Systems,
Decision and Control
M.M. Gupta and T. Yamakawa (Editors)

ON FUZZY PROBABILITY MEASURES

Heinz J. Skala

Universität Paderborn, Warburger Straße 100, FRG

Starting with Zadeh's (1968) attempt to give a precise meaning to the notions of a fuzzy event and a fuzzy probability we state very weak conditions on the fuzzy measurable space and on the fuzzy measure. Special attention is given to such fuzzy measures which are dominated by a linear functional. We continue the work of Klement et al. (1981) to find conditions which guarantee that a fuzzy measure has an integral representation. The main difference to their work is that we do not specify the σ-algebra in advance. Finally it is shown that the linear functionals can be rationalized in terms of coherent betting behaviour.

KEYWORDS: Fuzzy measure spaces, Fuzzy probabilities.

INTRODUCTION

When Zadeh (1968) introduced probabilities of fuzzy events he started from measurable functions. It was then natural to use the expected value of the membership function as measure of the fuzzy event. However, the question arises why to introduce a new concept when already everything is available for dealing with uncertainty (a σ-algebra and a probability measure).This I think was felt by subsequent authors when they proposed axioms in order to justify the "natural" definition of the probability of a fuzzy event. Let me only mention the papers by Klement et al. (1981) and Smets (1982). A good overview about related topics might be found in Höhle/Klement (1984). The axioms proposed so far give rise to a larger class of fuzzy probability measures than the ones originally considered by Zadeh. For example, let us only mention the requirement $m(f \vee g)+m(f \wedge g)=m(f)+m(g)$ as proposed by Klement et al. (1981). It is well known that any linear functional on a vector lattice satisfies it, however, there are usually a lot of other non-linear functionals which have this property. It is therefore interesting to find natural conditions which ensure that a certain functional is linear or at least dominated by a linear functional. In particular, linear functionals turn out to be rationalized by coherent betting behaviour. This in turn allows to create a theory of statistics based on probabilities of fuzzy events which is similar in gist to ordinary Bayesian statstics.

In the present paper we state very weak conditions on the fuzzy measurable space and on the fuzzy measure. These conditions are apparently the weakest ones that a functional must satisfy in order to qualify to measure some sort of vagueness. When looking for conditions which guarantee that a given functional is dominated (or dominates) one which can be represented as an integral with respect to a probability content we rely on a very effective version of the Hahn Banach theorem due to König (1978). We give some variants of it adapted for our purpose.

If certain additional assumptions are met it is even possible to obtain an integral representation with respect to a probability measure thus coming back to Zadeh's original proposal. However, neither the σ-algebra (as in the case of Klement et al.) nor the probability measure is pre assigned. They depend on the concrete situation. Without topological assumption it turns out that the crucial requirement is that the membership functions attain their supremum which has an obvious behavioural interpretation. Also note that Sugeno's (1974) set functions sometimes induce functionals with properties studied in our paper.

2. ON FUZZY MEASURABLE SPACES

Let X be an arbitrary set and let $\mathbb{F}$ denote a family of functions $f:X\to[0,1]$ closed under pointwise $\sup(f,g)=f\vee g$ and $\inf(f,g)=f\wedge g$. $(X,\mathbb{F})$ will be called a generalized fuzzy measurable space if the following holds:

(1) The constant functions $c(x)=c$, $c\in[0,1]$ belong to $\mathbb{F}$.

Sometimes it will be useful to have also other properties at our disposal (compare Klement et al. 1981).

(2) If $f\in\mathbb{F}$, then $1-f\in\mathbb{F}$.

(3) If all functions in the sequence $(f_n)_{n\in\mathbb{N}}$ belong to $\mathbb{F}$, then $\sup_{n\in\mathbb{N}} f_n \in\mathbb{F}$

(3') If all functions in the sequence $(f_n)_{n\in\mathbb{N}}$ belong to $\mathbb{F}$, then $\inf_{n\in\mathbb{N}} f_n\in\mathbb{F}$.

Examples:

1. The set of all functions $f:X\to[0,1]$ satisfies (1-3).
2. Let $\mathbb{A}$ be a σ-algebra on X and let $\mathbb{F}$ denote the set of all $\mathbb{A}$-measurable functions $f:X\to[0,1]$, then $(X,\mathbb{F})$ satisfies (1-3). This is the situation for which Klement et al. (1981) prove their main theorem.
3. Let X be a topological space and let $\mathbb{F}$ denote the set of all lower semicontinuous functions $f:X\to[0,1]$, then $(X,\mathbb{F})$ satisfies (1) and (3) but not necessarily (2). Analogously, if we consider the upper semicontinuous functions, then (1) and (3') is satisfied.

<u>Definition 1.</u> Let $(X,\mathbb{F})$ be a generalized fuzzy measurable space. A generalized fuzzy content is a mapping $m:\mathbb{F}\to[0,1]$ such that

(4) $m(0)=0$, $m(1)=1$,

(5) $f\leq g$ implies $m(f)\leq m(g)$,

(6) $m(\alpha g)=\alpha m(g)$, $\alpha\geq 0$,

(7) $m(f)\leq \sup_X f$

These are the apparently weakest conditions which guarantee that m is "well behaved". Moreover, the conditions are also well motivated. Remember e.g. that Zadeh (1968) defined the probability p(A) of a fuzzy event A by

(8) $p(A)=\int_X f_A(x)d\mu$

where μ is a probability measure on a σ-algebra $\mathfrak{A}$ on X and the membership function $f_A: X\rightarrow[0,1]$ is $\mathfrak{A}$-measurable.

3. MORE ON FUZZY CONTENTS

An interesting question is to state conditions which ensure that a fuzzy content m is dominated by a linear functional. In order to do so we rely on a very general version of the Hahn-Banach theorem due to König (1978). Let E be a real vector space, let E* denote the space of all real linear functionals on E, and let $E^{\#}$ consist of all real sublinear (i.e. homogeneous and subadditive) functionals.

Theorem 1. Let $\theta\in E^{\#}$ and let $\tau: T\rightarrow\mathbb{R}$, $\emptyset\neq T\subset E$, be such that $\tau\leq\theta|T$. If there are real numbers $\alpha,\beta>0$ such that for all $u,v\in T$ and $\varepsilon>0$ there exists an $x\in T$ with

(9) $\theta(x-\alpha u-\beta v)\leq\tau(x)-\alpha\tau(u)-\beta\tau(v)+\varepsilon$,

then there exists a $\varphi\in E^*$ such that $\varphi\leq\theta$ and $\tau\leq\varphi|T$.

Theorem 2. Let $(X,\mathbb{F})$ be a generalized fuzzy measurable space and let m be a generalized fuzzy content on $\mathbb{F}$. There exists a linear functional $\varphi\leq\sup$ on the space Bd(X) of bounded real functions on X such that $m(f)\leq\varphi(f)$ for all $f\in\mathbb{F}$ if there are real numbers $\alpha,\beta>0$ such that for all $f,g\in\mathbb{F}$ and $\varepsilon>0$ there exists an $h\in\mathbb{F}$ with

(10) $h(x)\leq\alpha f(x)+\beta g(x)+m(h)-\alpha m(f)-\beta m(g)+\varepsilon$ for all $x\in X$.

Proof. Observe that sup is a sublinear functional on Bd(X). Apply theorem 1 with the specifications $E=Bd(X)$, $\theta=\sup$, $T=\mathbb{F}$, $\tau=m$ and note that for $\theta=\sup$ the properties (9) and (10) are equivalent.

Remark. Remember, that to every linear functional $\varphi\leq\sup$ on Bd(X) there corresponds a probability content p_φ on X and vice versa. Thus the above result gives a sufficient condition in order that there exists a probability content p_φ on X such that

(11) $m(f)\leq\varphi(f)=\int f\,dp_\varphi$ for all $f\in\mathbb{F}$.

(For an integration theory with respect to a content, i.e. finitely additive measure, consult e.g. Dunford/Schwartz (1954).)
Very often we have more structural information about $\mathbb{F}$ and/or m. In particular in all our examples 1-3 $\mathbb{F}$ was convex.
For the following observe that the homogeneity of a functional m guarantees that it is concave if and only if it is super additive.

Theorem 3. Let $(X,\mathbb{F})$ be a generalized fuzzy measurable space and assume that $\mathbb{F}$ is convex. If the generalized fuzzy content m on $\mathbb{F}$ is concave (i.e., $m(\alpha f+(1-\alpha)g)\geq\alpha m(f)+(1-\alpha)m(g)$ for all $\alpha\in[0,1]$, $f,g\in\mathbb{F}$), then there exists a linear functional $\varphi\leq\sup$ on Bd(X) such that $m(f)\leq\varphi(f)$ for all $f\in\mathbb{F}$.

Proof. Take any $\alpha,\beta>0$, $\alpha+\beta=1$, and put $h=\alpha f+\beta g$, then it is easy to see that condition (9) is satisfied.

We are now going to deal with the problem to state conditions which ensure that m itself or its dominating linear functional φ can be represented by an integral with respect to a (σ-additive) probability measure. Obviously, the crucial properties are (3) and

(12) for all sequences $(f_n)_{n\in\mathbb{N}}$ in $\mathbb{F}$ such that $(f_n)_{n\in\mathbb{N}}\uparrow f\in\mathbb{F}$ it holds that $\varphi(f_n)\uparrow\varphi(f)$.

Note also the natural dual conditions, namely (3') and

(12') for all sequences $(f_n)_{n\in\mathbb{N}}$ in $\mathbb{F}$ such that $(f_n)_{n\in\mathbb{N}}\downarrow f\in\mathbb{F}$ it holds that $\varphi(f_n)\downarrow\varphi(f)$.

As, contrary to Klement et al. (1981), we do not assume these conditions from outright, an analogue of their main theorem will only hold in special cases. We shall now apply König's (1978) Hahn Banach theorem in the special case where X is a compact Hausdorff space.

Theorem 4. Let m be a generalized fuzzy content on $(X,\mathbb{F})$ where $\mathbb{F}$ consists of upper semicontinuous functions on a compact Hausdorff space X. There exists a Baire probability measure p_φ on X such that $m(f)\leq\int f\,dp_\varphi$ for all $f\in\mathbb{F}$ if there are real numbers $\alpha,\beta>0$ such that for all $f,g\in\mathbb{F}$ and $\varepsilon>0$ there exists an $h\in\mathbb{F}$ which satisfies (10).

Proof. By theorem 2 we obtain a linear functional $\varphi\leq\sup$ on Bd(X). To the restriction of φ to the subspace C(X) of continuous functions on X there corresponds a unique Baire probability measure p_φ according to the Riesz represen-

tation theorem. It is well known that for any upper semi-continuous function f it holds that $f=\inf\{g: g\in C(X),\ g\geq f\}$. Thus it is natural to define
$\int f dp_\varphi = \inf\{\int g\, dp_\varphi : g\in C(X),\ g\geq f\}$.
As obviously $\varphi(f)\leq \inf\{\int g dp_\varphi : g\in C(X),\ g\geq f\}$ it remains to show the converse inequality. For this note that $M=\{\inf(g_1,\ldots,g_n): n\geq 1,\ g_i\in C(X),\ g_i\geq f\}$ is downward directed. For any given $g'\geq f$, $g'\in C(X)$ we obtain by Dini's theorem

$$\inf_{g\in M} \sup(g-g')=\sup(f-g')\leq 0.$$

For any $\varepsilon>0$ we can therefore find a $g\in M$ such that $g\leq g'+\varepsilon$. As $\varphi(g)\leq\varphi(g')+\varepsilon$ and thus $\inf_{g\in M}\varphi(g)\leq\varphi(g')+\varepsilon$ for any given $g'\geq f$, $g'\in C(X)$ and all $\varepsilon>0$ we have $\inf_{g\in M}\varphi(g)\leq\varphi(f)$.

Corollary. Let $\mathbb{F}$ be a convex set of upper semicontinuous functions on a compact Hausdorff space X. If m on $\mathbb{F}$ is concave, then there exists a Baire measure p_φ on X such that $m(f)\leq\int f dp_\varphi$ for all $f\in\mathbb{F}$.

Remarks.

(i) Observe that the upper semicontinuous functions on a compact space attain their sup.

(ii) A similar theorem holds when $\mathbb{F}$ consists of lower semicontinuous functions.

(iii) The Baire measure p_φ extends in a unique way to a regular Borel measure.

(iv) With the obvious modifications we also get a proof of the above theorem for locally compact Hausdorff spaces.

We would like to note that super additive functionals very naturally occur in a variety of contexts. As an example we give a short proof of the well known fact that every 2-monotone capacity is dominated by a measure (in fact it is the lower envelope of the dominating measures).

Example. Let X be a compact Hausdorff space and denote by $\mathbb{K}$ the compact subset of X. A function $m:\mathbb{K}\to[0,1]$ is called a 2-monotone capacity if

(i) $m(\emptyset)=0$,

(ii) $K_1, K_2\in\mathbb{K}$ and $K_1\subset K_2$ implies $m(K_1)\leq m(K_2)$,

(iii) m is continuous on the right,

(iv) $m(K_1\cup K_2)+m(K_1\cap K_2)\geq m(K_1)+m(K_2)$.

Let $C^+(X)$ denote the convex cone of positive continuous functions on X. Choquet (1954) showed that by defining $\overline{m}: C^+(X)\to\mathbb{R}^+$ by

$$\overline{m}(f)=\int_0^\infty m(\{x\in X: f(x)\geq\gamma\})d\lambda$$

where $\gamma>0$ and λ denotes the Lebesgue measure one obtains a super additive functional $\overline{m}$ on $C^+(X)$. Note that $\overline{m}(f)\leq\sup f$ for all $f\in C^+(X)$. Using theorem 1 in the specification $T=C^+(X),\theta=\sup, \tau=\overline{m}, \alpha=\beta=1$ and putting $h=g+f$ one obtains $0\leq\overline{m}(h)-\overline{m}(g)-\overline{m}(f)$ by the super additivity of $\overline{m}$. Thus there exists a linear functional on $C(X)$ such that $\varphi\leq\sup$ and $\overline{m}\leq\varphi|C^+(X)$.
Observe that $m(K)=\inf\{\overline{m}(f):1_K\leq f\in C^+(X)\}$ and define the measure $p(K)$ of K by $p(K)=\inf\{\varphi(f):1_K\leq f\in C^+(X)\}$.

4. DOING WITHOUT TOPOLOGY

Recall theorem 4 where the topological assumption allowed us to state conditions ensuring that a generalized fuzzy content is dominated by an integral with respect to a genuine probability measure. One might suspect that this fact essentially depends on the special property of upper semicontinuous functions of attaining their supremum on a compact set. As in the context of fuzzy sets this amounts to the behavioural assumption of being able to specify a subset of maximal degree of membership we are going to exploit this property.

Lemma (Simons 1972) Let $Bd(X)$ be endowed with the usual sup norm $\|\cdot\|$. Suppose that for all $n\in\mathbb{N}$, $f_n\in Bd(X)$ and $\sup_{n\in\mathbb{N}}\|f_n\|<\infty$. Let there be a subset $Y\subset X$ such that whenever $\alpha_n\geq 0$ and $\sum_{n\in\mathbb{N}}\alpha_n=1$, there exists a $y\in Y$ such that

(13) $\sum_{n\in\mathbb{N}}\alpha_n f(y)=\sup_X(\sum_{n\in\mathbb{N}}\alpha_n f_n)$. Then

(14) $\sup_y \limsup_{n\to\infty} f_n(y)\geq\inf \sup_X(\mathrm{conv}\{f_n:n\in\mathbb{N}\})$

where conv denotes the convex hull.

Remark. Let $f_n\in\mathbb{F}$, $n\in\mathbb{N}$, where $\mathbb{F}$ is the generalized fuzzy measurable space of theorem 4. Then it is clear that $\sum_{n\in\mathbb{N}}\alpha_n f_n=1+\inf_k\sum_{n=1}^{k}\alpha_n(f_n-1)$. As every finite sum of upper semicontinuous functions is upper semicontinuous and the inf of upper semicontinuous functions is again upper semicontinuous, we see that condition (13) holds (with $Y=X$) as an upper semicontinuous function on a compact space X attains its sup.
For the following let $L(\mathbb{F})$ denote the subspace of $Bd(X)$ which is generated by a generalized fuzzy measurable space $\mathbb{F}$ and let $L_1(\mathbb{F})=\{f\in L(\mathbb{F}):0\leq f\leq 1\}$.

Theorem 5. Let m be a generalized fuzzy content on $(X,\mathbb{F})$ and assume that $\mathbb{F}$ is convex. Let there be real numbers $\alpha,\beta>0$, $\alpha+\beta=1$, such for all $f,g\in\mathbb{F}$ and $\varepsilon>0$ it holds that $0\leq m(\alpha f+\beta g)-\alpha m(f)-\beta m(g)+\varepsilon$. If, moreover, for any sequence $(f_n)_{n\in\mathbb{N}}$, $f_n\in L_1(\mathbb{F})$ it holds that every countable convex combination

$\sum_{n\in\mathbb{N}} \alpha_n f_n$, $\alpha_n \geq 0$, $\sum_{n\in\mathbb{N}} \alpha_n = 1$ attains its supremum on X (i.e. condition 13 holds), then there exists a σ-algebra $\sigma(\mathbb{F})$ making all functions $f\in\mathbb{F}$ measurable and a probability measure p_φ on $\sigma(\mathbb{F})$ such that $m(f)\leq\int f dp_\varphi$ for all $f\in\mathbb{F}$.

Proof. Applying theorem 2 the assumptions of the first part of theorem 5 provide us with a linear functional $\varphi\leq\sup$ on Bd(X) such that $m(f)\leq\varphi(f)$ for all $f\in\mathbb{F}$. In order to show that $L(\mathbb{F})$ is a vector lattice it suffices to verify $(f\wedge g)\in L(\mathbb{F})$ for any $f=\sum_{i=1}^{n}\alpha_i f_i$ and $g=\sum_{j=1}^{m}\beta_j g_j$, $\alpha_i,\beta_j\in\mathbb{R}^+$, f_i, $g_j\in\mathbb{F}$.

Let $a=\sum_{i=1}^{n}\alpha_i$, $b=\sum_{j=1}^{m}\beta_j$, $\bar{f}=\sum\frac{\alpha_i}{a}f_i$, and $\bar{g}-\sum\frac{\beta_j}{b}g_j$. By convexity $\bar{f},\bar{g}\in\mathbb{F}$. As $\mathbb{F}$ is a lattice containing the constants $c\in[0,1]$ also $\frac{a}{a+b}\bar{f}\wedge\frac{b}{a+b}\bar{g}\in\mathbb{F}$ and $(a+b)(\frac{a}{a+b}\bar{f}\wedge\frac{b}{a+b}\bar{g})\in L(\mathbb{F})$. Finally, we observe that for an arbitrary ordered vector space the equality $c(f\wedge g)=cf\wedge cg$, $c\in\mathbb{R}^+$, holds whenever either side of it exists.

Obviously $L(\mathbb{F})$ satisfies the Stone condition as it contains the constants. Thus, in order to show that φ, when restricted to $L(\mathbb{F})$, is an abstract integral we have to prove that for every decreasing sequence $(f_n)_{n\in\mathbb{N}}$ in $L(\mathbb{F})$ such that $(f_n)_{n\in\mathbb{N}}\downarrow 0$ it holds that $\varphi(f_n)\downarrow 0$.

Let $c=\sup_{n\in\mathbb{N}}\sup_X f_n$ then $\frac{1}{c}f_n\in L_1(\mathbb{F})$. As every countable convex combination $\frac{1}{c}\sum_{n\in\mathbb{N}}\alpha_n f_n$ attains its sup so does $\sum_{n\in\mathbb{N}}\alpha_n f_n$ and condition (13) is satisfied.

Observe that $\lim\sup_{n\to\infty} f_n=\inf f_n=0$ when $(f_n)_{n\in\mathbb{N}}\downarrow 0$.

Thus an application of the above lemma gives us

$0=\sup_X \inf_n f_n(x)\geq\inf_n \sup_X (\text{conv}\{f_n;\ n\in\mathbb{N}\})=\inf_n \sup_X f_n$.

As $\varphi(f_n)\leq\sup f_n$ and $\varphi(f_n)\geq 0$ for any $f_n\geq 0$ we have $\varphi(f_n)\downarrow 0$ whenever $(f_n)_{n\in\mathbb{N}}\downarrow 0$. By the Daniell-Stone theorem there exists therefore a smallest σ-algebra $\sigma(\mathbb{F})$ making all functions $f\in\mathbb{F}$ measurable and a probability measure p_φ on $\sigma(\mathbb{F})$ such that $m(f)\leq\int f dp_\varphi$ for all $f\in\mathbb{F}$.

5. BETTING ON FUZZY EVENTS

Remember that the concept of coherence is central for the Bayesian view of statistics. Without observations coherence simply means that an individual's degrees of belief in uncertain events should be such that it is impossible to make a Dutch book against him or her, i.e. a finite combination of bets such that the individual necessarily looses. It is a classical result of de Finetti that this is the case if and only if the degrees of belief are in accordance with a probability content. If countably many bets are allowed a corresponding result holds for probability measures. See in this connection Heath/Sudderth (1972) and Skala (1985).

We are now going to show that similar results hold in the context of fuzzy events. Let $\mathbb{F}$ be any set of fuzzy events on X. To every $f\epsilon\mathbb{F}$ a bookie assigns an odds functions $o_f:X\rightarrow[0,1]$. (Observe that a "crisp" odds function simply is a constant function.) If a gambler bets on f, then this (fuzzy) net outcome is $f-o_f$ if f occurs. Obviously it will not be advantageous to him to bet on f if $f(x)-o_f(x)<0$ for all $x\epsilon X$.

Theorem 6. Let $\mathbb{D}$ denote the set of differences $f-o_f$. Assume that there exist real numbers $\alpha,\beta>0$ such that for all $u,v\epsilon\mathbb{D}$ and $\varepsilon>0$ there is a $w\epsilon\mathbb{D}$ such that $w\leq\alpha u+\beta v+\varepsilon$ and $\sup u\geq 0$ for all $u\epsilon\mathbb{D}$. Then there exists a linear functional $\varphi\leq\sup$ on Bd(X) such that $\varphi(u)\geq 0$ for all $u\epsilon\mathbb{D}$ (i.e. $\varphi(f)\geq\varphi(o_f)$ for all $f\epsilon\mathbb{F}$).

Proof. This follows immediately from theorem 1 when specifying $\theta=\sup$ and $\tau=0$.

Remark. In the special case where $\mathbb{D}$ consists of all (finite) convex combinations this is essentially a result obtained by Heath/Sudderth (1978).

Especially, in the case of betting on fuzzy events one might be interested to ask for conditions implying that $\varphi(f)\geq\varphi(o_f)+\delta$ for all elements in $\mathbb{D}$.

Corollary. Let $\mathbb{D}$ be as above. Assume that there exist real numbers $\alpha,\beta>0$, $\alpha+\beta=1$, such that for all $u,v\epsilon\mathbb{D}$ and $\varepsilon>0$ there is a $w\epsilon\mathbb{D}$ such that $w\leq\alpha u+\beta v+\varepsilon$ and $\sup u\geq\delta$ for all $u\epsilon\mathbb{D}$. Then there exists a linear functional $\varphi\leq\sup$ on Bd(X) such that $\varphi(u)\geq\delta$ for all $u\epsilon\mathbb{D}$.

Proof. Put $\tau=\delta$.

For the sake of simplicity let us assume that with any $f_1,\ldots,f_n\ \epsilon\ \mathbb{F}$ also the bets on their convex combinations $\sum_{i=1}^{n}\alpha_i f_i$, $\alpha_i\epsilon[0,1]$, $\sum_{i=1}^{n}\alpha_i=1$, are available. Obviously a booki wants to avoid the existence of a bet on $f=\sum_{i=1}^{n}\alpha_i f_i$ such that $\inf(f-o_f)>0$ (or equivalently $\sup_X(o_f-f)<0$). This is achievable if and only if there exists a fuzzy content (i.e. a linear functional $\varphi\leq\sup$) such that $\varphi(o_f)\geq\varphi(f)$ for all convex combinations f. This is the rational behind fuzzy contents.

If X is a compact Hausdorff space even more can be said. Let $\mathbb{F}$ consist of lower semicontinuous functions f and assume that the corresponding odds o_f are upper semicontinuous, hence o_f-f is also upper semicontinuous.

Theorem 7. The following statements are equivalent:

(i) There exists a Baire measure p on X such that $\int(o_f - f)dp \geq 0$ for all $f \in \mathbb{F}$.

(ii) For every finite convex combination $\sum_{i=1}^{n} \alpha_i(o_{f_i} - f_i)$ there exists at least one $x \in X$ such that $\sum_{i=1}^{n} \alpha_i(o_{f_i}(x) - f_i(x)) \geq 0$.

Proof. Apply theorem 4 with the obvious modifications.

REFERENCES

Choquet, E.: Theory of capacities. Ann. Inst. Fourier 5 (1953/54) 131-295.

Dunford, N., Schwartz, J.T.: Linear Operators I, Wiley 1967.

Heath, D., Sudderth, W.: On a theorem of de Finetti, oddsmaking, and game theory. Ann. Math. Stat. 43 (1972) 2072-2077.

Heath, D., Sudderth, W.: On finitely additive priors, coherence, and extended admissibility. Ann. Stat. 6 (1978) 333-345.

Höhle, U., and Klement, E.P.: Plausibility measures. In Aspects of Vagueness, H.J. Skala, S. Termini, and E. Trillas (Eds.), Reidel (1984) 31-50.

Klement, E.P., Lowen, R. and Schwyhly, W.: Fuzzy probability measures. Fuzzy Sets and Systems 5 (1981) 21-30.

König, H.: Neue Methoden und Resultate aus Funktionalanalysis und konvexer Analysis. Oper. Res. Verf. 28 (1978) 6-16.

Simons, S.: A convergence theorem with boundary. Pacific J. Math. 40 (1972) 703-708.

Smets, Ph.: Probability of a fuzzy event: an axiomatic approach. Fuzzy Sets and Systems 7 (1982) 153-164.

Sugeno, M.: Theory of Fuzzy Integrals and its Applications. Ph. D. Thesis, Tokyo Institute of Technology, Tokyo 1974.

Zadeh, L.A.: Probability measures of fuzzy events. J. Math. Anal. Appl. 23 (1968) 421-427.

Fuzzy Logic in Knowledge-Based Systems,
Decision and Control
M.M. Gupta and T. Yamakawa (Editors)

ON PLAUSIBLE REASONING

S. T. Wierzchoń

Institute of Computer Science
Polish Academy of Sciences
00-901 Warszawa, PKIN
P.O. Box 22
Poland

This paper presents possible applications of the theory of discrete random sets (drs in short) in plausible and abductive reasoning as well as in analysis of experts' judgments. Special attention is focused on the properties of the set-theoretical sums and intersections of independent drs.

1. INTRODUCTION

The aim of the paper is to give a formal approach to the Dempster-Shafer theory of evidence and to study its applications in modelling different modes of reasoning.

We start by introducing basic notions from the theory of random sets. The discrete case is reviewied only. Such a line of presentation has some tutorial aspects and is quite sufficient for practical applications. A special attention is put into the set-theoretical unions and intersections of independent discrete random sets. From a theoretical point of view the resulting random sets give an account to so-called Sugeno g_λ measure [9], and from a practical point of view they are of use in the statistical theory of experts' judgments.

We propose a version of Bayes' theorem for Sugeno's measures; a counterpart of this theorem for probabilities is presented. An interesting feature of this last theorem is that it enables to compute *a posteriori* probability without referring to the *a priori* probabilities. Its applications to the problem of abductive reasoning are discussed.

2. DISCRETE RANDOM SETS

This section serves a concise introduction into the theory of random sets. The line of presentation follows that one proposed in [10].

2.1. Preliminary definitions

Let $X = \{x_1,\ldots,x_n\}$ be a discrete space, and let $D = \{D_1,\ldots D_d\}$ be its partition. Denote 2^X and M_D the algebras generated by X and D respectively.

For any $e \in E = \{0,1,2\}$ we define the sets A^e, $A \in M_D$, in the next way

$$A^0 = \{M \in 2^X : A \cap M = \emptyset\} \quad (1a)$$
$$A^1 = \{M \in 2^X : A \cap M^c = \emptyset\} \quad (1b)$$
$$A^2 = \{M \in 2^X : A \cap M^c \neq \emptyset \ \& \ A \cap M \neq \emptyset\} \quad (1c)$$

(Here M^c stands for the complement of M in X.) One observes that if $e \neq f$ then $A^e \cap A^f = \emptyset$. These definitions can be extended to subsets ε of E, namely

$$A^{\varepsilon} = \bigcup_{e\in\varepsilon} A^e \quad (1d)$$

We easily check that, for fixed A, the sets A^{ε}, $\varepsilon \subseteq E$, forms an algebra, i.e. for any α, $\beta \subseteq E$

$$A^{\alpha\cup\beta} = A^{\alpha} \cup A^{\beta},\ A^{\alpha\cap\beta} = A^{\beta} \cap A^{\beta},\ A^{E\setminus\alpha} = 2^X\setminus A^{\alpha},\ A^E = 2^X \quad (2)$$

In the sequel we will use the sets A^e, $A^{E\setminus e}$ with $e \in \{0,1\}$ only. The next properties will be of interest:

$$(\bigcup_{i=1}^{m} A_i)^e = \bigcap_{i=1}^{m} A_i^e \quad (3a)$$

$$(\bigcup_{i=1}^{m} A_i)^{E\setminus e} = \bigcup_{i=1}^{m} A_i^{E\setminus e} \quad (3b)$$

$$(\bigcap_{i=1}^{m} A_i)^e \supseteq \bigcup_{i=1}^{m} A_i^e \qquad (3c)$$

$$(\bigcap_{i=1}^{m} A_i)^{E\setminus e} \subseteq \bigcap_{i=1}^{m} A_i^{E\setminus e} \qquad (3d)$$

Consider now the class of sets generated by the finite intersections of the slots A^e and $A^{E\setminus e}$ where $e \in \{0,1\}$. Each such a nonempty intersection can be written in its canonical form

$$S^e = A_o^e \bigcap_{i=1}^{m} A_i^{E\setminus e} \qquad (4a)$$

where $A_o, A_1, \ldots A_m \in M_D$ are such that $A_o \cap A_i = \emptyset$, $i = 1\ldots m$, and $A_i \nsubseteq A_j$, $i \neq j$. Similarly we can consider (canonical) slots of the form

$$S = A_o^0 \bigcap_{i=1}^{m} A_i^{12} \cap B_o^1 \bigcap_{j=1}^{r} B_j^{02} \qquad (4b)$$

It is a simple exercise to verify that the class S_D^e (resp S_D) containing slots S^e (resp.S) is a semialgebra and that the class F_D^e (resp.F_D) generated by the finite unions of elements from S_D^e (resp.S_D) is an algebra.

2.2. Choquet Theorem

A drs can be viewed as a measurable mapping K from a probability space $(\Omega, \mathfrak{A}, Pr)$ to a measurable space $(2^X, F_D^e)$ where F_D^e is the already introduced algebra generated by the slots A^e, $A^{E\setminus e}$, $A \in M_D$. The probability law of K is the probability measure P^e defined by

$$P^e(F) = Pr(K^{-1}(F)), \qquad F \in F_D^e \qquad (5)$$

In many situations we define a probability P^e on $(2^X, F_D^e)$ immediately. This P^e is a probability law of some random set K, namely the canonical one.

We give a counterpart of the notion of distribution function for a drs K. For $A_o, A_1, \ldots, A_m$ we define recursively the function $R_m^e(A_o; A_1 \ldots A_m)$

$$R_o^e(A_o) = 1 - T^e(A_o) \qquad (6a)$$

$$R^e_m(A_o;A_1,\dots A_m) = R^e_{m-1}(A_o;A_1\dots A_{m-1})$$

$$- R^e_{m-1}(A_o\cup A_m;A_1\dots A_{m-1})$$

$$= \sum_{I\subseteq\{1\dots m\}} (-1)^{|I|} R^e(A_o \cup_o \bigcup_{i\in I} A) \qquad (6b)$$

Here $|I|$ = card (I). The set function T^e defined on M_D is said to be d-linear capacity (recall that d = card(D)) if (i) for any A, B in M_D such that $A \subset B$, $T^e(a) \le T^e(B)$, and (ii) $R^e_m \ge 0$, m=1, d-1. (For a general definition of Choquet capacity see e.g. [4].) Now let K be a canonical drs. For any A in M_D we define

$$T^e(A) = P^e(A^{E\setminus e}) = \begin{cases} P^o(K \cap A \neq \emptyset),\ e = 0 \\ P^1(K^c \cap A \neq \emptyset),\ e = 1 \end{cases} \qquad (7)$$

One can check that

$$R^e_m(A_o;A_1\dots A_m) = P^e(A^e_o \bigcap_{i=1}^{m} A^{E\setminus e}_i) \qquad (8)$$

is the probability that the drs K: (a) has the empty intersection with A_o and nonempty intersections with $A_1\dots A_m$ when e = 0, (b) contains A_o and does not contain any of the sets $A_1\dots A_m$ when e=1.

Referring to the notion of the canonical slot we verify that the assignment (8) is well defined. Further, employing the recurrent equation (6) we conclude that if T^e is a d-linear capacity on M_D such that $T^e(\emptyset) = 0$ and $0 \le T^e \le 1$, then the assignment (8) defines an additive measure on the semialgebra S^e_D. Proceding in a standard way we state that this measure can be uniquely extended to a probability measure P^e on the algebra F^e_D. In summary we get an extended version of Choquet's theorem, namely

Theorem 1 [10] Let X be a discrete space, let D be a partition of X and let T^e be a set function on M_D, the algebra of sets from D. Then there exists a probability measure on F^e_D satysfying the condition $P^e(A^{E\setminus e}) = T^e(A)$, for any $A \in M_D$ if and only if T^e is a d-linear capacity such that $T^e(\emptyset)=0$, $T^e \in [0,1]$ ■

Note that for e = 0 we get a version of standard formulation of Choquet theorem.

We can also define a probability measure P on F_D. To do this we introduce - after [9] - the notion of a d-planar capacity, i.e. the set function T defined on $M_D \times M_D$ and such that (i) T is monotone with respect to set inclusion in each of its arguments, (ii) the functions R_i^j (defined below) are nonnegative for i,j=0,1...d-1. The successive differences R_i^j are defined as follows

$$R_o^o\binom{B_o}{A_o} = 1 - T(A_o, B_o) \tag{9a}$$

$$R_i^j\binom{B_o;B_1\ldots B_j}{A_o;A_1\ldots A_i} =$$

$$R_{i-1}^j\binom{B_o;B_1\ldots B_j}{A_o;A_1\ldots A_{i-1}} - R_{i-1}^j\binom{B_o;B_1\ldots B_j}{A_o \cup A_i;A_1\ldots A_{i-1}}$$

$$= R_i^{j-1}\binom{B_o;B_1\ldots B_{j-1}}{A_o;A_1\ldots A_i} - R_i^{j-1}\binom{B_o \cup B;B_1\ldots B_{j-1}}{A_o;A_1\ldots A_i} \tag{9b}$$

The analogue of (8) is

$$P(A_o^o \bigcap_{i=1}^{r} A_i^{12} \cap B_o^o \bigcap_{j=1}^{s} B_j^{02}) = R_r^s\binom{B_o;B_1\ldots B_s}{A_o;A_1\ldots A_r} \tag{10}$$

When D = X, i.e. no approximation of the space X is involved, then $M_D = 2^X$ and $F_D^e = F_D$. In this case $P(F) = P^e(F)$, e = 0,1, $F \in F_D$.

2.3. Families of drs's

A finite family of drs's is viewed as a measurable mapping $K_1 \ldots K_m)$ from a probability space $(\Omega, \mathfrak{U}, Pr)$ to a measurable space

$\times_{i=1}^m (2^X, F_D^e)$. Assume for simplicity m=2.

A natural extension of the notion of the probability law to this two-dimensional case is

$$P_2^e(F) = Pr\{(K_1, K_2)^{-1})\}, \qquad F \in F_D^e \times F_D^e \tag{11}$$

Similary we define the d-linear capacity of order m = 2 as

$$T_2^e(A_1,A_2) = P_2^e(A_1^{E\setminus e} \times A_2^{E\setminus e}), \qquad A_1 \times A_2 \in M_D \times M_D \tag{12}$$

It is obvious that T_2^e defines a probability law of some drs $K=(K_1,K_2)$ if and only if (i) $T_2^e(\emptyset,\emptyset) = 0$, (ii) T_2^e is monotone with respect to set inclusion in each of its argument, and (iii) $0 \le T_2^e \le 1$. Following Section 2.1 we define a counterpart of successive differences (6); in particular

$$R_{st}^e({}^1A_o;A_1 \ldots {}^1A_s;{}^2A_o;{}^2A_1 \ldots {}^2A_t) =$$

$$P_2^e({}^1A_o^e \bigcap_{i=1}^{s} {}^1A_i^{E\setminus e} \times {}^2A_o^e \bigcap_{j=1}^{t} {}^2A_j^{E\setminus e}) \tag{13}$$

If

$$T_1^e(A_1,A_2) = T^e(A_1) \cdot T^e(A_2), \qquad A_1,A_2 \in M_D \tag{14}$$

where $T^e(A_1) = P_2^e(A_1^{E\setminus e} \times 2^X)$, $T^e(A_2) = P_2^e(2^X \times A_2^{Ee})$ we say that the drs's K_1, K_2 are independent.

3. CHARACTERISTICS OF DRS

Let K be a drs taking its values in 2^X and let D be an approximation of X. For any in M_D we define probabilities

$$\mathrm{Prob}(K \subseteq A) = P^o((A^c)^o) \tag{15}$$

$$\mathrm{Prob}(K \cap A \neq \emptyset) = P^o(A^{12}) \tag{16}$$

$$\mathrm{Prob}(K \supseteq A) = P^1(A^1) \tag{17}$$

and assume that Prob(K = ∅) = 0. In this case these probabilities correspond to the next notions used in Shafer's theory of evidence [8]: belief function - Bel, plausibility function - Pl, and commonality function - Q.

Let us show that f(A) = Prob(K ⊆ A) is really a belief function. It is evident that f(∅) = 0, f(X) = 1. Referring to (3a) we state

that if $A \subseteq B$ then $A^o \supseteq B^o$, hence $f(A) \le f(B)$. Lastly, for any sequence $A_1, \ldots A_n$ of sets in M_D we have - cf properties (3c) and (3a)

$$f(A_1 \cup \ldots \cup A_n) = P^o((A_1^c \cap \ldots \cap A_n^c)^o) \geq P^o(\bigcup_{i=1}^{n} (A_i^c)^o)$$
$$= \sum_{I \subseteq N} (-1)^{|I|+1} P^o(\bigcap_{i \in I} (A_i^c)^o)$$
$$= \Sigma_{I \in N} (-1)^{|I|+1} P^o((\bigcap_{i \in I} A_i)^o)$$
$$= \Sigma_{I \in N} (-1)^{|I|+1} f(\bigcap_{i \in I} A). \quad (18)$$

Here N stands for the family of subsets of the set { 1...n } without the empty set. According to Theorem 2.1 in [8] we state that f(A) defined above is really a belief function. Hence we will write Bel(A) instead of f(A).

Remark. When Prob(K = Ø) > 0 we define Bel(A) as the conditional probability Prob(K ⊆ Ø|K ≠ Ø).

As $P^o(A^{12}) = 1 - P^o(A^o) = 1 - Bel(A^c)$ we state that Prob(K∩A ≠ Ø) qualifies as a plausibility function.

Noticing that the canonical slots underlying the algebra F_D^o are of the form

$$(A^c)^o \bigcap_{D_i \subseteq A} D_i^{12} \quad (19a)$$

we define

$$m(A) = P^o((A^c)^o \bigcap_{D_i \subseteq A} D_i^{12}), \qquad A \in M_D \quad (20)$$

Employing the recurrent equation (6) one verifies that

$$\sum_{A \in M_D} m(A) = 1 \quad (21)$$

$$Bel(A) = \sum_{B \subseteq A} m(B) \quad (22)$$

$$Pl(A) = \sum_{B \cap A \neq \emptyset} m(B) \quad (23)$$

It is important to note that the belief function Bel discussed above is so-called belief function carried by partition *D* . When *D* equals to X then Bel is a "usual" belief function. In this case the canonical slots (19) are of the form

$$(A^c)^0 \bigcap_{x_i \in A} \{x_i\}^{12} = (A^c)^0 \cap A^1 \tag{19b}$$

and the quantity m(A) can be viewed as Prob(K = A). Since (19b) can be written in an equivalent form $A^1 \bigcap_{x_i \in A^c} \{x\}^{02}$ we state that

$$Q(A) = \text{Prob}(K \supseteq A) = \sum_{A \subseteq B} m(B) \tag{24}$$

It is interesting to observe that m-values can be obtained from the appropriate Q-values as follows

$$m(A) = \sum_{B \supseteq A} (-1)^{|B-A|} Q(B) \tag{25}$$

Emploing the recurrent equations (6) to the right hand side of (24) we get a characterization of Q in terms of Bel

$$Q(A) = \sum_{B \subseteq A} (-1)^{|B|} \text{Bel}(B^c) \tag{26}$$

This relation is still valid in a general case when $D \neq X$; similarly the relation (24) remains valid in this general case with m-values as defined in the equation (20).

4. UNIONS AND INTERESECTIONS OF INDEPENDENT DRS'S

In this section we introduce two important, from a practical standpoint, types of drs's. From a theoretical point of view both the types lead to the notion of so-called g_λ - measure [9].

Let us make a next remark first. To this moment *D* was viewed as a partition of the base space X. This partition can be termed as the disjunctive covering of X. We can also use so-called conjunctive covering of X, i.e a family *C* of subsets of X such that C ∪ C' = X for any C, C' in *C* and $\bigcap_{C \in \mathcal{C}} C_i = \emptyset$. When *C* is a conjunctive covering of X then $\mathcal{C}^c = \{C^c: C \in \mathcal{C}\}$ is a disjunctive covering of X.

4.1 Intersections of independent drs's.

Let $\mathcal{C} = \{C_1,\dots,C_d\}$ be a conjunctive covering of X, and let $\mathcal{M}_C$ be a family of subsets of X obtained as finite intersections of elements from $\mathcal{C}$. Let $K_1,\dots,K_d$ be a family of independent drs's such that $\mathrm{Prob}(K_i \supseteq C_i) = 1$, i = 1,...,d, i.e all values of i-th drs's contain the subset C_i. Assume also that we know the next probabilities

$$p_i = \mathrm{Prob}(K_i \in (C_i{}^c)^0) = \mathrm{Prob}(K_i = C_i),$$
$$q_i = \mathrm{Prob}(K_i \in (C_i{}^c)^1) = \mathrm{Prob}(K_i = X),$$
$$r_i = \mathrm{Prob}(K_i \in (C_i{}^c)^2).$$

Of course $p_i + q_i + r_i = 1$ for i = 1,...,d. When $r_i = 0$ then K_i can be viewed as a counterpart of Shafer's simple support function (see [8]). Define K to be the intersection of all K_i's . As

$$\{A\} = (A^c)^0 \cap A^1 = \bigcap_{j\in J_A} (C_j{}^c)^1 \bigcap_{i\in J\setminus J_A} (C_i{}^c)^0 \tag{27}$$

where $J_A = \{j: C_j{}^c \subseteq A\}$, $J = \{1,\dots,d\}$ we find that for any $A \in \mathcal{M}_D$

$$\mathrm{Prob}(K = A) = \prod_{j\in J_A} q_j \prod_{i\in J\setminus J_A} p_i \tag{28}$$

Another, easy computable characteristics of the drs K are

$$\mathrm{Prob}(K \in A^e) = \begin{cases} \prod_{j\in J_A} p_j & \text{for } e = 0 \\ \prod_{j\in J_A} q_j & \text{for } e = 1 \end{cases} \tag{29}$$

$$\mathrm{Prob}(K \cap A \neq \emptyset) = 1 - \mathrm{Prob}(K \in A^0) = 1 - \prod_{j\in J_A} p_j \tag{30}$$

Assume now that $r_i = 0$, i = 1,...,d and that

$$C_i = \{x_i\}^c, \quad i = 1,\dots,d. \tag{31}$$

Under these assumptions (28) can be written in the form

$$\mathrm{Prob}(K = A) = (\prod_{j=1}^{d} p_j)^{1-|A|} \prod_{x_i \in A} \mathrm{Prob}(K = \{x_i\}) \qquad (32)$$

Denoting

$$\mathrm{Prob}(K = \emptyset) = \prod_{j=1}^{d} p_j = a \qquad (33)$$

we compute the "density" function of the conditional drs

$$m(A) = \mathrm{Prob}(K = A)/\mathrm{Prob}(K \neq \emptyset) = \mathrm{Prob}(K = A)/(1 - a) \qquad (34)$$

This m-function possesses the property

$$m(A) = [(1 - a)/a]^{|A|-1} \prod_{x_i \in A} m(\{x_i\}) =$$

$$= b^{|A|-1} \prod_{x_i \in A} m(\{x_i\}) \qquad (35)$$

One recognizes that the belief function corresponding to the conditional drs is Sugeno's g_λ measure with the parameter $\lambda = b$. Recall that a set function $g\colon 2^X \longrightarrow [0,1]$ qualifies as a g_λ measure if it admits the next properties [9]

(G1) $g(\emptyset) = 0$, $g(X) = 1$,

(G2) $g(A \cup B) = g(A) + g(B) + \lambda g(A)g(B)$, $A \cap B = \emptyset$, $\lambda > -1$.

A very nice feature of such a measure is that it is determined uniquely by the set of its "densities" $g_i = g(\{x_i\})$, $i = 1,\ldots,n$ according to the rule

$$g(A) = \lambda^{-1}[\prod_{x_i \in A} (1 + \lambda g_i) - 1] \qquad (36)$$

and the parameter λ is found by solving the equation $g_\lambda(X) = 1$. Of course we can extend the formula (36) to the general situation by setting $g_i = g(C_i^{\,c})$, $i = 1,\ldots,d$. It is obvious that the plausibility function induced by the m-function (35) is also a g_λ measure with the parameter $\lambda = -b/(b - 1)$ - see [12].

4.2 Unions of independent drs's.

Suppose now $D = \{D_1,\ldots,D_d\}$ is a disjunctive partition of the base space X. The counterparts of the drs's K_i introduced above are the drs's L_i with the next properties.

(i) $\mathrm{Prob}(L_i \subseteq D_i) = 1$,
(ii) $u_i = \mathrm{Prob}(L_i \in D_i^1) = \mathrm{Prob}(L_i = D_i)$,
$v_i = \mathrm{Prob}(L_i \in D_i^0) = \mathrm{Prob}(L_i = \emptyset)$,
$w_i = \mathrm{Prob}(L_i \in D_i^2)$.

(Again $u_i + v_i + w_i = 1$.) For the drs L being the sum of L_i's we find

$$\mathrm{Prob}(L = A) = \prod_{i \in I_A} u_i \prod_{j \in I \setminus I_A} v_j, \qquad A \in M_D \tag{37}$$

where M_D is the algebra generated by the sets from D, $I = \{1, \dots, d\}$, $I_A = \{i \in I: D_i \subseteq A\}$. Another important characteristics of L are

$$\mathrm{Prob}(L \in A^e) = \begin{cases} \prod_{i \in I_A} v_i & \text{when } e = 0 \\ \prod_{i \in I_A} u_i & \text{when } e = 1 \end{cases} \tag{38}$$

$$\mathrm{Prob}(L \cap A \neq \emptyset) = 1 - \prod_{i \in I_A} v_i \tag{39}$$

In particular, when

$$D_i = \{x_i\} \tag{40}$$

we state that

$$\mathrm{Prob}(L = A) = \left(\prod_{j=1}^{d} v_j\right)^{|A|-1} \mathrm{Prob}(L = \{x_i\}) \tag{41}$$

Again, denoting

$$z = \mathrm{Prob}(L = \emptyset) = \prod_{j=1}^{d} v_j \tag{42}$$

and defining the conditional m-function

$$m_U(A) = \mathrm{Prob}(L = A)/\mathrm{Prob}(L \neq \emptyset) = \mathrm{Prob}(L = A)/(1 - z) \tag{43}$$

we assert that the belief function corresponding to this drs is a Sugeno's measure with the parameter $\lambda = (1-z)/z$.

4.3 Applications of drs's to analysis of experts' judgments.

The drs's introduced in Section 4.1 can be used in the combination of disconfirmatory pieces of evidence [12], while the drs's introduced in Section 4.2 apper to be a useful tool in the statistical theory of experts judgments.

Suppose $A = \{a_1, \ldots, a_k\}$ is a set of alternatives and $R = [r_{ij}]_{k\times k}$ is a matrix characterizing a preference relation over A given by an expert. Hence $r_{ij} \in \{0, 1\}$ and $r_{ij} = 1$ means e.g. that i-th alternative is not worse than j-th alternative.

Let $\mathcal{A} = \{(i,j): 1 \le i \le k, 1 \le j \le k\}$. The matrix R can be identified with a subset Y_R of $\mathcal{A}$ constructed such that $(i,j) \in Y_R$ if $r_{ij} = 1$. Now as expert is a human being only, he is fallible and it is convenient to treat his judgments as a drs.

Having n matrices obtained from n experts we are interested in finding a group opinion. To do this we refer to the notion of the distance (in the sense of Kemeny-Snell) between preferences that can be written as (see [5] for details)

$$d(A_1, A_2) = \mu(A_1 \Delta A_2) \tag{44}$$

where Δ stands for the set-theoretical symmetric difference and μ is a finitely additive measure over $\mathfrak{U}$, the algebra of subsets of $\mathcal{A}$. If A_1 is a drs, say Y_R, and A_2 is a subset of $\mathcal{A}$, say A, then (44) reads as

$$d(Y_R, A) = \sum_{i,j=1}^{k} \mu(\{i,j\}) \,|Prob((i,j) \in Y_R) - \chi_A(\{i,j\})| \tag{45}$$

where χ_A stands for the characteristic function of the set A. The group opinion is found by solving the next optimization problem (see [5] for details).

$$\min_{A\in\mathfrak{U}} \sum_{i=1}^{n} d^p(Y_{R(i)}, A) \tag{46}$$

It has been shown that under sufficiently general assumptions there exists a solution $M(p) \in \mathfrak{U}$ to the problem (46). This M(p) is said to be the mean set of order p characterizing answers of experts i = 1, n. M(1) is referred to as the median while M(2) as the set-theoretical mean-value.

In the theory of independent binary comparisons the experts' judgments can be described as the Bernoulli random variable. Leting z_{ij} to denote a result of comparison of a pair (a_i, a_j) and assuming that $\{z_{ij}\}$ are mutually independent, we state that the answer given by an expert can be characterized by the drs with independent elements. More precisely

$$\mathrm{Prob}(Y_R = B) = \prod_{i,j=1}^{k} h[(i,j), p_{ij}(Y_R)] \tag{47}$$

where

$$h[(i,j), p_{ij}(Y_R)] = \begin{cases} p_{ij}(Y_R) & \text{when } (i,j) \in B \\ 1 - p_{ij}(Y_R) & \text{otherwise} \end{cases} \tag{48a}$$

$$p_{ij}(Y_R) = \mathrm{Prob}\{(i,j) \in Y_R\} \tag{48b}$$

It is interesting to notice that (47) corresponds to (25) under the assumption that $Q(A \cup B) = Q(A)Q(B)$.

When judgments are performed by n experts we may be interested e.g. in testing the hypothesis

$$(\forall B \in \mathcal{U})\ \mathrm{Prob}(Y_{R(1)} = B) = \ldots = \mathrm{Prob}(Y_{R(n)} = B) \tag{49}$$

Details concerning this problem can be found in [5].

5. BAYES' THEOREM FOR SUGENO'S MEASURE

Let G be a Sugeno's g_λ measure over 2^X. It was shown in [11] that this g induces on 2^X a unique probability measure

$$P(A) = \frac{\ln(1 + \lambda G(A))}{\ln(1 + \lambda)} \tag{50}$$

According to our earlier discussion this G can be viewed as a belief function derived from a conditional drs K' defined by its density function (43). This remark applies to the case of $\lambda > 0$; when $\lambda < 0$ we obtain a plausibility function. Knowing the value of λ we easily recover the density function of unconditional drs underlying K' or L'.

Suppose now that G is defined over $2^{X \times Y}$. Then for $A \subseteq X$, $B \subseteq Y$

$$G_X(A|B) = G(A \times Y \cap X \times B)/G(X \times B)$$
$$= [(\lambda_o + 1)^{P(A|B)} - 1]/\lambda_o \qquad (51)$$

where

$$\lambda_o = G(B) = (1 + \lambda)^{P(B)} - 1 \qquad (52)$$

defines conditional Sugeno's measure over 2^X. Here P is the probability measure induced by G and $P(\cdot|B)$ is a conditional measure derived from P. There holds the next

<u>Theorem 2</u> [11]. Let $\{G(\cdot|y_j), j = 1,...,m\}$, $m = |Y|$, be a collection of the conditional Sugeno's measures derived from an unknown Sugeno's measure G_{XY}. Then the *a posteriori* measure G_Y: $2^Y \longrightarrow [0, 1]$ is characterized by its densities of the form

$$g(y_j|A) = G(\{y_j\}|A) = \lambda G(A|y_j)/\lambda \qquad (53)$$

and the parameter

$$\lambda = \prod_{j=1}^{m} [1 + \lambda_j G(A|y_j)] - 1 \qquad (54)$$

Here λ_j stands for the parameter characterizing j-th conditional measure. ■

Noticing that Sugeno's measure tends to a probabbility measure as its parameter λ tends to zero we find the next probabilistic counterpart of the above theorem.

<u>Theorem 3</u> [12]. Let $\{P(\cdot|y_j), j = 1,...,m\}$ be a collection of conditional probabilities derived from an unknown probability measure P_{XY}. Then the *a posteriori* probability measure P_Y over 2^Y is characterized by its probability mass function of the form

$$p(y_j|A) = P(\{y_j\}|A) = \Big(\sum_{t=1}^{m} [P(A|y_t)b_j/P(A|y_j)b_t]\Big)^{-1} \qquad (55)$$

where

$$b_t = [1 - \sum_{x \in X} p^2(x|y_t)]/2 \qquad (56)$$

■

Notice that b_t can be viewed as a measure of "informativity" of the t-th probability distribution function. When P_t is the uniform probability distribution then b_t takes its maximal value $b_t = (n-2)/2n$ and when P_t is a Dirac measure then $b_t = 0$.

6. ABDUCTIVE REASONING

Casual links are typically represented by statements of the form

IF h THEN r (57)

where h denotes a hypothesis and r stands for a result. By abduction we understand the next pattern of reasoning [1]

$$\frac{\text{IF h THEN r} \quad \text{r}}{\text{h}} \qquad (58)$$

i.e. abduction tells us that one of hypothetical explanation for r (within the frames of acculul ated knowledge) is h.

Because of the lack of complete information we assume that all the knowledge about a problem under considerations can be represented by means of the conditional statements (57) and the credibility of each statement is assessed by a probability. Hence the formula

$$\mathrm{Prob}(r|h) \qquad (59)$$

denotes the probability of a result r given hypothesis h.

Let R and H denote fixed and finite sets of results and hypothesis, respectively. Let $m = |R|$ and $n = |H|$. Our knowledge about the problem can be represented concisely by the set of conditional probabilities $\mathrm{Prob}(r_j|h_i)$ and abduction can be stated as "Given result r and a set of the conditional probabilities of the form (59) find $\mathrm{Prob}(h_i|r)$". $\mathrm{Prob}(h_i|r)$ is known as the posterior probability and usually it is computed by using Bayes' theorem. Such a procedure is viewed as the belief revision and to perform it we need prior opinion about h_i's. This section is addressed to the case when this prior opinion is not available. To solve it we refer to Theorem 3. The rule (55) seems to be reasonable in the sense that when determining the strength of our guess, $Prob(h_i|r)$, we are forced to utilize the whole knowledge we have at our disposal. Note also the next properties of our rule.

(i) When $Prob(r|h_i)$ increases (resp. decreases) then $Prob(h_i|r)$ increases (resp. decreases).

(ii) When b_i decreases (resp. increases) then $Prob(h_i|r)$ increases (resp. decreases). In other words, the more informative rule we have, the clearer explanation is generated.

(iii) Suppose that our knowledge changes and we add a new hypothesis, h_{n+1}, to the existing set H. Then

$$Prob_{new}(h_i|r) = \left(\frac{1}{Prob_{old}(h_i|r)} + \frac{Prob(r|h_{n+1})b_i}{Prob(r|h_i)b_{n+1}} \right)^{-1} \qquad (60)$$

To perform a logical analysis of (55) assume that m = n = 2. In this case the equation (55) reduces to the form

$$Prob(h_1|r_1) = \frac{Prob(r_2|h_2)}{Prob(r_2|h_1) + Prob(r_2|h_2)} \qquad (61)$$

It appears that (61) imitates quite well the problem of determining validity of $r_j \longrightarrow h_i$ from $h_i \longrightarrow r_j$ where $\longrightarrow$ stands for the logical implication.

Denote r_1 as x, r_2 as ¬x and similarly h_1 as y and h_2 as ¬y. Suppose now that $Prob(h_1|r_1)$ models the degree of truth of the formula $x \longrightarrow y$ and $Prob(r_2|h_2)$ models the degree of truth of the formula $\neg y \longrightarrow \neg x$ equivalent from the logical standpoint to previous one. The same interpretation applies to $Prob(h_2|r_1)$ and $Prob(r_2|h_1)$. The analogies among logical conclusions and conclusions drawn with the help of the rule (61) can be seen by comparing tables 1 and 2.

Table 1

IF		THEN	
$\neg y \longrightarrow \neg x$	$y \longrightarrow \neg x$	$x \longrightarrow y$	$x \longrightarrow \neg y$
T	F	T	F
F	T	F	T
T	T	T	T
F	F	?	?

Table 2

IF		THEN	
$Prob(r_2 \mid h_2)$	$Prob(r_2 \mid h_1)$	$Prob(h_1 \mid r_1)$	$Prob(h_2 \mid r_1)$
1	0	1	0
0	1	0	1
1	1	0.5	0.5
0	0	?	?

Consider for example third line of Table 2. It models situation when we know that the result r_2 is true (abbreviated as T in Table 1) under hypothesis h_1 as well as under hypothesis h_2. In the probabilistic setting this means that $Prob(r_1 \mid h_i) = 0$, $i = 1, 2$. Hence, observing r_1 we cannot say about the truthness of h_1 and h_2. This is confirmed by the logical analysis (see 3-rd line in Table 1): knowing that $x \longrightarrow y$ and $x \longrightarrow \neg y$ are true and knowing that x is true both y or $\neg y$ may be true. Note also that the assumption that (simultaneously) $\neg y \longrightarrow \neg x$ and $y \longrightarrow \neg x$ are false is unrealistic and nothing can be said about the truthness of $x \longrightarrow y$ and $x \longrightarrow \neg y$. This is modelled by the fourth line in tables 1 and 2.

In summary the rule (61) seems to be a useful tool in the analysis of casual links among entities. As our knowledge concerning these links is not complete, we hope that (61) is appropriate under the Closed World Assumption suggested by Reiter [7]. It is worth noticing that (61) is a special case of the general formula (53) what offers its applications in less restrictive situations, e.g. when the strength of a given link is assessed in terms of an interval probability.

7. DEMPSTER'S RULE OF COMBINATION AND AN UNCERTAIN LOGIC

Let K_1, K_2 be two drs's defined over (M, F_X) and let m_1, m_2 denote their density functions. It can be verified that

$$m(A) = \text{Prob}(K_1 \cap K_2 = A \mid K_1 \cap K_2 \neq \emptyset) = \frac{\sum\{m_1(B)m_2(C) \mid B, C \in M,\ B \cap C = A\}}{\sum\{m_1(E)m_2(F) \mid E, F \in M,\ E \cap F \neq \emptyset\}} \tag{62}$$

The equation (62) is known as the Dempster's rule of combination [8].

Now let **P** be a finite set of propositions such that (i) if $p \in \mathbf{P}$ then $\neg p \in \mathbf{P}$, (ii) if $p, q \in \mathbf{P}$ then $p \wedge q \in \mathbf{P}$. Here $\neg$ and $\wedge$ stand for the logical negation and conjunction, respectively. A belief function defined on the Boolean algebra, **P**, is such that (a) $Bel(\mathbf{0}) = 0$, where **0** is the ever false proposition, (b) $Bel(\mathbf{1}) = 1$, where **1** is the ever true proposition, and (c) if p entails q (i.e. $p \longrightarrow q \equiv \mathbf{1}$) then $Bel(q) \geq Bel(p)$. The proposition p with $0 < Bel(p) < 1$ is said to be the uncertain proposition [6]. One should note that under this general setting $Bel(p) = 0$ does not mean that $Bel(\neg p) = 1$ while $Bel(p) = 1$ does mean that $Bel(\neg p) = 0$. This shows that in the framework of the Dempster-Shafer theory we can model situations when an agent has no opinion on the truth/falsity of the statement p as well as situations when the agent is certain that p is true.

Referring to the fundamental properties of belief function we have the next assessment true for any $p, q, r \in \mathbf{P}$

$$\begin{aligned} Bel(p \vee q) &= Bel[(p \vee q \vee r) \wedge (p \vee q \vee \neg r)] \\ &\geq Bel(p \vee q \vee r) + Bel(p \vee q \vee \neg r) - 1 \\ &\geq Bel(q \vee r) + Bel(p \vee \neg r) - 1 \end{aligned} \tag{63}$$

Under the logical setting this inequality gives an account to the resolution principle in the logic with uncertain premises (see also [2]), namely

$$\frac{Bel(q \vee r) = a,\ Bel(p \vee \neg r) = b}{Bel(p \vee q) \geq \max(0,\ a+b-1)} \tag{64}$$

Particularly, with $q \equiv \mathbf{0}$ we get a counterpart of the *Modus Ponens*. The rule (64) does not seem to be quite satisfactory, however. The manifold application of the rule decreases the lower bound to zero.

This inconvenience can be avoided with the help of the Dempster's rule of combination. Restricting ourselves to the *Modus Ponens* only we observe that the rule $p \longrightarrow q$ induces its own frame $\mathfrak{F} = \{(p,q), (p,\neg q), (\neg p,q), (\neg p,\neg q)\}$, where e.g. (p,q) means "p and (simultaneously) q are true". In such a context an expert has an ability to assign different portions of belief to subsets of $\mathfrak{F}$, i.e. he is able to determine his own interpretation of the truth of the rule. Under the logical setting it is appropriate to use the next allocation

$$m_{p\to q}(A) = \begin{cases} b & \text{if } A = \mathfrak{F}\setminus\{(p,\neg q)\} \\ 1 - b & \text{if } A = \mathfrak{F} \\ 0 & \text{otherwise} \end{cases} \tag{65}$$

Observe that $\mathfrak{F}\setminus\{(p,\neg q)\}$ is the set of all possibilities for which $p \longrightarrow q$ remains true; hence $Bel(p \longrightarrow q) = b$. Similarly, the allocation of probability that correspond to $Bel(p) = a$ may be of the form

$$m(A) = \begin{cases} a & \text{if } A = \{(p,q), (p,\neg q)\} \\ 1 - a & \text{if } A = \mathfrak{F} \\ 0 & \text{otherwise} \end{cases} \tag{66}$$

Applying the Dempster's rule of combination we get the next pattern of reasoning

$$\frac{Bel(p) = a,\ Bel(p \longrightarrow q) = b}{Bel(q) = ab} \tag{67}$$

As $ab \geq \max(0, a+b-1)$ this new result is consistent with the more general formula (64).

8. FINAL REMARKS

The theory of random sets seems to be a useful tool in a fuzzy analysis as well as in modelling various modes of reasoning. The first problem is explored by I.R. Goodman (see e.q. [3]). In [13] I present other applications of this theory in modelling some concepts from the theory of fuzzy sets (see also [14]). This approach exploits the optimization problem suggested in Section

4.3. The paper [15] describes a practical application of the ideas from section 7.

9. REFERENCES

[1] Charniak, E. and McDermott, D., Introduction to Artificial Intelligence (Addison Wesley, New York, 1985)

[2] Dubois, D. and Prade, H., Busefal No 27 (1986) 120

[3] Goodman, I.R., Fuzzy sets as equivalence classes of random sets, in: Yager, R.R., (ed.), Recent Developments in Fuzzy Sets and Possibility Theory (Pergamon Press, New York, 1982) pp. 327-343

[4] Matheron, G., Random Sets and Integral Geometry (J. Willey & Sons, New York, 1975)

[5] Orlov, A.I., Stability in social and economical models, in Russian, (Наука Publishing House, Moscow, 1979)

[6] Prade, H., IEEE Trans. on Pattern Analysis and Machine Intelligence, PAMI-7 (1985) 260

[7] Reiter, R., On closed world data bases, in: Gaillaire, H.,and Minker, J., (eds.), Logic and Data Bases (Plenum Press, N.Y., 1978)

[8] Shafer, G. A Mathematical Theory of Evidence (Princeton, N.J., 1976)

[9] Sugeno, M., Theory of fuzzy integral and its applications, Ph. D. thesis (Tokyo Institute of Technology, Tokyo, 1974)

[10] Vorobiov, O.,Ju., A Mean-in-Measure Modelling, in Russian, (Наука Publishing House, Moscow, 1985)

[11] Wierzchoń, S.T., The use of Dempster,s upper and lower probabilities in decision-making, in: Albrycht, J., and Wiśniewski, J., (eds.), Proc. of the Polish Symp. on Interval and Fuzzy Mathematics, (Institute of Mathematics, Poznań, Poland, 1985) pp. 229-243

[12] Wierzchoń, S.T., An inference rule based on Sugeno measure, in: Bezdek, J.C., (ed.) The Analysis of Fuzzy Information (CRC Press, Boca Raton, 1987)

[13] Wierzchoń, S.T., Random sets and belief structures, in preparation

[14] Wierzchoń, S.T., Deduction methods for fuzzy logics, in print

[15] Wierzchoń, S.T., and Mostowska, M., An application of the Dempster-Shafer theory to medical diagnosis, in print

Fuzzy Logic in Knowledge-Based Systems,
Decision and Control
M.M. Gupta and T. Yamakawa (Editors)

FUZZY NUMBERS FOR EXPERT SYSTEMS

J. J. Buckley

Mathematics Department
University of Alabama at Birmingham
Birmingham, AL 35294 USA

W. Siler
Carraway Medical Center
Birmingham, AL 35294 USA

This paper discusses operations, both set theoretic and algebraic, on fuzzy numbers needed in applications programs in fuzzy expert systems. We argue that the user must select a proper t-norm, and co-t-norm, to obtain desirable results when combining fuzzy numbers with an AND (intersection) or an OR (union). Algebraic operations on fuzzy numbers should not be performed in the abstract, as if they exist on their own, but instead in applications fuzzy numbers are values of fuzzy variables. We argue that prior associations between these fuzzy variables must be accounted for in order to obtain correct results in algebraic manipulations of fuzzy numbers.

Fuzzy Numbers, Possibility Distributions, Expert Systems.

1. INTRODUCTION

The purpose of this paper is to discuss operations on fuzzy numbers that may occur in fuzzy expert systems (FES). We first consider set theoretic operations on fuzzy numbers. Fuzzy numbers are fuzzy sets and therefore we may AND (intersect) and OR (union) two fuzzy numbers. We next discuss algebraic operations on fuzzy numbers which includes: (1) comparisons (inequalities); (2) sum and product of positive fuzzy numbers; and (3) the extended max and min of discrete fuzzy numbers in [0,1]. Throughout our discussion on the algebraic operations we argue that prior associations between the fuzzy variables, whose "values" are the fuzzy numbers, must be taken into account in order to obtain correct answers. Our FES shell FLOPS ([3],[14]) allows for fuzzy numbers and discrete fuzzy sets as data types and therefore we must be able to manipulate fuzzy numbers within the system in any applications program.

Any symbol with a "bar" on top represents a fuzzy subset of the real numbers. So $\overline{A},\overline{B},\cdots,\overline{M},\overline{N},\cdots$ are all fuzzy subsets of the reals. The membership function for any fuzzy set $\overline{A}$ will be denoted by $\mu(x|\overline{A})$. The extension of max and min are written as $\overline{\max}$ and $\overline{\min}$, respectively.

Let $\overline{M},\overline{N},\overline{O},\overline{P},\overline{Q},\cdots$ be fuzzy numbers. We will consider two basic types of fuzzy numbers. The first is a flat fuzzy number denoted by (a/b,c/d) where the membership function is: (1) zero outside [a,d]; (2) one on [b,c]; (3) continuous and monotonically increasing from zero to one on [a,b]; and (4) continuous and monotonically decreasing from one to zero on [c,d]. We usually use straight line segments on [a,b] and [c,d] and then the graph is a trapezoid with base [a,d] and a triangle if $b = c$. In this paper (a/b,c/d) will always represent a trapezoidal fuzzy number. If $a \geq 0$, then the fuzzy number is positive.

The flat, or triangular, fuzzy number will quite often arise from subjective expert opinion. An expert may believe that a future interest rate will be approximately between 8% and 10% and then we might use (0.07/0.08,0.10/0.11) for this future rate [5]. The other type of fuzzy number has its membership function ([2],[6],[14]) given by

$$\mu(x) = \begin{cases} \exp\left(\lambda\left[\dfrac{x-\theta}{\sigma}\right]^2\right), & -k\sigma + \theta \leq \mu \leq \theta + k\sigma, \\ 0, & \text{otherwise}, \end{cases} \tag{1}$$

where λ is a constant that makes $\mu(x)$ equal to some specific value when x is a certain number of σ-values from the central value θ. For example, $\lambda = \ln(0.5)$ makes $\mu(x) = 0.5$ for $x = \theta \pm \sigma$. The fuzzy number in equation (1) will arise from data obtained from certain measurements. Suppose that in an image processing problem we want a fuzzy number $\overline{A}$ for the area of a region. We count the number of pixels in the region which becomes θ and σ^2 is then our estimate of the error in θ ([2],[6],[14]). This second type of fuzzy number is positive when $\theta - k\sigma \geq 0$. Usually we would take $k = 3$ in equation (1).

A discrete fuzzy number is just the discrete version of (a/b,c/d) or equation (1). A discrete fuzzy number of type (a/b,c/d) is positive for only a finite number of values in [a,d] with its values strictly increasing (decreasing) on [a,b] ([c,d]) and always one between b and c. Similarly, a discrete fuzzy number of the second type has its values computed from equation (1) but for only a finite number of x values between $-k\sigma + \theta$ and $k\sigma + \theta$.

Real valued fuzzy variables will be denoted by X,Y,Z,$\cdots$. We may define an abstract algebra of fuzzy numbers just like that for real numbers but in applications fuzzy numbers do not appear "alone" but as "values" of fuzzy variables. Suppose we have a real valued fuzzy variable X whose "values" are given by fuzzy number $\overline{M}$. Therefore, $\overline{M}$ becomes a possibility distribution for X and

$$\text{Poss}[X = x] = \mu(x|\overline{M}). \tag{2}$$

When we have two fuzzy variables X and Y we will always associate $\overline{M}$ with X and $\overline{N}$ with Y. The algebra of the fuzzy numbers $\overline{M},\overline{N},\cdots$ must relate back to associations (if any) between their fuzzy variables. We will say that X and Y are positively associated if large (small) values of X are associated with large (small) values of Y. X and Y are negatively associated when large (small) values of X are associated with small (large) values of Y. When there is no association between X and Y we say X and Y are non-interactive (independent). We argue in Section III that the association between X and Y is important in determining $\overline{M} + \overline{N}$, $\overline{MN}$, $\overline{\max}(\overline{M},\overline{N})$ and evaluating $\overline{M} \leq \overline{N}$, $\overline{M} < \overline{N}$, etc. Examples of positively, and negatively, associated fuzzy variables are given in Section III. We will sometimes say $\overline{M}$ and $\overline{N}$ have an association which means, more precisely, that X and Y have the association.

The joint distribution of X and Y is given by

$$\Pi(x,y) = \min\left[\mu(x|\overline{M}),\mu(y|\overline{N})\right], \tag{3}$$

when X and Y are non-interactive and

$$\Pi(x,y) = \min\left[\mu(x|\overline{M}),\mu(y|\overline{N}),\mu(x,y|\overline{R})\right], \tag{4}$$

when X and Y interact [15]. The interaction between X and Y is measured by the fuzzy relation $\overline{R}$ whose values are

$$x\overline{R}y = \mu(x,y|\overline{R}). \tag{5}$$

Let us now introduce some other notation that will be used in this paper. T denotes any t-norm and C represents any co-t-norm. The Lukasiewicz AND and OR are

$$\text{LAND}(\alpha,\beta) = \max(\alpha + \beta - 1,\ 0), \tag{6}$$

$$\text{LOR}(\alpha,\beta) = \min(\alpha + \beta,\ 1). \tag{7}$$

The fuzzy set whose membership function is always zero is ϕ and $\mathcal{U}$ in the fuzzy set whose membership function is always one. If [ ] denotes a statement, then v[ ], a number in [0,1], is the value of the statement. For example, when we evaluate $\overline{M} \leq \overline{N}$ we get the value $v[\overline{M} \leq \overline{N}]$, in [0,1], for a measure of how much $\overline{M}$ is less than or equal to $\overline{N}$. For continuous (non-discrete) fuzzy numbers $\overline{M}$ and $\overline{N}$ we will use

$$v[\overline{M} = \overline{N}] = \sup_x\{\min\left[\mu(x|\overline{M}),\mu(x|\overline{N})\right]\}, \tag{8}$$

for a standard measure of equality. This is simply the height of the intersection of $\overline{M}$ and $\overline{N}$. Equation (8) may be used to compare any two fuzzy sets.

The next section discusses set theoretic operations on fuzzy numbers followed by a more detailed discussion on the algebraic operations on fuzzy numbers. The last section contains a brief summary and our conclusions.

2. SET OPERATIONS ON FUZZY NUMBERS

Since fuzzy numbers are fuzzy sets we may AND (intersect) and OR (union) two fuzzy numbers. Therefore, we wish to compute

$$\overline{P} = \overline{M} \text{ AND } \overline{N}, \tag{9}$$

and

$$\overline{Q} = \overline{M} \text{ OR } \overline{N}. \tag{10}$$

In general we would compute $\overline{P}$ and $\overline{Q}$ as

$$\mu(x|\overline{P}) = T\left[\mu(x|\overline{M}),\mu(x|\overline{N})\right], \tag{11}$$

$$\mu(x|\overline{Q}) = C\left[\mu(x|\overline{M}),\mu(x|\overline{N})\right], \tag{12}$$

for t-norm T and co-t-norm C. The selection of T and C will depend on $\overline{M}$ and $\overline{N}$. Sometimes we would use LAND for T and other times we would employ min for T. In certain situations one should use LOR for C and in other cases one will employ max for C.

Let us consider a few examples to illustrate the choice of T and C. Other applications are in Section 3A. We use the usual negation so (not $\overline{M}$) is

$$\mu\left[x|(\text{not } \overline{M})\right] = 1 - \mu(x|M). \tag{13}$$

Now compute

$$\overline{P} = \overline{M} \text{ AND } (\text{not } \overline{M}), \tag{14}$$

and

$$\overline{Q} = \overline{M} \text{ OR } (\text{not } \overline{M}). \tag{15}$$

Using LAND for T and LOR for C we get $\overline{P} = \phi$ and $\overline{Q} = \mathcal{U}$. This result appears intuitively correct. If instead one uses min for T and max for C, then $\overline{P} \neq \phi$ and $\overline{Q} \neq \mathcal{U}$.

In Section 3A we will define fuzzy sets $(< \overline{M})$, $(\leq \overline{M})$, $(\geq \overline{M})$ and $(> \overline{M})$. For example. $(< \overline{M})$ is the numbers fuzzily less than $\overline{M}$ and $(\geq \overline{M})$ is the numbers fuzzily greater than or equal to $\overline{M}$. Now compute

$$\overline{P}_1 = (< \overline{M}) \text{ AND } (\geq \overline{M}), \tag{16}$$

and

$$\overline{P}_2 = (\overline{M}) \text{ AND } (\geq \overline{M}). \tag{17}$$

Using LAND $\overline{P}_1 = \phi$ but T = min does not produce $\overline{P}_1 = \phi$. If min = T, then $\overline{P}_2 = \overline{M}$ but LAND = T does not give $\overline{M}$ for $\overline{P}_2$. We may also wish to evaluate

$$\overline{Q}_1 = (< \overline{M}) \text{ OR } \overline{M}, \tag{18}$$

and

$$\overline{Q}_2 = (\geq \overline{M}) \text{ OR } \overline{M}. \tag{19}$$

With LOR = C we get $\overline{Q}_1 = (\leq \overline{M})$ and max = C gives $\overline{Q}_2 = (\geq \overline{M})$. We therefore see that we may have to pick our t-norm, or co-t-norm, to obtain a desirable result. A FES shell should make available to the user different t-norms and co-t-norms in order to perform necessary set theoretic operations on fuzzy numbers.

We have no theorem to tell the reader how to pick a correct t-norm or co-t-norm. However, we do feel that for AND: (1) if you wish to discount the overlap (if any) between $\overline{M}$ and $\overline{N}$ as much as possible then use a small t-norm like LAND; and (2) if you want to weight the overlap as much as possible use min. For OR: (1) to weight the overlap as much as possible have LOR = C; and (2) to discount overlap as much as possible employ a small co-t-norm such as max. These principles are also illustrated in the difference between the exclusive OR (EOR) and the inclusive OR (IOR).

Consider the statement

$$\overline{A} = (\overline{M} \text{ OR } \overline{N}). \tag{20}$$

Do we mean: (1) $\overline{A} \approx \overline{M}$, or $\overline{A} \approx \overline{N}$, or $\overline{A}$ is approximately equal to both $\overline{M}$ and $\overline{N}$; or (2) $\overline{A} \approx \overline{M}$, or $\overline{A} \approx \overline{N}$, but $\overline{A}$ is not approximately equal to both $\overline{M}$ and $\overline{N}$? The first case is IOR and the second case is EOR. We would consider using LOR for IOR and max for EOR. As an example let $\overline{M} = (1/2,2/3)$ and $\overline{N} = (2/3,3/4)$. Then $\overline{M}$ LOR $\overline{N} = (1/2,3/4)$ but employing max for C $\overline{M}$ OR $\overline{N}$ has a "dip" in its membership function on [2,3]. If $\overline{A} = (2/2.5,2.5/3)$ the value of equation (20) would be one for IOR but 0.75 for EOR. The values of $\overline{A} = \overline{M}$ and $\overline{A} = \overline{N}$ are both 0.75 so the IOR gives 0.75 but we get one for $\overline{A}$ approximately equal to both $\overline{M}$ and $\overline{N}$. In IOR we increased the overlap as much as possible with LOR and decreased the overlap in EOR with max.

3. ALGEBRA OF FUZZY NUMBERS

This section is broken down into three subsections: (1) comparison of fuzzy numbers (inequalities); (2) addition and multiplication of positive fuzzy numbers; and (3) the extended max and min of discrete fuzzy numbers in [0,1].

3.A. INEQUALITIES

We require a method of evaluating $\overline{M} \leq \overline{N}$, $\overline{M} < \overline{N}$, $\overline{M} = \overline{N}$, $\overline{M} > \overline{N}$ and $\overline{M} \geq \overline{N}$ for fuzzy numbers because such statements may appear within rules in a FES ([2],[6]). We first discuss a set theoretic procedure for comparing non-interactive continuous fuzzy numbers. We then introduce a second method, for interactive continuous fuzzy numbers or for discrete fuzzy numbers, of comparing fuzzy variables.

When we were designing our FES shell FLOPS the procedures known to us for evaluating $\overline{M} < \overline{N}$ and $\overline{M} \leq \overline{N}$ produced the same answer [9]. We therefore proceeded to construct our own methods of computing $\overline{M} \leq \overline{N}$ and $\overline{M} < \overline{N}$.

We began with defining fuzzy sets $(< \overline{N})$, and $(> \overline{N})$, of numbers fuzzily less than, or greater than, $\overline{N}$. The conditions we placed on $(< \overline{N})$ and $(> \overline{N})$ were: (1) $(< \overline{N})$ and $(> \overline{N})$ should be subsets of (not $\overline{N}$); (2) $(< \overline{N})$ and $(> \overline{N})$ should be disjoint; and (3) the union of $(< \overline{N})$ and $(> \overline{N})$ should equal (not $\overline{N}$). It was clear to us that $(< \overline{N})$ would be the "left side" of (not $\overline{N}$) and $(> \overline{N})$ would be the "right side" of (not $\overline{N}$). Therefore, if $\overline{N} = (a/b,c/d)$, then

$$(<\overline{N}) = \begin{cases} (\text{not } \overline{N}) & \text{on } (-\infty, b], \\ 0, & \text{otherwise}, \end{cases} \tag{21}$$

and

$$(>\overline{N}) = \begin{cases} (\text{not } \overline{N}) & \text{on } [c, +\infty), \\ 0, & \text{otherwise}. \end{cases} \tag{22}$$

For the second type of fuzzy number (equation (1)) we have $(<\overline{N}) = (\text{not } \overline{N})$ on $(-\infty, \theta]$ and zero otherwise and $(>\overline{N}) = (\text{not } \overline{N})$ on $[\theta, +\infty)$ and zero otherwise. Using some measure of equality (say equation (8)) we define

$$v[\overline{M} < \overline{N}] = v[\overline{M} = (<\overline{N})], \tag{23}$$

and

$$v[\overline{M} > \overline{N}] = v[\overline{M} = (>\overline{N})]. \tag{24}$$

When $\overline{M}$ and $\overline{N}$ are both fuzzy numbers derived from measurements (equation (1)) it is not difficult to obtain closed form solutions, in terms of θ and σ, for equations (23) and (24).

To obtain the other two inequalities we defined the fuzzy sets

$$(\leq \overline{N}) = (<\overline{N}) \text{ OR } (\overline{N}), \tag{25}$$

$$(\geq \overline{N}) = (>\overline{N}) \text{ OR } (\overline{N}), \tag{26}$$

of numbers fuzzily less than or equal, greater than or equal, to $\overline{N}$. To evaluate equations (25) and (26) we chose LOR to give the overlap between $(<\overline{N})$ and $\overline{N}$, and $(>\overline{N})$ and $\overline{N}$, maximum weight. When $\overline{N} = (a/b, c/d)$ we obtain

$$(\leq \overline{N}) = \begin{cases} 1, & \text{on } (-\infty, b], \\ \overline{N}, & \text{on } [b, +\infty), \end{cases} \tag{27}$$

and

$$(\geq \overline{N}) = \begin{cases} \overline{N}, & \text{on } (-\infty, c], \\ 1, & \text{on } [c, +\infty). \end{cases} \tag{28}$$

If we had used max for OR in equations (25) and (26) we would get an undesirable "dip" in $(\leq \overline{N})$ on $[a, b]$ and an undesirable "dip" in $(\geq \overline{N})$ on $[c, d]$. For the second type of fuzzy number we get $(\leq \overline{N}) = \overline{N}$ on $[\theta, +\infty)$ and equal to one otherwise and $(\geq \overline{N}) = \overline{N}$ on $(-\infty, \theta]$ and equal to one otherwise. We therefore define

$$v[\overline{M} \leq \overline{N}] = v[\overline{M} = (\leq \overline{N})], \tag{29}$$

and

$$v[\overline{M} \geq \overline{N}] = v[\overline{M} = (\geq \overline{N})]. \tag{30}$$

Equations (29) and (30) are also easily evaluated when $\overline{M}$ and $\overline{N}$ are given by equation (1).

Using the fuzzy sets $(\geq \overline{M})$, $(\leq \overline{N})$, $(> \overline{M})$ etc. we can construct fuzzy intervals such as

$$\overline{S} = (\geq \overline{M}) \text{ AND } (\leq \overline{N}), \tag{31}$$

and

$$\overline{V} = (> \overline{M}) \text{ AND } (< \overline{N}). \tag{32}$$

In order to maximize the overlap we would employ min to evaluate equations (31) and (32).

Although our method of comparing fuzzy numbers was intuitively clear to us, it is not new. Sanchez [13] also considers the fuzzy sets $(\leq \overline{N})$, $(\geq \overline{N})$ and fuzzy intervals. If $\overline{N} = (a/b, b/d)$, then his membership function for $(\leq \overline{N})$ is 0.5 at $x = b$ where as our value will be one for $x = b$.

Our method of evaluating $\overline{M} \leq \overline{N}$ and $\overline{M} < \overline{N}$ turns out to be equivalent to one of Dubois and Prade's methods ([1],[8]). Suppose $\overline{M}$ and $\overline{N}$ are two triangular fuzzy numbers and let $\mu(x|(< \overline{N}))$ be the membership function for $(< \overline{N})$. Then the value of $\overline{M} < \overline{N}$ is equal to

$$\sup\{\min[\mu(x|\overline{M}), \mu(x|(< \overline{N}))]\}. \tag{33}$$

Equation (33) is equivalent to

$$\Pi_M(]N, +\infty)), \tag{34}$$

given in [8], which is their measure of $\overline{M} < \overline{N}$.

The above procedure for comparing fuzzy numbers is not applicable to discrete fuzzy numbers because we will usually obtain zero for our measure of equality (equation (8)). If $\overline{M}$ and $\overline{N}$ are discrete, then quite often their intersection is ϕ and therefore $v[\overline{M} = \overline{N}]$, from equation (8), is zero. Therefore, for discrete $\overline{M}$ and $\overline{N}$ and continuous interacting $\overline{M}$ and $\overline{N}$, we define

$$v[\overline{M} < \overline{N}] = \sup\{\Pi(x,y)\,|\,x < y\}, \tag{35}$$

$$v[\overline{M} \leq \overline{N}] = \sup\{\Pi(x,y)\,|\,x \leq y\}, \tag{36}$$

$$v[\overline{M} = \overline{N}] = \sup\{\Pi(x,y)\,|\,x = y\}, \tag{37}$$

$$v[\overline{M} \geq \overline{N}] = \sup\{\Pi(x,y)\,|\,x \geq y\}, \tag{38}$$

$$v[\overline{M} > \overline{N}] = \sup\{\Pi(x,y)\,|\,x > y\}. \tag{39}$$

We notice that $v[\overline{M} = \overline{N}]$ from equation (37) will be the same as equation (8) for non-interactive continuous $\overline{M}$ and $\overline{N}$. We do not use equations (35) through (39) for continuous non-interactive $\overline{M}$ and $\overline{N}$ because: (1) equations (35) and (36) will give the same result; (2) equations (38) and (39) will produce the same answer; and (3) the two methods of computing $v[\overline{M} < \overline{N}]$, equations (23) and (35), give different results, •••, and the two methods of finding $v[\overline{M} > \overline{N}]$, equations (24) and (39), produce different answers. In FLOPS we now use the previous method for comparing non-interactive continuous $\overline{M}$ and $\overline{N}$, especially for the fuzzy numbers given in equation (1).

Let us consider an example of two strongly positively associated variables whose joint discrete distribution is given in Table II with marginals shown in Table I. Using the joint in Table II we see that $v[\overline{M} < \overline{N}] = 0$, $v[\overline{M} \leq \overline{N}] = 0$, $v[\overline{M} = \overline{N}] = 0$, $v[\overline{M} \geq \overline{N}] = 1$ and $v[\overline{M} > \overline{N}] = 1$. If we had erroneously assumed no interaction, so the joint is the minimum of the marginals, we obtain $v[\overline{M} < \overline{N}] = 0.50$, $v[\overline{M} \leq \overline{N}] = 0.75$, $v[\overline{M} = \overline{N}] = 0.75$, $v[\overline{M} \geq \overline{N}] = 1$ and $v[\overline{M} \geq \overline{N}] = 1$.

Table I

Discrete Fuzzy Numbers in the Example in Section 3A.

Table Ia

Discrete Fuzzy Number $\overline{M}$

x	2.5	3	3.5	4	4.5	5	5.5
$\mu(x\|M)$	0.25	0.50	0.75	1.00	0.75	0.50	0.25

Table Ib

Discrete Fuzzy Number $\overline{N}$

x	1.5	2	2.5	3	3.5	4	4.5
$\mu(x\|N)$	0.25	0.50	0.75	1.00	0.75	0.50	0.25

Table II

Strongly Positively Associated Variables in the Example in Section 3A.

$\overline{N}$ \ $\overline{M}$	2.5	3	3.5	4	4.5	5	5.5
4.5	0	0	0	0	0	0	0.25
4	0	0	0	0	0	0.50	0
3.5	0	0	0	0	0.75	0	0
3	0	0	0	1.00	0	0	0
2.5	0	0	0.75	0	0	0	0
2.	0	0.50	0	0	0	0	0
1.5	0.25	0	0	0	0	0	0

With no prior information on associations we would default to the non-interactive method since it appears to give good results for negatively associated variables. Assuming non-interaction will, as the above example shows, give some incorrect answers if the variables are strongly positively associated. Therefore, a FES shell should allow the user different methods of comparing fuzzy numbers, especially for discrete and positively associated fuzzy numbers.

3.B. ADDITION AND MULTIPLICATION

Let X and Y be two fuzzy variables with possibility distributions $\overline{M}$ and $\overline{N}$, respectively, which are positive fuzzy numbers. We wish to find the possibility distribution of X + Y and XY which will be $\overline{M} + \overline{N}$ and $\overline{MN}$, respectively. In this section we will discuss how to add and multiply $\overline{M}$ and $\overline{N}$ depending on prior associations between X and Y. For simplicity we will assume $\overline{M}$ and $\overline{N}$ are discrete positive fuzzy numbers. Similar results hold for positive continuous fuzzy numbers.

Let us first consider adding $\overline{M}$ and $\overline{N}$. If $\overline{P}$ is to be the sum of $\overline{M}$ and $\overline{N}$, then the extension principle gives ([11],[12])

$$\mu(z|\overline{P}) = \sup\{T(\mu(x|\overline{M}),\mu(y|\overline{N}))\,|x + y = z\}, \qquad (40)$$

for $\overline{P}$, where T is any t-norm. The classical principle employs min for T. If $\Pi(x,y)$ is the joint possibility distribution for X and Y, whose marginals are $\overline{M}$ and $\overline{N}$, then we also have

$$\mu(z|\overline{P}) = \sup\{\Pi(x,y)\ |x + y = z\}, \tag{41}$$

for $\overline{P}$. When X and Y are non-interactive (independent) then equation (41) reduces to the classical extension principle but when X and Y interact equation (40) will not necessarily give the sum of $\overline{M}$ and $\overline{N}$.

First assume that X and Y are strongly negatively associated with the joint given in Table IVa whose marginals are shown in Table III. Then X + Y can take on only the value of six so that $\mu(z|\overline{P}) = 1$ when $z = 6$ and is zero otherwise. We are using equation (41) to find $\overline{P}$. Table IVb gives an example of X and Y negatively associated (not maximally negatively associated) and now the sum is given in Table V. This sum can also be obtained using the extension principle if we have LAND for T. If X and Y are positively associated it appears that the classical extension principle (equation (40) with min = T) works well for finding $\overline{M} + \overline{N}$.

Table III

Discrete Fuzzy Numbers in the Example in Section 3B.

Table IIIa

Discrete Fuzzy Number $\overline{M}$

x	0.5	1	1.5	2	2.5	3	3.5
$\mu(x\|\overline{M})$	0.25	0.50	0.75	1.00	0.75	0.50	0.25

Table IIIb

Discrete Fuzzy Number $\overline{N}$

x	2.5	3	3.5	4	4.5	5	5.5
$\mu(x\|\overline{N})$	0.25	0.50	0.75	1.00	0.75	0.50	0.25

Table IV

Negative Associations between X and Y in the Example in Section 3B.

Table IVa

Maximum Negative Association

$\overline{M}$ \ $\overline{N}$	2.5	3	3.5	4	4.5	5	5.5
3.5	0.25	0	0	0	0	0	0.
3	0	0.50	0	0	0	0	0
2.5	0	0	0.75	0	0	0	0
2	0	0	0	1.0	0	0	0
1.5	0	0	0	0	0.75	0	0
1	0	0	0	0	0	0.50	0
0.5	0	0	0	0	0	0	0.25

Table IVb

Negative Association

$\overline{M}$ \ $\overline{N}$	2.5	3	3.5	4	4.5	5	5.5
3.5	0.25	0.25	0	0	0	0	0
3	0.25	0.50	0.25	0.25	0	0	0
2.5	0	0.50	0.75	0.50	0.25	0.25	0
2	0	0.25	0.75	1.00	0.75	0.50	0
1.5	0	0.25	0.50	0.50	0.75	0.50	0
1	0	0	0	0.25	0.25	0.50	0.25
0.5	0	0	0	0	0	0.25	0.25

Table V

The Sum of $\overline{M}$ and $\overline{N}$ given in Table IVb

z	4.5	5	5.5	6	6.5	7	7.5
$\mu(z\|\overline{P})$	0.25	0.5	0.75	1.0	0.75	0.50	0.25

However, we hardly ever know the joint distribution (exact association) of X and Y but we quite often know of positive or negative association. Consider a process with two outputs X and Y with possibility distributions $\overline{M}$ and $\overline{N}$, respectively. The outputs are <u>competitive</u> (competing for scarce resources) when X and Y are negatively associated and <u>complementary</u> if X and Y are positively associated. For competitive outputs if X gets large then Y tends to be small and vice versa. Competitive outputs will have small dispersion (support) for the combined output $\overline{M} + \overline{N}$ [10]. An analogous situation occurs in statistics when $\text{var}(X + Y) < \text{var}(X) + \text{var}(Y)$ for negatively correlated random variables. Complementary outputs tend to have large dispersion (support) for the total output $\overline{M} + \overline{N}$ [10]. In statistics $\text{var}(X + Y) > \text{var}(X) + \text{var}(Y)$ if X and Y are positively correlated random variables. Examples of competitive outputs abound in industry where a firm produces multiple products all utilizing the same limited resources. Complementary outputs occur when one output is a byproduct of the other output (sawdust and lumber, slag and iron, etc.). If we have no information as to the association between X and Y then we suggest defaulting to the usual extension principle for finding $\overline{M} + \overline{N}$.

Similar results hold for finding $\overline{MN}$ the possibility distribution for XY. If $\overline{Q} = \overline{MN}$ then the extension principle gives

$$\mu(z|\overline{Q}) = \sup\{T(\mu(x|\overline{M}),\mu(y|\overline{N}))\,|xy = z\}, \tag{42}$$

for a t-norm T and the joint distribution method is

$$\mu(z|\overline{Q}) = \sup\{\Pi(x,y)\,|xy = z\}. \tag{43}$$

If X and Y are non-interactive, or positively associated, equation (42) gives good results for $\overline{M}\cdot\overline{N}$ when T = min. However, when X and Y are negatively associated we need to use equation (43) to find $\overline{Q}$. The dispersion (support) of $\overline{Q}$ will be smaller when X and Y are negatively associated than when they are positively associated.

Although we usually do not know the exact association between X and Y to find $\overline{MN}$ we sometimes can easily determine whether it is positive or negative. For example, if X is the price of a product and Y is its weakly demand, then X and Y are negatively associated which would be used to find the possibility distribution of weekly revenue XY. Also, if X is a future interest rate and Y is the future investment, then X and Y are positively associated which is taken into account in finding the possibility distribution of the future return XY [5]. With no information about the association between X and Y we would default to the classical extension principle to determine $\overline{MN}$.

The arithmetic of fuzzy numbers is usually not performed within a production rule. That is, $\overline{M} + \overline{N}$ and $\overline{MN}$ are preliminary calculations whose results may be input into a pattern within a rule. The discussion above implies that we need new methods of calculating $\overline{M} + \overline{N}$ and $\overline{MN}$ when these fuzzy numbers are "values" of negatively associated fuzzy variables. Various methods of performing fuzzy arithmetic must be made available to the user of a FES shell.

3.C. MAX AND MIN

In a FES all data items and rules have attached to them confidence values (cf-values). In FLOPS ([3],[14]) these cf-values are single numbers in [0,1]. We would like to allow the cf-values to be fuzzy. A fuzzy cf-value will be a discrete fuzzy number in [0,1].

It is, of course, possible to consider continuous fuzzy numbers $\overline{M} = (a/b,c/d)$ in [0,1] for cf-values, but the added accuracy for all numbers between a and d to be cf-values does not seem warranted. We would think possible cf-values to be 0.0,0.1,•••,0.9,1.0, or 0.00,0.01,•••,0.99,1.00, which implies fuzzy cf-values to be discrete fuzzy numbers.

Two examples of fuzzy cf-values are presented in Table VI. For the fuzzy cf-value $\overline{M}$ the confidence values are 0.5, 0.6, 0.7 and 0.8 but our confidence in these values is given by 0.50, 1,00, 0.67 and 0.33, respectively.

Table VI

Discrete Fuzzy Confidence Values in Section 3C.

Table VIa

Discrete Fuzzy Confidence Value $\overline{M}$

x	0.50	0.6	0.7	0.8
$\mu(x\|\overline{M})$	0.50	1.00	0.67	0.33

Table VIb

Discrete Fuzzy Confidence Value $\overline{N}$

x	0.50	0.6	0.7	0.8	0.9
$\mu(x\|\overline{N})$	0.25	0.50	0.75	1.00	0.50

Given an initial data base we must process the fuzzy cf-values through the fuzzy rules to obtain a fuzzy set of conclusions. The structure of a rule is [2]

$$\text{If } [\ ] * \cdots * [\ ], \text{ then } \cdots, \tag{44}$$

where each [ ] is a pattern and $*$ stands for AND or OR. When the left hand side of the rule is evaluated each pattern will produce a fuzzy cf-value. We now must evaluate

$$\overline{\alpha}_1 * \overline{\alpha}_2 * \cdots \overline{\alpha}_n = \overline{\beta}, \tag{45}$$

where $\overline{\alpha}_i$ is the fuzzy cf-value for the ith pattern and $\overline{\beta}$ will be the fuzzy cf-value for the rule antecedent. For simplicity let us assume the rule has only two patterns so we must compute

$$\overline{\alpha}_1 * \overline{\alpha}_2 = \overline{\beta}, \tag{46}$$

for $*$ = AND or $*$ = OR. For notational convenience also let the two patterns be called $\mathcal{P}_1$ and $\mathcal{P}_2$ so our simple rule is

$$\text{If } \mathcal{P}_1 * \mathcal{P}_2, \text{ then } \cdots. \tag{47}$$

Let us first consider $*$ = OR so

$$\overline{\beta} = \overline{\alpha}_1 \text{ OR } \overline{\alpha}_2, \tag{48}$$

with $\overline{\alpha}_i$ a discrete fuzzy number in [0,1]. It is natural to employ max for OR in equation (48) so that

$$\overline{\beta} = \overline{\max}(\overline{\alpha}_1, \overline{\alpha}_2). \tag{49}$$

The usual method of finding $\overline{\max}$ is to use the extension principle so we have

$$\mu(z|\overline{\beta}) = \sup\{T(\mu(x|\overline{\alpha}_1), \mu(y|\overline{\alpha}_2)) \,|\, \max(x,y) = z\}, \tag{50}$$

where T is a t-norm. However, this method does not allow for any prior associations between $\mathcal{P}_1$ and $\mathcal{P}_2$. We have argued ([4],[7]), when the cf-values are single numbers in [0,1], that prior associations between the patterns in a rule antecedent must be taken into account in determining final antecedent confidence. Now we argue that the joint distribution $\Pi(x,y)$ should be used to find $\overline{\beta}$. $\Pi(x,y)$ is the joint distribution of the fuzzy variables, which give the confidence in $\mathcal{P}_1$ and $\mathcal{P}_2$, whose possibility distributions are $\overline{\alpha}_1$ and $\overline{\alpha}_2$, respectively. Therefore

$$\mu(z|\overline{\beta}) = \sup\{\Pi(x,y) \,|\, \max(x,y) = z\}. \tag{51}$$

First suppose $\mathcal{P}_1$ and $\mathcal{P}_2$ are strongly negatively associated because: (1) a high cf-value for $\mathcal{P}_1$ implies a low cf-value for $\mathcal{P}_2$; and (2) a low cf-value for $\mathcal{P}_1$ implies a high cf-value for $\mathcal{P}_2$. In other words, if $\mathcal{P}_1$ tends to be true (false) then $\mathcal{P}_2$ tends to be false (true) as in $\mathcal{P}_1 = [\overline{A} \leq (1/2, 2/3)]$ and $\mathcal{P}_2 = [\overline{A} \geq (2/3, 3/4)]$ for fuzzy number $\overline{A}$. An example of this situation is given in Table VII. Table VII is the joint distribution $\Pi(x,y)$ of negatively associated fuzzy variables whose marginals are $\overline{\alpha}_1 = \overline{M}$ and $\alpha_2 = \overline{N}$ from Table VI. Using the joint (equation (51)) the $\overline{\max}$, or $\overline{\beta}$ is shown in Table VIII.

Table VII

Joint Distribution $\Pi(x,y)$ for Negatively Associated Fuzzy Variables in the Example in Section 3C.

$\overline{\alpha}_1 = \overline{M}$				
0.9	0.5	0	0	0
0.8	0.5	1.0	0	0
0.7	0	0.75	0.67	0
0.6	0	0	0.50	0.33
0.5	0	0	0	0.25
$\alpha_2 = \overline{N}$	0.5	0.6	0.7	0.8

Table VIII

$\overline{MAX} = \overline{\beta}$ of Variables in Table VII.

z	0.5	0.6	0.7	0.8	0.9
$\mu(z\|\overline{\beta})$	0	0	0.75	1.0	0.50

Notice that $\overline{\beta}$ agrees with the "larger" $\overline{M}$ at 0.7, 0.8 and 0.9 but is zero for 0.5, 0.6. This is because to obtain $z = 0.5$ or 0.6 <u>both</u> $\mathcal{P}_1$ and $\mathcal{P}_2$ must have low confidence but to obtain $z = 0.8$ or 0.9 either $\mathcal{P}_1$ or $\mathcal{P}_2$ can have high confidence. We conclude that the usual $\overline{\max}$, equation (50) with min for T, does not work for strongly negatively associated variables.

When $\mathcal{P}_1$ and $\mathcal{P}_2$ are non-interactive (independent) then equation (51) reduces to equation (50), with min for T, because the joint $\Pi(x,y)$ is just the minimum of the marginals. When $\mathcal{P}_1$ and $\mathcal{P}_2$ are positively associated it appears that the usual $\overline{\max}$ works well to find the final confidence in $\mathcal{P}_1$ OR $\mathcal{P}_2$.

We next consider the case where $*$ = AND so that

$$\overline{\beta} = \overline{\alpha}_1 \text{ AND } \overline{\alpha}_2. \tag{52}$$

Now we would consider employing min for AND so we have

$$\overline{\beta} = \overline{\min}(\overline{\alpha}_1, \overline{\alpha}_2), \tag{53}$$

for $\overline{\alpha}_i$ discrete fuzzy numbers in [0,1]. The common procedure for finding $\overline{\min}$ is

$$\mu(z|\overline{\beta}) = \sup\{T(\mu(x|\overline{\alpha}_1), \mu(y|\overline{\alpha}_2)) \,|\, \min(x,y) = z\}, \tag{54}$$

with min for T. Employing the joint we also have

$$\mu(z|\overline{\beta}) = \sup\{\Pi(x,y) \,|\, \min(x,y) = z\}. \tag{55}$$

When $\mathcal{P}_1$ and $\mathcal{P}_2$ are non-interactive (independent) the joint procedure reduces to the extension principle for finding $\overline{\beta}$. If $\mathcal{P}_1$ and $\mathcal{P}_2$ are positively associated it again appears that the classical extension principle method (equation (54) with min = T) works well for calculating the confidence of $\mathcal{P}_1$ AND $\mathcal{P}_2$. However, when $*$ = AND we would probably not have a situation where $\mathcal{P}_1$ and $\mathcal{P}_2$ are negatively associated because then the rule would never fire. If $\mathcal{P}_1$ and $\mathcal{P}_2$ are negatively associated, then our confidence in $\mathcal{P}_1$ AND $\mathcal{P}_2$ would always be low because when $\mathcal{P}_1$ tends to be true (false), $\mathcal{P}_2$ tends to be false (true). If our confidence in $\mathcal{P}_1$ AND $\mathcal{P}_2$ is low, then our confidence in the rule antecedent is also low and this rule would probably never execute. It makes little sense to write rules that never fire so $\mathcal{P}_1$ and $\mathcal{P}_2$ in $\mathcal{P}_1$ AND $\mathcal{P}_2$ should not be (strongly) negatively associated.

If we are to allow fuzzy confidence values in fuzzy expert systems, then new procedures ($\overline{\max}$ for negatively associated fuzzy confidence values) need to be developed to process these fuzzy cf-values through the system to obtain a fuzzy set of conclusions.

4. SUMMARY AND CONCLUSIONS

This paper discusses operations, both set theoretic and algebraic, on fuzzy numbers needed in applications programs in fuzzy expert systems. We argued that a number of t-norms, and co-t-norms, should be made available to the user of a FES shell in order to perform the AND (intersection) and the OR (union) of fuzzy numbers. We presented a general principle, relating to how much weight one wishes to place on the overlap of the fuzzy numbers, to help a user select a proper t-norm and co-t-norm.

In applications fuzzy numbers do not exist "alone" but exist as "values" of fuzzy variables. Therefore, fuzzy numbers are also interpreted as possibility distributions. We argued that prior associations between the fuzzy variables should be employed in computing the results of algebraic operations performed on their fuzzy (values) numbers. We presented examples showing that for (strongly) positively, and negatively, associated variables the classical extension principle does not produce correct answers. The analogy in statistics is that to find the distribution of a function of random variables one must use their joint distribution which will contain all correlations between the variables. One should not attempt to draw an analogy between fuzzy numbers and real numbers because no association exists between real numbers. Associations may exist between fuzzy numbers when they are properly considered as "values" of fuzzy variables. When designing a FES shell one should make available a number of different methods for manipulating, and comparing, fuzzy numbers especially taking into account the extreme cases of strongly positively, or negatively, associated fuzzy variables.

REFERENCES

1. G. Bortolan and R. Degani, "A Review of Some Methods for Ranking Fuzzy Subsets," Fuzzy Sets and Systems 15(1985) 1-19.
2. J. J. Buckley, W. Siler and D. Tucker, "Fuzzy Expert System", Fuzzy Sets and Systems 20(1986) 1-16.
3. J. J. Buckley, W. Siler and D. Tucker, "FLOPS, A Fuzzy Expert System: Applications and Perspectives", in C. V. Negoita and H. Prade (eds.), Fuzzy Logics in Knowledge Engineering (Verlag TUV Rheinland, Koln, W.-Germany) 1986, 256-274.
4. J. J. Buckley and W. Siler, "Fuzzy Operators for Possibility Interval Sets", Fuzzy Sets and Systems 22(1987) To Appear.
5. J. J. Buckley, Fuzzy Mathematics of Finance, Fuzzy Sets and Systems 22(1987). To Appear.
6. J. J. Buckley and W. Siler, Echocardiogram Analysis Using Fuzzy Numbers and Relations, Fuzzy Sets and Systems. To Appear.
7. J. J. Buckley and W. Siler, "Managing Uncertainty in a Fuzzy Expert System", Int. J. Man-Machine Studies. Submitted.
8. D. Dubois, "Ranking Fuzzy Numbers in the Setting of Possibility Theory", Information Sciences 30(1983) 183-224.
9. D. Dubois and H. Prade, Fuzzy Sets and Systems (Academic Press, N. Y.) 1980.
10. D. Dubois and H. Prade, "Additions of Interactive Fuzzy Numbers", IEEE Trans. on Automatic Control 26(1981) 926-936.

11. D. Dubois and H. Prade, Inverse Operations for Fuzzy Numbers, in: E. Sanchez and M. M. Gupta (eds.), Fuzzy Information, Knowledge Representation and Decision Analysis (Pergamon Press, N. Y.) 1983, 391-396.

12. D. Dubois and H. Prade, "Fuzzy-set-theoretic Differences and Inclusions and their Use in the Analysis of Fuzzy Equations", Control and Cybernetics 13(1984) 129-146.

13. E. Sanchez, "Medical Applications with Fuzzy Sets", in A. Jones, A. Kaufmann and H.-J. Zimmermann (eds.) Fuzzy Sets Theory and Applications (D. Reidel) 1985, 331-348.

14. W. Siler and D. Tucker, FLOPS, A Fuzzy Logic Production System User's Manual, Kemp-Carraway Heart Institute, Birmingham, AL 1986.

15. L. A. Zadeh, "The Concept of a Linguistic Variable and its Application to Approximate Reasoning - I", Information Sciences 8(1975) 199-249.

Fuzzy Logic in Knowledge-Based Systems,
Decision and Control
M.M. Gupta and T. Yamakawa (Editors)
Elsevier Science Publishers B.V. (North-Holland), 1988

A DEEPER LOOK ON FUZZY NUMBERS FROM A THEORETICAL AS WELL AS FROM A PRACTICAL POINT OF VIEW

Etienne E. Kerre
Seminar for Mathematical Analysis
State University of Gent, Belgium.

Antoine van Schooten
Seminar and Laboratory for Computer Science
State University of Gent, Belgium.

The first part of this paper is a serious attempt to grasp the different existing definitions of a fuzzy number. Starting from the most general concept i.e. a fuzzy set on the reals, several supplementary conditions are imposed in order to obtain smaller classes having a richer algebraic structure. The second part of the paper treats fuzzy numbers from a practical point of view. It describes a computer representation that is suitable for expert systems dealing with fuzzy information.

Keywords : fuzzy numbers, fuzzy arithmetic, approximate reasoning.

1. INTRODUCTION

The concept of a fuzzy number may be seen as one of the most fundamental concepts in fuzzy set theory. However, a walk through fuzzy litterature reveals the existence of many different definitions for this fundamental notion. We mention explicitly : the definition of Mizumoto and Tanaka simply imposing a convexity condition, the different concepts of Dubois and Prade using a monotony property and several forms of continuity and thirdly the more mathematically founded approach of Rodabaugh a.o. defining a fuzzy number as an equivalence class of decreasing fuzzy sets on R, satisfying some boundary condition. Anyway this is a very dangerous situation, since the corresponding calculus of fuzzy numbers highly depends on the definition choosen.

In the first part of this paper we give a theoretical, consistent description of fuzzy numbers and their properties. Moreover an attempt is made to link the different existing definitions. We start from a very general concept, namely the so-called fuzzy quantity as introduced by Dubois and Prade. Imposing supplementary conditions leads successively to the notion of a fuzzy interval, a fuzzy closed interval and finally a fuzzy number. On the basis of Zadeh's extension principle, unary as well as binary algebraic operations between fuzzy quantities are defined. The structure-properties of these operations are indicated; special attention will be called to the solution of questions such as : do we obtain a fuzzy closed interval by adding two fuzzy closed intervals?

In the second part of this paper, we present a computer-representation for a fuzzy quantity.
Using this representation and referring to articles by Dong, Shah,

Wong and Dubois, Prade, we show that arithmetic operations on fuzzy quantities and especially on fuzzy numbers can be carried out with a (very) good performance. Moreover, the suitability of our representation of fuzzy quantities for their use in fuzzy expert systems is outlined : operations such as union (OR), intersection (AND), compatibility and modus ponens can be done in a very efficient way. An example of a fuzzy production rule will also be given.

2. DEFINITIONS

We start from the most recent definitions of Dubois and Prade [1], moving from the most general to the more specific notions by successive adding of supplementary conditions.

(1) A *fuzzy quantity* is a fuzzy set on the real line R.

We will make no distinction between a fuzzy set and its membership function, hence a fuzzy quantity is simply an element of $[0,1]^R$.

It's well-known that a fuzzy quantity Q may be interpreted : Q(x) is the possibility that the value of the fuzzily known quantity Q is equal to x. For every $\alpha\in]0,1]$ we define the *α-level* (or α-cut) of Q as $Q_\alpha=\{x|x\in R \text{ and } Q(x)\geqslant\alpha\}$.

The 1-level of Q is called the *kernel* of Q :

$$\ker(Q)=\{x|x\in R \text{ and } Q(x)=1\}.$$

For every $\alpha\in[0,1[$ we may define the *strong α-level* (or strong α-cut) of Q as $Q_{\bar{\alpha}}=\{x|x\in R \text{ and } Q(x)>\alpha\}$.

The strong 0-level of Q is called the *support* of Q :

$$\text{supp}Q=\{x|x\in R \text{ and } Q(x)>0\}.$$

(2) A fuzzy quantity Q is called *non-negative (non-positive)* iff $\text{supp}Q\subseteq[0,+\infty[$ $(\text{supp}Q\subseteq]-\infty,0])$.

(3) A fuzzy quantity Q satisfying $\ker(Q)\neq\phi$ is called a *normalized fuzzy quantity*. Every element of ker(Q) is called a *modal value* of Q.

(4) A convex fuzzy quantity Q is called a *fuzzy interval;* it satisfies the convexity condition

$$(\forall(x_1,x_2)\in R^2)(\forall\lambda\in[0,1])(Q(\lambda x_1+(1-\lambda)x_2)\geqslant\min(Q(x_1),Q(x_2))).$$

(5) An upper-semicontinuous fuzzy interval is called a *fuzzy closed interval.*

(6) A *fuzzy real number* or shortly a fuzzy number is a fuzzy closed interval with a unique modal value and bounded support. Some authors replace the condition bounded support by compact support. As a compact set in R being equivalent to a closed and

bounded set, these authors only keep some kind of "flying" numbers (a simple triangular fuzzy number for example has no compact support!).

3. PROPERTIES OF (STRONG) α-LEVELS

The α-levels play a central role in the description of and the operations on fuzzy sets especially on fuzzy quantities. In the following we give an extensive list of their properties. Let X be an arbitrary set and let I denote the unit interval [0,1].

3.1. Decomposition of a fuzzy set in terms of its (strong) α-levels.

For $A \in I^X$ holds

$$A = \bigcup_{\alpha \in ran(A) \setminus \{0\}} \alpha A_\alpha \qquad A = \bigcup_{\alpha \in ran(A) \setminus \{1\}} \alpha A_{\overline{\alpha}}$$

where $\cup$ denotes Zadeh's union using supremum and αA_α stands for :

$$\alpha A_\alpha : X \rightarrow [0,1]$$
$$x \mapsto \alpha, \text{ if } A(x) \geqslant \alpha$$
$$x \mapsto 0, \text{ if } A(x) < \alpha.$$

3.2. Inclusion and equality in terms of (strong) α-levels

For $A \in I^X, B \in I^X$ we have :

$$A \subseteq B \Leftrightarrow (\forall \alpha \in]0,1])(A_\alpha \subseteq B_\alpha) \Leftrightarrow (\forall \alpha \in [0,1[)(A_{\overline{\alpha}} \subseteq B_{\overline{\alpha}})$$

$$A = B \Leftrightarrow (\forall \alpha \in]0,1])(A_\alpha = B_\alpha) \Leftrightarrow (\forall \alpha \in [0,1[)(A_{\overline{\alpha}} = B_{\overline{\alpha}}).$$

3.3. Zadeh's extended operations in terms of (strong) α-levels

Let J be an arbitrary index set and $(\forall i \in J)(A_i \in I^X)$.

Then we obtain for $\alpha \in]0,1]$

$$(\bigcap_{i \in J} A_i)_\alpha = \bigcap_{i \in J} (A_i)_\alpha, \ (\bigcup_{i=1}^{n} A_i)_\alpha = \bigcup_{i=1}^{n} (A_i)_\alpha, \ (\bigcup_{i \in J} A_i)_\alpha \supseteq \bigcup_{i \in J} (A_i)_\alpha$$

$$(coA)_\alpha \neq coA_\alpha, \ (coA)_\alpha = co(A_{\overline{1-\alpha}})$$

and for $\alpha \in [0,1[$

$$(\bigcup_{i \in J} A_i)_{\overline{\alpha}} = \bigcup_{i \in J} (A_i)_{\overline{\alpha}}, \ (\bigcap_{i=1}^{n} A_i)_{\overline{\alpha}} = \bigcap_{i=1}^{n} (A_i)_{\overline{\alpha}}, \ (\bigcap_{i \in J} A_i)_{\overline{\alpha}} \subseteq \bigcap_{i \in J} (A_i)_{\overline{\alpha}}$$

3.4. Convexity condition in terms of (strong) α-levels

Let $Q\in I^{R}$ (R may be replaced by an arbitrary linear space). Then :

$$Q \text{ is convex} \Leftrightarrow (\forall\alpha\in]0,1])(Q_\alpha \text{ is convex})$$
$$\Leftrightarrow (\forall\alpha\in[0,1[)(Q_{\overline{\alpha}} \text{ is convex})$$

Since the convex sets in R are precisely the intervals in R, we obtain the following characterization of a fuzzy interval :

$$Q \text{ is a fuzzy interval} \Leftrightarrow (\forall\alpha\in]0,1])(Q_\alpha \text{ is an interval})$$
$$\Leftrightarrow (\forall\alpha\in[0,1[)(Q_{\overline{\alpha}} \text{ is an interval}).$$

3.5 Structure of (strong) α-levels

The set of (strong) α-levels form a descending chain with respect to inclusion, i.e. for $A\in I^X$ we have :

$$\alpha_1\leqslant\alpha_2 \Rightarrow A_{\alpha_2}\subseteq A_{\alpha_1} \text{ and } A_{\overline{\alpha_2}}\subseteq A_{\overline{\alpha_1}}.$$

Moreover $A_{\overline{\alpha}}\subseteq A_\alpha$, $\forall\alpha\in]0,1[$.

3.6. Cartesian product of fuzzy sets in terms of (strong) α-levels

Let $A_1\in[0,1]^X$, $A_2\in[0,1]^X$. Then :

$$(A_1xA_2)_\alpha=(A_1)_\alpha x(A_2)_\alpha,\ \alpha\in]0,1]$$

$$(A_1xA_2)_{\overline{\alpha}}=(A_1)_{\overline{\alpha}}x(A_2)_{\overline{\alpha}},\ \alpha\in[0,1[,$$

where $A_1xA_2(x,y)=\min(A_1(x),A_2(y))$, $\forall(x,y)\in X^2$.

3.7. Direct and inverse image in terms of (strong) α-levels.

Let $f\in Y^X$, $g\in Z^{XxY}$, $A\in[0,1]^X$, $B\in[0,1]^Y$. Then we have the following relations for the direct and inverse images under f :

$$f(A_\alpha)\subseteq(f(A))_\alpha \text{ and } f^{-1}(B_\alpha)=(f^{-1}(B))_\alpha,\ \alpha\in]0,1]$$
$$f(A_{\overline{\alpha}})=(f(A))_{\overline{\alpha}} \text{ and } f^{-1}(B_{\overline{\alpha}})=(f^{-1}(B))_{\overline{\alpha}},\ \alpha\in[0,1[$$

$$g(A_\alpha,B_\alpha)\subseteq(g(A,B))_\alpha,\ \alpha\in]0,1]$$

$$g(A_{\overline{\alpha}},B_{\overline{\alpha}})=(g(A,B))_{\overline{\alpha}},\ \alpha\in[0,1[.$$

It's quite interesting to mention here a necessary and sufficient condition to obtain the reverse inclusion for the direct images [2] :

$$(\forall\alpha\in]0,1])(g(A_\alpha,B_\alpha)=(g(A,B))_\alpha)$$
$$\Leftrightarrow(\forall z\in Z)(\sup_{g(x,y)=z}\min(A(x),B(y)) \text{ is attained}).$$

The proofs of these useful properties are quite simple. As an example we show the characterization of convex fuzzy sets in terms of the strong α-levels. Let $Q\in I^X$ be convex, $\alpha\in[0,1[$, $\lambda\in[0,1]$, $x_1\in Q_{\bar{\alpha}}$, $x_2\in Q_{\bar{\alpha}}$. Then we have $Q(x_1)>\alpha, Q(x_2)>\alpha$ and hence, from the convexity of Q : $Q(\lambda x_1+(1-\lambda)x_2)>\alpha$ i.e. $\lambda x_1+(1-\lambda)x_2\in Q_{\bar{\alpha}}$. Conversely, let $x_1\in X, x_2\in X$ and putting $\min(Q(x_1),Q(x_2))=\alpha$ we obtain for every $\varepsilon\in]0,\alpha[$: $Q(x_1)>\alpha-\varepsilon, Q(x_2)>\alpha-\varepsilon$ and hence from the convexity of $Q_{\overline{\alpha-\varepsilon}}$: $\lambda x_1+(1-\lambda)x_2\in Q_{\overline{\alpha-\varepsilon}}$ i.e. $Q(\lambda x_1+(1-\lambda)x_2)>\alpha-\varepsilon$, $\forall\varepsilon\in]0,\alpha[$. Taking limits for $\varepsilon\to 0$, we obtain $Q(\lambda x_1+(1-\lambda)x_2)\geqslant\alpha$ or, replacing α, the convexity of Q.

4. SOME TOPOLOGICAL REFLECTIONS

4.1. Level characterization of upper- and lower-semicontinuous fuzzy quantities

Let (X,τ) be a topological space and R being equipped with the euclidean topology $\tau_{|\cdot|}$ derived from the absolute value mapping $|\cdot|$. It's wellknown that a subbase for $\tau_{|\cdot|}$ is given by $\{]\alpha,+\infty[\ |\alpha\in R\}\cup\{]-\infty,\alpha[\ |\alpha\in R\}$. The general form for continuity of an X-R mapping f is given as :

$$f \text{ is continuous}\Leftrightarrow(\forall O_2\in\tau_{|\cdot|})(f^{-1}(O_2)\in\tau)$$
$$\Leftrightarrow(\forall F_2\in\tau'_{|\cdot|})(f^{-1}(F_2)\in\tau')$$

where $\tau',\tau'_{|\cdot|}$ denote the class of closed sets corresponding with $\tau,\tau_{|\cdot|}$.

Using the subbase one easily obtains the following equivalent forms :

$$f \text{ is continuous}\Leftrightarrow(\forall\alpha\in R)(f^{-1}(]\alpha,+\infty[)\in\tau)$$
$$\text{and } (\forall\alpha\in R)(f^{-1}(]-\infty,\alpha[)\in\tau)$$

$$\Longleftrightarrow (\forall \alpha \in R)(f^{-1}([\alpha,+\infty[) \in \tau')$$

$$\text{and } (\forall \alpha \in R)(f^{-1}(]-\infty,\alpha] \in \tau').$$

Splitting these conjunctions leads to the notions of upper- and lower-semicontinuity :

$$f \text{ is upper-semicontinuous} \Longleftrightarrow (\forall \alpha \in R)(f^{-1}(]-\infty,\alpha[) \in \tau)$$

$$\Longleftrightarrow (\forall \alpha \in R)(f^{-1}([\alpha,+\infty[) \in \tau')$$

$$f \text{ is lower-semicontinuous} \Longleftrightarrow (\forall \alpha \in R)(f^{-1}(]\alpha,+\infty[) \in \tau)$$

$$\Longleftrightarrow (\forall \alpha \in R)(f^{-1}(]-\infty,\alpha]) \in \tau').$$

In particular, if $X=R$ and $\text{range}(f) \subseteq [0,1]$, i.e. if we restrict ourselves to fuzzy quantities Q, we obtain :

$$Q \text{ is upper-semicontinuous} \Longleftrightarrow (\forall \alpha \in]0,1])(Q_\alpha \text{ is closed})$$

$$Q \text{ is lower-semicontinuous} \Longleftrightarrow (\forall \alpha \in [0,1[)(Q_{\bar{\alpha}} \text{ is open}).$$

4.2. Level characterization of a fuzzy closed interval

From 4.1. and 3.4. one immediately deduces the following characterization of fuzzy closed intervals :

Q is a fuzzy closed interval $\Longleftrightarrow (\forall \alpha \in]0,1])(Q_\alpha$ is a closed interval)

Q is a fuzzy real number $\Rightarrow (\forall \alpha \in]0,1])(Q_\alpha$ is a compact interval)

but not the converse implication.

4.3. Link between upper- (lower-) semicontinuity and right- (left-) continuity

Generally there is no correspondence between these notions in case of arbitrary R-R functions. However, if we only consider monotone (not necessarily strict) R-R mappings, we can establish the following connections. Let Q be a fuzzy quantity, then we have :

if Q is increasing :

Q is upper-semicontinuous $\Longleftrightarrow$ Q is right-continuous

Q is lower-semicontinuous $\Longleftrightarrow$ Q is left-continuous

if Q is decreasing :

Q is upper-semicontinuous $\Longleftrightarrow$ Q is left-continuous

Q is lower-semicontinuous $\Longleftrightarrow$ Q is right-continuous.

4.4. Link between convexity and monotony

It's easily proved that every decreasing or increasing fuzzy quantity Q is convex. Hence every Rodabaugh and Lowen fuzzy number constitutes a convex fuzzy quantity [3]. The converse however does not hold. We have to distinguish between : sup Q is attained or sup Q is not attained. If sup Q is attained then we have :

$$Q \text{ is convex} \Rightarrow (\forall a \in \mathbb{R})(a = \sup Q \Rightarrow Q_{/]-\infty,a]} \text{ is increasing and } Q_{/[a,+\infty[} \text{ is decreasing}).$$

Hence triangular numbers, trapezoïdal numbers and L-R numbers of Dubois and Prade [4] are special cases of fuzzy intervals.
If sup Q is not attained then we have [5] :

$$Q \text{ is convex and } Q \text{ is upper-semicontinuous} \Rightarrow Q \text{ is monotone.}$$

5. The extension principle of Zadeh

It's well-known that an ordinary X-Y mapping f can be extended in a natural way into a $P(X)$-$P(Y)$ mapping, also denoted f, using the notion of direct image :

$$f : P(X) \to P(Y)$$
$$A \mapsto \{f(x) \mid x \in A\}, \ \forall A \in P(X),$$

where $P(X)$ denotes the power class of X.
Zadeh's extension principle [6] generalizes this construction. Every X-Y mapping f can be extended into a I^X-I^Y mapping as follows :

$$f : I^X \to I^Y$$
$$A \mapsto f(A), \ \forall A \in I^X$$

where f(A) denotes the extended direct image of A under f :

$$f(A) : Y \to [0,1]$$
$$y \mapsto \sup_{f(x)=y} A(x), \ \forall y \in \mathrm{ran}(f)$$
$$y \mapsto 0 \qquad , \ \forall y \in Y \setminus \mathrm{ran}(f).$$

Especially, if $X=X_1 x \ldots x X_n$ and $A=A_1 x \ldots x A_n$ where $A_i \in I^{X_i}$, we obtain :

$$f(A_1,\ldots,A_n)\cdot y = \sup_{f(x_1,\ldots,x_n)=y} \min(A_1(x_1),\ldots,A_n(x_n)),\ \forall y \in ran(f).$$

6. UNARY OPERATIONS ON FUZZY QUANTITIES

6.1. Definitions

Due to the extension principle every unary operation on $\mathbb{R}$ gives rise to a corresponding unary operation on the set of fuzzy quantities $I^{\mathbb{R}}$. We will consider explicitly : the opposite function $T(x)=-x$, the scalar multiplication $H_\lambda(x)=\lambda x$, the inverse function $O(x)=\frac{1}{x}$, the absolute value function $|.|(x)=|x|$ and the exponential function $\exp(x)=e^x$. If f denotes one of these unary operations on $\mathbb{R}$, we obtain a fuzzy quantity f(Q) for every fuzzy quantity Q as follows :

$$f(Q) : \mathbb{R} \to [0,1]$$

$$y \mapsto \sup_{f(x)=y} Q(x) \quad \text{if } f^{-1}(\{y\}) \neq \phi$$

$$y \mapsto 0 \quad \text{if } f^{-1}(\{y\}) = \phi.$$

The corresponding results are summarized in table 1.

6.2. Properties

(P.1) The opposite of a fuzzy real number is a fuzzy real number.

(P.2) The scalar product of a fuzzy real number and a real number different from zero is a fuzzy real number.

(P.3) The inverse of a fuzzy real number is in general no fuzzy real number.

(P.3') The inverse of a non-negative or a non-positive fuzzy closed interval is a fuzzy closed interval.

(P.4) The absolute value of a fuzzy real number is a fuzzy real number.

(P.5) The exponential of a fuzzy real number is a (non-negative) fuzzy real number.

f	f(Q)	f(Q).y
T	-Q	$Q(-y)$, $\forall y \in R$
$H_\lambda, \lambda \in R^*$	λQ	$Q(\frac{y}{\lambda})$, $\forall y \in R$
O	$\frac{1}{Q}$	$Q(\frac{1}{y})$, $\forall y \in R^*$ 0, $y=0$
$\lvert . \rvert$	$\lvert Q \rvert$	$\sup(Q(y),Q(-y))$, $\forall y \in [0,+\infty[$ 0, $\forall y \in]-\infty,0[$
exp	e^Q	$Q(\ln y)$, $\forall y \in]0,+\infty[$ 0, $\forall y \in]-\infty,0]$

table 1

As an example we prove (P.3'). Consider a non-negative fuzzy closed interval Q. Then from the continuity of O and the upper-semicontinuity of Q the chain rule garantuees the upper-semicontinuity of $\frac{1}{Q}$. The convexity condition for $\frac{1}{Q}$ reads :

$$\frac{1}{Q}(\lambda x_1+(1-\lambda)x_2) \geqslant \min(\frac{1}{Q}(x_1),\frac{1}{Q}(x_2)), \forall (x_1,x_2) \in R^2,\ \forall \lambda \in [0,1].$$

If $x_1<0$ or $x_2<0$, then $\frac{1}{x_1}<0$ and $\frac{1}{x_2}<0$ and hence $Q(\frac{1}{x_1})=0$ or $Q(\frac{1}{x_2})=0$ and thus the convexity condition. This is even true if $x_1=0$ or $x_2=0$. Now consider $x_1>0$ and $x_2>0$ and $\lambda \in [0,1]$ such that : $0<x_1 \leqslant \lambda x_1+(1-\lambda)x_2 \leqslant x_2$. Hence :

$$0<\frac{1}{x_2} \leqslant \frac{1}{\lambda x_1+(1-\lambda)x_2} \leqslant \frac{1}{x_1}.$$

Applying the convexity condition for Q we obtain :

$$Q(\frac{1}{\lambda x_1+(1-\lambda)x_2}) \geqslant \min(Q(\frac{1}{x_1}),Q(\frac{1}{x_2}))$$

hence the convexity condition for $\frac{1}{Q}$.

7. BINARY OPERATIONS ON FUZZY QUANTITIES

7.1. Definitions

Due to the extension principle every binary operation on R can be extended into a binary operation on the set of fuzzy quantities $F(R)$. Let f be a R^2-R function and $Q_1 \in F(R), Q_2 \in F(R)$. Then $f(Q_1,Q_2)$ is defined as :

$$f(Q_1,Q_2): R \to [0,1]$$

$$z \mapsto \sup_{f(x,y)=z} \min(Q_1(x),Q_2(y)),\ \forall z \in \mathrm{ran}(f)$$

$$z \mapsto 0,\ \forall z \in R \setminus \mathrm{ran}(f).$$

As examples of binary operations on R we consider : the sumfunction (S), the productfunction (P), the differencefunction (V), the quotientfunction (Q), the maximum- and minimumfunction (max,min). The corresponding extensions are denoted as :

$$Q_1 \oplus Q_2, Q_1 \odot Q_2, Q_1 \ominus Q_2, Q_1 \oslash Q_2, \widetilde{\max}(Q_1,Q_2), \widetilde{\min}(Q_1,Q_2).$$

The results are summarized in table 2.

The values indicated in the third column are obtained by straightforward manipulation of the definition; more details can be found in [7].

7.2. The α-level representation of fuzzy binary operations

The key for this representation consists of the sufficient condition as formulated by Nguyen [2] :

If f is continuous

then $(\forall \alpha \in]0,1])((f(Q_1,Q_2))_\alpha = f((Q_1)_\alpha,(Q_2)_\alpha))$.

f	$f(Q_1,Q_2)$	$f(Q_1,Q_2).z$
S	$Q_1 \oplus Q_2$	$\sup_{x\in R} \min(Q_1(x),Q_2(z-x)$, $\forall z\in R$
P	$Q_1 \odot Q_2$	$\sup_{x\in R^*} \min(Q_1(x),Q_2(\frac{z}{x}))$, $\forall z\in R^*$ $\max(Q_1(0),Q_2(0))$, $z=0$
V	$Q_1 \ominus Q_2$	$\sup_{x\in R} \min(Q_1(x+z),Q_2(x))$, $\forall z\in R$
Q	$Q_1 \oslash Q_2$	$\sup_{x\in R} \min(Q_1(x.z),Q_2(x))$, $\forall z\in R$
max	$\widetilde{\max}(Q_1,Q_2)$	$\sup_{\max(x,y)=z} \min(Q_1(x),Q_2(y))$, $\forall z\in R$
min	$\widetilde{\min}(Q_1,Q_2)$	$\sup_{\min(x,y)=z} \min(Q_1(x),Q_2(y))$, $\forall z\in R$

table 2

Since S,P,V,Q,max,min are continuous, we may write :

$$(Q_1 \oplus Q_2)_\alpha = (Q_1)_\alpha + (Q_2)_\alpha$$

$$(Q_1 \odot Q_2)_\alpha = (Q_1)_\alpha \cdot (Q_2)_\alpha$$

$$(Q_1 \ominus Q_2)_\alpha = (Q_1)_\alpha - (Q_2)_\alpha$$

$$(Q_1 \oslash Q_2)_\alpha = (Q_1)_\alpha : (Q_2)_\alpha$$

$$(\widetilde{\max}(Q_1,Q_2))_\alpha = \max((Q_1)_\alpha,(Q_2)_\alpha)$$

$$(\widetilde{\min}(Q_1,Q_2))_\alpha = \min((Q_1)_\alpha,(Q_2)_\alpha).$$

The second members of these equalities are defined as follows :

Let $A\subseteq R$, $B\subseteq R$ and $*$ a binary operation on R, then we have :

$$A*B=\{c \mid (\exists a\in A)(\exists b\in B)(c=a*b)\}.$$

Since the α-levels of a fuzzy real number (even of a closed fuzzy interval) are precisely compact intervals in R, we can use the interval calculus of Moore [8] to calculate the second members in the last equalities. For the sake of completeness we mention explicitly :

$$[a,b]+[c,d]=[a+c,b+d]$$

$$[a,b]-[c,d]=[a-d,b-c]$$

$$[a,b].[c,d]=[\min(ac,ad,bc,bd),\max(ac,ad,bc,bd)]$$

$$[a,b]:[c,d]=[a,b].[\frac{1}{d},\frac{1}{c}] \text{ if } 0\notin[c,d]$$

$$\max([a,b],[c,d])=[\max(a,c),\max(b,d)]$$

$$\min([a,b],[c,d])=[\min(a,c),\min(b,d)]$$

where $[a,b]$ and $[c,d]$ denote arbitrary compact intervals in R.

7.3. Properties

7.3.1. Properties of fuzzy addition

(P.1) Q_1 is a fuzzy real number and Q_2 is a fuzzy real number $\Rightarrow Q_1 \oplus Q_2$ is a fuzzy real number.

Q_1 is a fuzzy closed interval and Q_2 is a fuzzy closed interval $\Rightarrow Q_1 \oplus Q_2$ is a fuzzy closed interval.

(P.2) $(Q_1 \oplus Q_2) \oplus Q_3 = Q_1 \oplus (Q_2 \oplus Q_3)$.

(P.3) $Q_1 \oplus 0 = 0 \oplus Q_1 = Q_1$, where 0 denotes the mapping $x \mapsto 0$, $\forall x \in R^*, 0 \mapsto 1$.

(P.4) Most fuzzy real numbers have no symmetrical element w.r.t. addition i.e. $Q \oplus (-Q) \neq 0$, for most Q's.

(P.5) The opposite of the sum of two fuzzy real numbers equals the sum of their opposites.

(P.6) Fuzzy addition is commutative.

7.3.2. Properties of fuzzy multiplication.

(P.1) Q_1 is a fuzzy real number and Q_2 is a fuzzy real number $\Rightarrow Q_1 \odot Q_2$ is a fuzzy real number.

Q_1 is a fuzzy closed interval and Q_2 is a fuzzy closed interval $\Rightarrow Q_1 \odot Q_2$ is a fuzzy closed interval.

(P.2) $(Q_1 \odot Q_2) \odot Q_3 = Q_1 \odot (Q_2 \odot Q_3)$.

(P.3) $Q_1 \odot 1 = 1 \odot Q_1 = Q_1$, where 1 denotes the mapping

$x \mapsto 0, \forall x \in R \setminus \{1\}, 1 \mapsto 1$.

(P.4) Most fuzzy real numbers have no inverse element w.r.t. multiplication i.e. $Q \odot \frac{1}{Q} \neq 1$ for most Q's.

(P.5) $Q_1 \odot Q_2 = (-Q_1) \odot (-Q_2)$

$-(Q_1 \odot Q_2) = (-Q_1) \odot Q_2 = Q_1 \odot (-Q_2)$.

N.B. Multiplication $\odot$ is not necessarily distributive w.r.t. addition $\oplus$. We only have :

$Q_1 \odot (Q_2 \oplus Q_3) \subseteq (Q_1 \odot Q_2) \oplus (Q_1 \odot Q_3)$,

for Q_1, Q_2 and Q_3 fuzzy real numbers.

8. COMPUTER-REPRESENTATION OF FUZZY QUANTITIES

The practical calculation of the sum of two fuzzy real numbers for example, needs a lot of computer time especially the execution of Zadeh's extension principle is very time-consuming. For this reason several authors developed approximate methods to represent and to operate on fuzzy quantities. During the building of an expert system in order to detect coronary diseases, we have had a very good experience with a special type of fuzzy quantity, i.e. piecewise linear fuzzy quantities with bounded support. This type of fuzzy quantity shows a sufficient degree of generality to deal with the problems met during the creation of an expert system that requires the manipulation of imprecise and incomplete information. On the other hand this type of fuzzy quantity is still simple enough to be presented in a very simple and compact way and to permit a sufficiently fast execution of the operations between them. Moreover they allow a very simple representation in expert system friendly languages such as LISP.
In the following we only consider *piecewise linear fuzzy quantities with bounded support*.

8.1. List-representation of a piecewise linear fuzzy quantity with bounded support

Our practical type of fuzzy quantity Q may be represented by a (finite) list of ordered pairs (x_i, α_i) where $x_i \in R$ and $\alpha_i \in [0,1]$:

list representation of $Q = lr(Q) = ((x_1\ \alpha_1)(x_2\ \alpha_2) \ldots (x_n\ \alpha_n))$

where $(x_i)_{i=1}^{n}$ denotes a non-decreasing sequence.

Every x_i is called a breakpoint for Q. An explicit expression for a piecewise linear fuzzy quantity with bounded support is given as :

$$Q(x)=0, \ \forall x \in]-\infty, x_1[\cup]x_n, +\infty[$$

$$Q(x)=\max\{\alpha_i \mid (x\ \alpha_i) \text{ appears in } lr(Q)\}, \ \forall x \in \{x_1\} \cup \ldots \cup \{x_n\}$$

$$Q(x)=\alpha_i+(\alpha_{i+1}-\alpha_i)\frac{x-x_i}{x_{i+1}-x_i}, \ \forall x \in]x_i, x_{i+1}[, \ x_i<x_{i+1}.$$

The definition of Q in the breakpoints guarantees the upper-semi-continuity of Q. To illustrate the definition let's give an example.

Example 1.

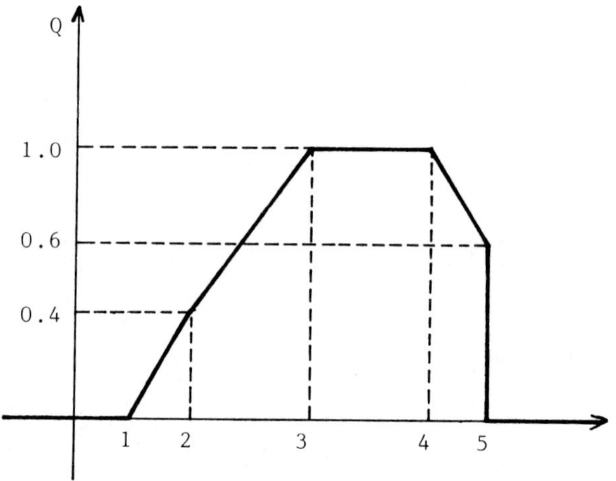

lr(Q)=((1 0.0)(2 0.4)(3 1.0)(4 1.0)(5 0.6)(5 0.0))

A breakpoint x may appear one, two or three times in the list representation, according to the following situation :

case 1 : Q is continuous at x. In this case we obtain a list of the form $(\ldots(x_{i-1}\alpha_{i-1})(x\alpha_i)(x_{i+1}\alpha_{i+1})\ldots)$ where $x_{i-1}<x<x_{i+1}$.

case 2 : (Q is left-continuous at x and Q is not right-continuous at x)

or (Q is right-continuous at x and Q is not left-continuous at x)

In this case we obtain a list of the form :

$(\ldots(x_{i-1}\alpha_{i-1})(x\alpha_i)(x\alpha_{i+1})(x_{i+2}\alpha_{i+2})\ldots)$ with $Q(x)=\max(\alpha_i,\alpha_{i+1})$

i.e. $\lim\limits_{\substack{z\to x\\ <}} Q(z)=\alpha_i$ and $\lim\limits_{\substack{z\to x\\ >}} Q(z)=\alpha_{i+1}$.

case 3 : Q is not left-continuous at x and Q is not right-continuous at x.

In this case the list representation of Q looks as follows :

$$(\ldots(x_{i-1}\alpha_{i-1})(x\alpha_i)(x\alpha_{i+1})(x\alpha_{i+2})(x_{i+3}\alpha_{i+3})\ldots)$$

i.e.

$$\lim_{\substack{z\to x\\ <}} Q(z)=\alpha_i,\quad Q(x)=\alpha_{i+1},\quad \lim_{\substack{z\to x\\ >}} Q(z)=\alpha_{i+2}.$$

The latter situation occurs particularly in the list-representation of a crisp number.

Example 2.

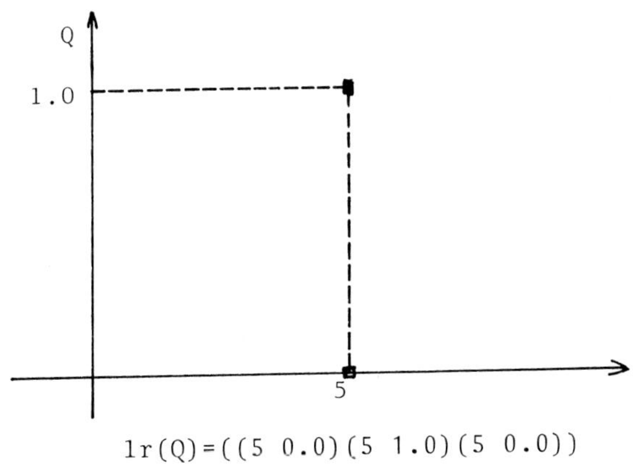

lr(Q)=((5 0.0)(5 1.0)(5 0.0))

Q = crisp number 5.

In PASCAL programming language implemented on an ordinary PC one could use the following data structure :

```
TYPE

   UnitInterval=0..1;
```

```
ListRepresentation =^Pairlist;
Pairlist = RECORD
   x:Real;
   y:ListRepresentation
   END;
```

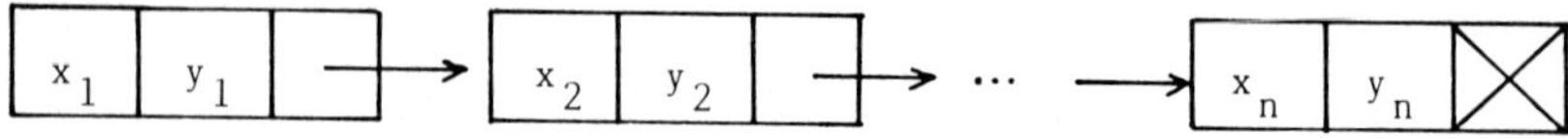

8.2. Level representation of a piecewise linear fuzzy quantity with bounded support

As outlined in 7.2, we may use the α-level representation to perform the operations on fuzzy quantities. However since our practical numbers are not necessarily convex, the corresponding α-levels are not necessarily intervals but only closed (due to the upper-semicontinuity) and bounded (due to the bounded support) subsets of R that can be written as finite unions of compact intervals (due to the piecewise linearity). Let Q be represented by the following list :

$$lr(Q)=((x_1\alpha_1)(x_2\alpha_2)\ldots(x_n\alpha_n))$$

Putting $\{\lambda_1,\ldots,\lambda_m\}=\{\alpha_1\}\cup\{\alpha_2\}\cup\ldots\cup\{\alpha_n\}$, where $0=\lambda_1<\lambda_2<\ldots<\lambda_m$ we obtain the following level representation (or cut representation) for Q :

$$((\lambda_1(a_1^1b_1^1)(a_2^1b_2^1)\ldots(a_{n_1}^1\,b_{n_1}^1))$$
$$(\lambda_2(a_1^2b_1^2)(a_2^2b_2^2)\ldots(a_{n_2}^2\,b_{n_2}^2))$$
$$\ldots$$
$$(\lambda_m(a_1^mb_1^m)(a_2^mb_2^m)\ldots(a_{n_m}^m\,b_{n_m}^m))),$$

where $Q_{\lambda_j}=[a_1^j,b_1^j]\cup[a_2^j,b_2^j]\cup\ldots\cup[a_{n_j}^j,b_{n_j}^j]$, $\forall j\in\{2,\ldots,m\}$

$$\overline{Q}_{\overline{\lambda}_1}=[a_1^1,b_1^1]\cup[a_2^1,b_2^1]\cup\ldots\cup[a_{n_1}^1,b_{n_1}^1].$$

($\overline{Q}_{\overline{\lambda}_1}$ = the closure of $Q_{\overline{\lambda}_1}$).

Example 3.

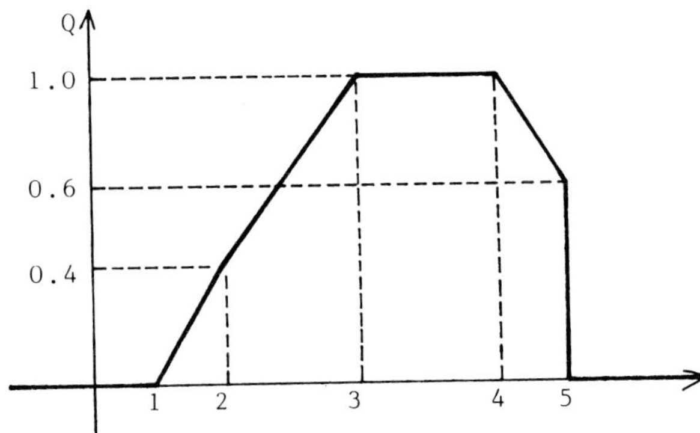

cut representation of Q=cr(Q)=((0.0(1 5))
(0.4(2 5))
(0.6(2.5 5))
(1.0(3 4)))

In PASCAL the level representation can be represented by the following datastructure :

```
TYPE
 UnitInterval = 0..1;
 IntervalList =^Intervals;
 Intervals = RECORD
  a,b:Real;
  next:IntervalList
  END;
 LevelRepresentation =^Alfacut;
  Alfacut = RECORD
   Level:UnitInterval;
   Intervals:IntervalList;
   next:LevelRepresentation
   END;
```

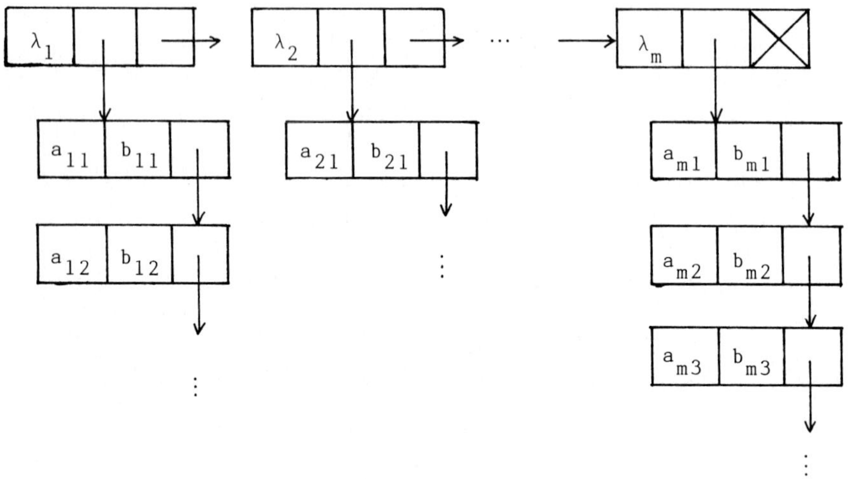

9. Algorithm for the implementation of binary operations on piecewise linear fuzzy quantities with bounded support

Let X_1, X_2, Y be three sets, f a $X_1 x X_2$-Y mapping, $Q_1 \in I^{X_1}, Q_2 \in I^{X_2}$. Due to the extension principle we obtain a fuzzy set $Q=f(Q_1, Q_2)$ on Y. A good approximation for Q is obtained in the following way :

(1) Determine the levelrepresentation of Q_1, starting from the listrepresentation of Q_1 as described in 8. So we obtain :

$((\alpha_1 (a_1^1 b_1^1)(a_2^1 b_2^1) \ldots (a_{n_1}^1 b_{n_1}^1))$

$(\alpha_2 (a_1^2 b_1^2)(a_2^2 b_2^2) \ldots (a_{n_2}^2 b_{n_2}^2))$

...

$(\alpha_k (a_1^k b_1^k)(a_2^k b_2^k) \ldots (a_{n_k}^k b_{n_k}^k)))$.

(2) Determine the levelrepresentation of Q_2 :

$((\beta_1(c_1^1 d_1^1)(c_2^1 d_2^1)\ldots(c_{m_1}^1 d_{m_1}^1))$

$(\beta_2(c_1^2 d_1^2)(c_2^2 d_2^2)\ldots(c_{m_2}^2 d_{m_2}^2))$

...

$(\beta_l(c_1^l d_1^l)(c_2^l d_2^l)\ldots(c_{m_l}^l d_{m_l}^l)))$.

(3) Calculate $L=\{\lambda_1,\ldots,\lambda_s\}=\{z|(\exists i\in\{1,\ldots,k\})(z=\alpha_i)\}$
$\cup\{z|(\exists i\in\{1,\ldots,l\})(z=\beta_i)\}$.

(4) Determine the missing λ_j-level of Q_1 and Q_2 for some $\lambda_j\in L$. Then apply the rules from 7.2 to determine the λ_j-level of Q. If both fuzzy quantities Q_1, Q_2 are convex, then the λ_j level of each number is an interval and so Moore's interval analysis can be used. If on the contrary one of them is not convex, the corresponding λ_j-level consists of a finite union of compact intervals; on each of these intervals Moore's calculus of intervals can be applied.

(5) Repeat (4) for every remaining value $\lambda_j\in L$.

Example 4.

Consider Q_1 and Q_2 given as :

lr(Q_1)=((9 0.0)(10 1.0)(12 0.0))

lr(Q_2)=((1 0.0)(2 0.4)(3 1.0)(4 1.0)(5 0.6)(5 0.0))

cr(Q_1)=((0.0(9 12))(1.0(10 10)))

cr(Q_2)=((0.0(1 5))(0.4(2 5))(0.6(2.5 5))(1.0(3 4)))

cr($Q_1 \oplus Q_2$)=

((0.0(10 17))(0.4(11.4 16.2))(0.6(11.93 15.8))(1.0(13 14)))

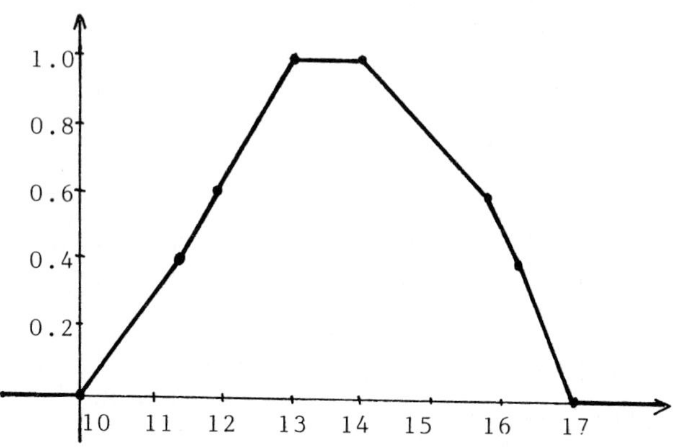

$cr(Q_1 \odot Q_2) =$

((0.0(9 60))(0.4(18.8 56))(0.6(22.4 54))(1.0(30 40)))

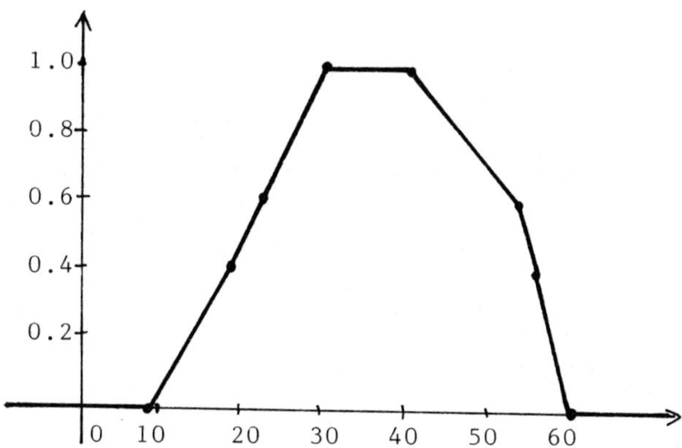

$cr(Q_1 \ominus Q_2) =$

((0.0(4 11))(0.4(4.4 9.2))(0.6(4.6 8.467))(1.0(6 7)))

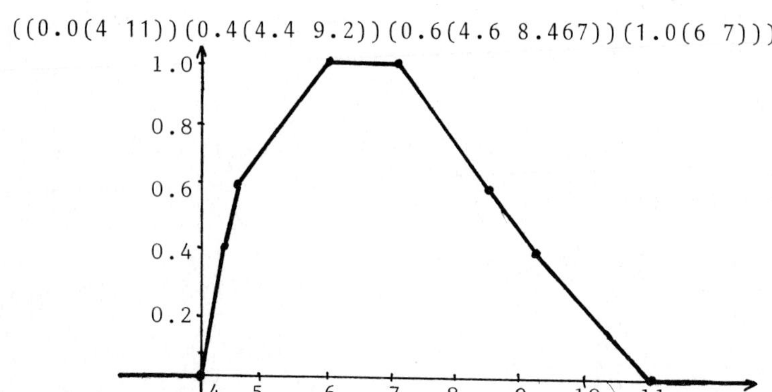

$cr(Q_1 \odot Q_2) =$

((0.0(1.8 12))(0.4(1.88 5.6))(0.6(1.92 4.629))(1.0(2.5 3.33)))

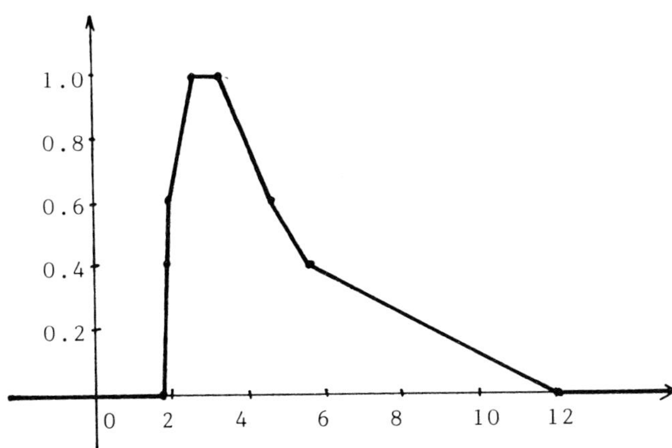

10. COMPARISON WITH EXISTING METHODS

As mentioned earlier, there exist already several methods to obtain a practical representation of fuzzy quantities. Hereafter the most important ones are considered with respect to four criteria : generality, implementation facilities, performance, deviation from the exact result.

(1) The method of Baas and Kwakernaak (1979) [9]
This method is based on nonlinear programming techniques.
generality : arbitrary fuzzy sets
implementation : very difficult
performance : bad
result : exact.

(2) The method of Dubois and Prade (1979) [10]
This method consists of a "partition" of the fuzzy quantity in non-increasing and non-decreasing parts; the operation is carried out on each of these parts separately.
generality : piecewise continuous fuzzy quantities and non-decreasing algebraic operations
implementation : difficult
performance : moderate
result : good.

(3) FWA algorithm of Dong and Wong (1985) [11,12]
These authors use a discretization of the grade of membership [0,1] and apply Moore's interval analysis.
generality : "arbitrary" fuzzy quantities
implementation : rather easy
performance : very good
result : good.

(4) The method of discretization
Some authors simply consider a discrete number of points in R and make use of a straightforward application of Zadeh's extension principle on that discrete domain.
generality : arbitrary fuzzy quantities
implementation : very easy
performance : very bad and strongly dependent on the discrete domain choosen
result : bad.
(5) The L-R representation of fuzzy quantities [10]
This method is restricted to a special kind of fuzzy quantities that may be characterized by at most four parameters.
generality : moderate due to the very special type of fuzzy quantity
implementation : very easy
performance : very good
result : good.
(6) Our method of piecewise linear fuzzy quantities with bounded support
generality : general enough to deal with the practical representation problems in expert systems
implementation : very easy
performance : very good
result : good.
This method agrees very well with the method of Dong and Wong. There are two main differences : firstly the admissible fuzzy quantities and secondly the choice of the α-levels. In our model this choice is indicated by the fuzzy quantities considered and hence not arbitrary as in Dong and Wong.

11. APPLICATION TO APPROXIMATE REASONING TECHNIQUES IN FUZZY EXPERT SYSTEMS

Let's consider the canonical form of a general rule in a fuzzy expert system :

IF (X_1 is F_1) and (X_2 is F_2) and ... and (X_n is F_n)
THEN Y is G

where :

X_i a variable with domain U_i, $i \in \{1,\dots,n\}$,
Y a variable with domain V,
F_i a fuzzy set in U_i, $i \in \{1,\dots,n\}$,
G a fuzzy set in V.

Using the fuzzy antecedent matching technique we are able to determine the truth-value of the proposition X_i is F_i; this truth-value may be a fuzzy set in [0,1], it will be denoted as $T(X_i \text{ is } F_i)$. By means of Zadeh's extension principle we obtain the following rules from fuzzy logic :

$T(P \text{ and } Q)=\widetilde{\max}(0,T(P)\oplus T(Q)\ominus 1)$

$T(P \text{ or } Q)=\widetilde{\min}(1,T(P)\oplus T(Q))$

$T(\text{if } P \text{ then } Q)=\widetilde{\min}(1,1\ominus T(P)\oplus T(Q))$

$T(Q)=\widetilde{\max}(0,T(\text{if } P \text{ then } Q)\oplus T(P)\ominus 1)$ (fuzzy modus ponens).

Hence, in order to evaluate a general rule as above we have to execute the following algorithm :

(1) Determine $T(X_i \text{ is } F_i)$ for every $i\in\{1,\ldots,n\}$, using fuzzy antecedent matching :
$T(X_i \text{ is } F_i/X_i \text{ is } H_i)(t)=\sup\{H_i(x)\,|\,x=F_i(t)\}$, $\forall t\in[0,1]$
where $H_i\in[0,1]^{U_i}$.

(2) Combine all $T(X_i \text{ is } F_i)$, $i\in\{1,\ldots,n\}$, by and's using the above formula for conjunction.

(3) Apply the fuzzy modus-ponensrule to obtain $T(Y \text{ is } G)$.

(4) If necessary, for example if there is more than one rule with the same consequent Y is G, then apply the so-called "evidence combination" rule, using disjunction.

In this algorithm only fuzzy addition, fuzzy substraction, fuzzy maximum and minimum appear. As outlined above, these operations can easily be performed using our practical representation of fuzzy quantities.

Note. After the preparation of this paper, we became informed about a new book on fuzzy numbers written by Gupta and Kaufmann [13]. At a first glance this book gives an extensive treatment of the arithmetic with fuzzy numbers, especially from a practical point of view. So we can highly recommend it to the practitioners of fuzzy numbers.

REFERENCES

[1] D. Dubois, H. Prade, Théorie des possibilités. Masson (Paris) 1985.

[2] H.T. Nguyen, A note on the extension principle for fuzzy sets, J. Math. Anal.Appl.64 (1978) 369-380.

[3] S. Rodabaugh, Fuzzy addition in the L-fuzzy real line, Fuzzy Sets and Systems 8 (1982) 39-52.

[4] D.Dubois, H. Prade, Fuzzy real algebra, some results, Fuzzy Sets and Systems 2 (1979) 327-348.

[5] L. Biacino, A. Lettieri, Some properties on fuzzy numbers and the triangular inequality (1986) preprint.

[6] L. Zadeh, The concept of a linguistic variable and its application to approximate reasoning I, Information Sciences, 8 (1975) 199-249.

[7] E. Kerre, Vage verzamelingen met toepassingen in de informatica, lecture notes, State University of Gent, (Gent) 1985, private publication, 270 p. (in Dutch).
[8] R. Moore, Interval Analysis, Prentice Hall, Englewood Cliffs, (N.Y.) 1966.
[9] S. Baas, H. Kwakernaak, Rating and ranking of multiple-aspect alternatives using fuzzy sets, Automatica, 13 (1979) 47-58.
[10] D. Dubois, H. Prade, Fuzzy Sets and Systems : Theory and Applications, Academic Press (New York) 1980.
[11] W. Dong, F. Wong, Fuzzy weighted averages and the implementation of the extension principle, Fuzzy Sets and Systems (1986) (to appear).
[12] W. Dong, H. Shah, F. Wong, Fuzzy computations in risk and decision analysis, Civ. Eng. Syst. 2 (1985) 201-208.
[13] A. Kaufmann, M. Gupta, Introduction to Fuzzy Arithmetic, Van Nostrand Rheinhold (New York) 1986.

Fuzzy Logic in Knowledge-Based Systems,
Decision and Control
M.M. Gupta and T. Yamakawa (Editors)
Elsevier Science Publishers B.V. (North-Holland), 1988

ON THE METHODOLOGY OF SOLVING FUZZY RELATIONAL EQUATIONS AND ITS IMPACT ON FUZZY MODELLING

Siegfried Gottwald[1] and Witold Pedrycz[2]

[1]Logic Group, Department of Philosophy, Karl Marx University, Leipzig, G.D.R.

[2]Department of Automatic Control and Computer Sci., Silesian Technical University, Gliwice, Poland
and
Research & Development Centre of Electronic Medical Instruments, Zabrze, Poland

The approach toward fuzzy model building which uses fuzzy relational equations presents a series of theoretical problems. A first one is that of the solvability of systems of fuzzy relational equations - and the connected one of the evaluation of approximate solutions. These problems are discussed from a general point of view regarding fuzzy modelling. For applicational reasons - e.g. distinguishing direct and inverse use of fuzzy models - here some different points have to be considered. We survey essential ideas of earlier solutions, formulate other ones and new problems.

Key words: fuzzy relational equations, fuzzy control, fuzzy models, solvability indices, confidence intervals

1. INTRODUCTION

Describing fuzzy models by their input - output behaviour is a very convenient tool in fuzzy modelling. This approach immediately leds to fuzzy relational equations - usually to systems of them - and to the necessity to solve them. Solving such systems of equations is a complicated task not only for numerical reasons but also because of the fact that in essential cases only approximate solutions do exist. Therefore, a detailled discussion of the solvability behaviour of systems of fuzzy relational equations and of factors which influence or change it becomes interesting.

The choice of the fuzzy operators used in the equations is one point for discussions. Another one is the change from sup-min resp. inf-max compositions to other, perhaps more complicated ones, e.g. to the convex combination of such simple kinds of compositions as done in recent work of OHSATO / SEKIGUCHI /12/,/13/ which also can be extended to systems of such equations.

In all those cases, assuming solvability of the system under consideration, extremal solutions of the systems are available from corresponding extremal solutions of the single equations. Additionally, in interesting special cases of non-solvability of such systems those constructions give best possible approximate solutions. Hence, a "brute-force" methodology can be taken into account which starts with considering such a combination of solutions of the single equations of the system as an a priori approximate solution. If doing so, it becomes essential to discuss the quality of an approximate solution. Such a discussion can begiven by using global, but also by using local solvability - or better: solution - indices. Such indices can also be used to define analogues to confidence intervals (with respect to each point of the corresponding universe of discourse) for the actual approximate solution. Such an approach provides means for evaluating the quality of the actual approximate solution at hand, of the fuzzy model, and hence also for ways to look for improved models.

Of course, the content of this paper does not exhaust the entire research on this matter, instead it presents first steps into a new period for fuzzy modelling by indicating fundamental problems required to be solved, and by additionally providing the reader with some algorithms.

2. PRELIMINARY NOTIONS: COMPARISON OF TWO FUZZY SETS. INVERSE PROBLEM

First some notations and notions used later on. By capital letters we denote fuzzy sets and fuzzy relations defined in finite universes of discourse, e.g. $A : \underline{A} \rightarrow [0,1]$ or $R : \underline{A} \times \underline{B} \rightarrow [0,1]$. To the fundamental concepts of fuzzy sets theory such as union and intersection we add a definition of fuzzy equality of two fuzzy sets (relations). The idea is of basic character and comes from a logical resp. set-theoretical point of view for which a further extension is also rather straightforward. Remember that (crisp) sets C and D defined in an universe of discourse $\underline{A}$ are called equal each other iff C is contained in D and D is contained in C; or in other words: iff for each point x of $\underline{A}$ it holds true that $x \in C$ iff $x \in D$. More formally, one requires satisfaction of the two conditions $C \subseteq D$ and $D \subseteq C$, or of the condition $\forall x (x \in C \text{ iff } x \in D)$.

Modelling the connectives involved in "$\subseteq$", "and" and "iff" in our fuzzy set theoretic setting via any residuation operator φ (known also as φ-operator) for implication and a continuous t-norm $\underline{t}$ for conjunction, a (local) degree to which two fuzzy sets C and D are equal each other at a point a_n of $\underline{A}$ is given as a number $\|C = D\|(a_n)$ defined as e.g. in /4/ by

$$\|C = D\|(a_n) =_{def} (C(a_n) \varphi D(a_n)) \, \underline{t} \, (D(a_n) \varphi C(a_n)) . \tag{1}$$

Going one step further, one usually likes to have a number expressing a (unique) degree to which C and D are equal each other in a global sense. For this one has to aggregate the partial evaluations (1) of equality over the whole universe $\underline{A}$.

There is no unique way to perform this task. Perhaps the most evident ways are to take an optimistic or a pessimistic form of aggregation. In the optimistic case one prefers to modelize the degree of the statement "C and D are equal each other" by the maximal value of $\|C = D\|(x)$ over $\underline{A}$, viz.

$$\|C = D\| =_{def} \max_{x \in \underline{A}} \|C = D\|(x) . \tag{2}$$

Note that the max-operator here corresponds to the existential quantifier $\exists$; so one takes only one selected element in $\underline{A}$ in which C and D are equal at the highest grade. On the opposite, i.e. pessimistic pole one uses the formula

$$\|C = D\| =_{def} \min_{x \in \underline{A}} \|C = D\|(x) . \tag{3}$$

This degree of equality has a pessimistic character. Obviously, the min-operator corresponds to the universal quantifier $\forall$. Generalized quantifiers like "almost everywhere" instead of these "sometimes" or "always" versions applied to the localized equality values of C, D would give global evaluations in between the optimistic and the pessimistic point of view.

Another possibility to get an intermediate value of global equality not so likely to cause overestimation resp. underestimation as (2) or (3) is to take e.g. the average value of all the local equality degrees, i.e. to consider

$$\|C = D\| =_{def} \sum_{x \in \underline{A}} \|C = D\|(x) / \mathrm{card}(\underline{A}) \ . \tag{4}$$

In the sequel we will also utilize some characteristics that can be deduced from histograms of the values of equality index (1) computed for each element of the universe of discourse.

An interesting problem that arises in relation to such an equality index (1) is related to the so-called inverse problem, cf. /16/. In concise formulation we can state it as follows:

for a given fuzzy set C and a certain value v determine such a fuzzy set D for which the inequality

$$\|C = D\|(x) \geq v \tag{5}$$

holds true for all points x of the universe of discourse.

In other words, one is interested to get a fuzzy set D such that C and D are equal each other to a degree not lower than v (the equality of C and D again here thought in the pointwise sense).

Omitting a detailled discussion which has been performed in /16/ it is worthy to recall only some main results. At the beginning notice that (5) usually does not have a unique solution. Instead, one can indicate a certain range of the unit interval in which the values of the membership function of D may lie to fulfill (5). Thus, the resulting fuzzy set D can in fact be interpreted as an interval-valued fuzzy set, a so-called Φ-fuzzy set, cf. /19/.

It is remarkable to notice that replacing in (5) the inequality sign by equality may cause that there is no solution. Thus, our (5) forms an appropriate problem statement.

Few facts now will be summarized to give properties of the fuzzy sets forming a solution of (5). We start with rewriting (1) in a slightly modified form utilizing properties of the φ-operator which, starting from a continuous t-norm $\underline{t}$ may be defined as

$$u \varphi w =_{def} \sup \{ z \in [0,1] \mid u \, \underline{t} \, z \leq w \} \ . \tag{6}$$

Straightforward calculations give

$$\|C = D\|(x) = \begin{cases} C(x) \varphi D(x) \ , & \text{if } D(x) < C(x) \\ D(x) \varphi C(x) \ , & \text{if } D(x) > C(x) \\ 1 \ , & \text{if } D(x) = C(x) \ . \end{cases} \tag{7}$$

Let us denote by $D^{\wedge}(a)$ the set of values of a membership function $D(x)$ at point a satisfying (5), i.e.

$$D^{\wedge}(a) =_{def} \{ D(a) \in [0,1] \mid \|C = D\|(a) \geq v \} \ .$$

Thus, if we put $v = 0$ then obviously $D(a)$ may be any of the values from the unit interval, therefore in this case $D^{\wedge}(a) = [0,1]$. On the opposite side, putting $v = 1$ reduces $D^{\wedge}(a)$ to exactly one point: $D^{\wedge}(a) = \{C(a)\}$. For intermediate situations, viz. for $v \in (0,1)$, $D^{\wedge}(a)$ is derived by solving the two inequalities

$$C(a) \varphi D(a) \geq v \ , \tag{8}$$

$$D(a) \varphi C(a) \geq v \ . \tag{9}$$

Solving the first of them we get a subinterval $[d_1, C(a)]$ while the second leads us to a subinterval $[C(a), d_2]$. Finally, $D^*(a)$ is created by taking the union of both these intervals:

$$D^*(a) = [d_1, C(a)] \cup [C(a), d_2] = [d_1, d_2] .$$

This set of solutions $D^*(a)$ will be called "equality interval" or "confidence interval" for the fuzzy set D (at point a) induced by v for the given fuzzy set C.

This inverse construction will be of considerable use for determining resulting fuzzy sets for fuzzy models constructed in the framework of fuzzy relational equations. To have a clear picture what goes on in this area and which form of solution can be obtained there, we start a more thorough discussion of this topic in Section 6.

3. FUZZY RELATIONAL EQUATIONS AND THEIR SOLUTIONS

In a general setting, by a fuzzy relational equation we mean an equation describing relationships between fuzzy sets in two (possibly different) spaces. Such a form of relationship is represented by a fuzzy relation. More precisely, let be given two fuzzy sets $A : \underline{A} \rightarrow [0,1]$ and $B : \underline{B} \rightarrow [0,1]$ as well as a fuzzy relation $R : \underline{A} \times \underline{B} \rightarrow [0,1]$. Then a fuzzy relational equation is put down generally as

$$B = \Theta(R, A)$$

where Θ is a suitable operator producing a fuzzy set B out of a fuzzy set A and a fuzzy relation R. Some types of fuzzy relational equations studied in research which was performed until now are the following:

- fuzzy relational equations with sup-$\underline{t}$ composition

$$B = A \,\square\, R , \tag{10}$$

$$B(b) = \max_{a \in \underline{A}} (A(a) \,\underline{t}\, R(a,b)) , \tag{11}$$

  here Θ corresponds in crisp case to taking the full image of a set under a relation (or mapping) which often is written as R"A too;
- fuzzy relational equations with inf-$\underline{s}$ composition

$$B = A \vartriangle R , \tag{12}$$

$$B(b) = \min_{a \in \underline{A}} (A(a) \,\underline{s}\, R(a,b)) , \tag{13}$$

  which in a suitable sense is dual to the foregoing case;
- adjoint fuzzy relational equations /15/ with inf-φ composition

$$B = A \circledphi R , \tag{14}$$

$$B(b) = \min_{a \in \underline{A}} (A(a) \,\varphi\, R(a,b)) . \tag{15}$$

Besides those types, which might be viewed as of a basic nature, one can consider some types of a more complex form which sometimes are formed on the basis of the types given above. Of such possibilities we recall two:

- a convex combination of (10) and (12) as proposed in /12/

$$B = \lambda \cdot (A \,\square\, R_1) + (1 - \lambda) \cdot (A \vartriangle R_2) , \tag{16}$$

$$\begin{aligned} B(b) = {} & \lambda(b) \cdot \max_{a \in \underline{A}} (A(a) \,\underline{t}\, R_1(a,b)) + \\ & + (1 - \lambda(b)) \cdot \min_{a \in \underline{A}} (A(a) \,\underline{s}\, R_2(a,b)) \end{aligned} \tag{17}$$

with $\lambda: \underline{B} \rightarrow [0,1]$ and the t-norm $\underline{t}$ as well as the t-conorm $\underline{s}$ in /12/ taken as max, min only;

- fuzzy relational equations with equality operator /3/

$$B = A \circledequal R \,, \tag{18}$$

$$B(b) = \max_{a \in \underline{A}} ((A(a) \varphi R(a,b)) \wedge (R(a,b) \varphi A(a))) \,. \tag{19}$$

For all the basic forms of equations (10) - (15) the family of solutions, if solutions do exist at all, has been characterized and their extremal (maximal or minimal) elements have been obtained. Moreover, for systems

$$B_i = A_i \,\square\, R \,, \qquad i = 1(1)N \,, \tag{20}$$

$$B_i = A_i \,\vartriangle\, R \,, \qquad i = 1(1)N \,, \tag{21}$$

$$B_i = A_i \,\circled{\varphi}\, R \,, \qquad i = 1(1)N \tag{22}$$

of such equations the relevant results are also available - mainly under the condition that they have an exact solution at all. To have an overall picture of the results they are collected in Table 1.

Type of equation	Interpretation	Solution of a single equation (in case of solvability)	Solution of a system of equations (in case of solvability)
$B = A \,\square\, R$ sup - $\underline{t}$ composition	the (fuzzified) full image of a fuzzy set A under a fuzzy relation R; o r for each point b of B there is a point a of A such that b can be reached from point a	$\hat{R} = A \circled{\varphi} B$ $\hat{R} = \sup \mathcal{R}'$	$\hat{R} = \bigcap_{i=1}^{N} (A_i \circled{\varphi} B_i)$ $\hat{R} = \bigcap_{1 \le i \le N} \sup \mathcal{R}'_i$
$B = A \vartriangle R$ inf - $\underline{s}$ composition	dual to the first case in sense that $B = A \vartriangle R$ iff $\overline{B} = \overline{A} \,\square\, \overline{R}$	$\check{R} = A \circled{\beta} B$ $\check{R} = \inf \mathcal{R}''$	$\check{R} = \bigcup_{i=1}^{N} (A_i \circled{\beta} B_i)$ $\check{R} = \bigcup_{1 \le i \le N} \inf \mathcal{R}''_i$
$B = A \circled{\varphi} R$ inf - φ composition	each point b of B can be reached from every Point a of A	$\tilde{R} = A \circled{t} B$ $\tilde{R} = \inf \mathcal{R}'''$	$\tilde{R} = \bigcup_{i=1}^{N} (A_i \circled{t} B_i)$ $\tilde{R} = \bigcup_{1 \le i \le N} \inf \mathcal{R}'''_i$

Table 1. Basic types of fuzzy relational equations, their (logical) interpretations, and solutions

Concerning the notation used in this table we add two explanations. First, by $\mathcal{R}'$, $\mathcal{R}''$, $\mathcal{R}'''$ we denote respectively the sets of solutions of the systems (20),(21),(22) of fuzzy relational equations. Secondly, as a dual to the φ-operator (6) we use the β-operator defined for all $u,w \in [0,1]$ as:

$$u \,\beta\, w =_{def} \inf \{ z \in [0,1] \mid u \,\underline{s}\, z \geq w \} \,. \tag{23}$$

At present, for the general case it seems impossible to give simple and easy-to-check conditions for the solvability of a system of equations. This was one of the reasons for the discussion of degrees of solvability e.g. in /4/. Here we will present now only conditions for the solvability of single equations; cf. Table 2. The case of systems of fuzzy relational equations we consider again in the next section.

Type of equation	Condition
$B = A \,\square\, R$	$(\forall b \in \underline{B})(\exists a \in \underline{A})(A(a) \geq B(b))$
$B = A \,\Delta\, R$	$(\forall b \in \underline{B})(\exists a \in \underline{A})(A(a) \leq B(b))$
$B = A \circledcirc_{\varphi} R$	no restrictive condition in case $\underline{t}$ is strongly monotonuous in both arguments

Table 2. Necessary and sufficient conditions that a single equation has a solution

4. APPROXIMATE SOLUTIONS FOR SYSTEMS OF FUZZY RELATIONAL EQUATIONS - WAYS TOWARD A MORE FLEXIBLE TREATMENT OF FUZZY MODELS

As it becomes quite clear from the discussions in the previous sections, the results e.g. of Table 1 have a significant value only in case that solutions really exist, i.e. that not only approximate solutions do exist. If this true solvability is not the case - and it is this more uncomfortable situation one usually mets in practice - then one has to think about other possibilities to overcome the problem of the non-existence of (true) solutions.

A simple, and perhaps for the practitioner the most obvious, way out is to use the formulas which describe solutions - in case of solvability - also if there does not exist a solution of the system of fuzzy relational equations to be considered - and then to check the quality of the "approximate solution" got this way. If this quality is not too bad, i.e. if the results/outputs B_i given for the corresponding data/inputs A_i by the approximate solution are not "too far" from the results/outputs B_i one intends to get from the data (A_i, B_i) constituting the intended model and hence the considered system of fuzzy relational equations, in such a case at least an approximate fuzzy model is realized and can be used within some quality bounds. To go such a way is in accordance with fuzzy modelling at all, because any fuzzy model aims at a rough, i.e. approximate description of real processes, situations, and the like.

But having taken this point of view one can move one step further: instead of having proven a formula to give a solution in case of solvability, one can start from a formula which one guesses to describe a solution - of course, if there are some acceptable reasons for such a guess. And, indeed, for some classes of fuzzy relational equations such acceptable guesses are available. To present some basic ones let us distinguish for relational equations $B = \Theta(A,R)$ two different types:

- such an equation will be called to be of sup-type in case one has

$$B(b) = \Theta(A,R)(b) = \sup_{a \in \underline{A}} \mathfrak{G}(A(a), R(a,b)) \tag{24}$$

where the term $\mathfrak{G}$ is built up using the membership degrees $A(a), R(a,b)$ and combining them e.g. by a t-norm, a φ-operator or some suitable other kind of "simple" operator;

- such an equation will be called to be of inf-type in case one has

$$B(b) = \Theta(A,R)(b) = \inf_{a \in \underline{A}} G(A(a),R(a,b)) \tag{25}$$

with term G choosen correspondingly.

To support the acceptability of the guesses we will consider the following facts which should be taken into account.
Fact 1: The union of any two solutions of a fuzzy relational equation of sup-type is again a solution of this equation - and hence this equation has a greatest solution.
Fact 2: The intersection of any two solutions of a fuzzy relational equation of inf-type is again a solution of this equation - and hence this equation has a smallest solution.
Fact 3: Any system of fuzzy relational equations of sup-type has in case of solvability as greatest solution the intersection of all the greatest solutions of its single equations.
Fact 4: Any system of fuzzy relational equations of inf-type has in case of solvability as smallest solution the union of all the smallest solutions of its single equations.

It is interesting to mention that those types of behaviour one also can find with some mixed types of fuzzy relational equations. To have an example we refer to the convex combination form (16),(17) of fuzzy relational equations discussed by OHSATO / SEKIGUCHI /12/,/13/ for $A \,\square\, R_1$ as sup-min composition and $A \,\triangle\, R_2$ as inf-max composition. Indeed, in the paper /13/ a solvability behaviour of such equations is proven which combines facts 1 and 2: the type (16) of equations has an extremal solution $(\check{R}_1,\hat{R}_2)$ in the sense that $R_1{}' \subseteq \check{R}_1$ and $R_2{}' \subseteq \hat{R}_2$ for all solutions $(R_1{}',R_2{}')$ of this equation. Extending those results one can prove that for solutions (R_1',R_2') and (R_1'',R_2'') of that equation also $(R_1' \cup R_1'', R_2' \cap R_2'')$ is a solution. And extending facts 3 and 4 one can prove that an extremal solution of a system of such equations (16) is determined - in case of its existence - in its "first coordinate" as intersection of the "first coordinates" of the extremal solutions of the single equations, and in its "second coordinate" as union of the "second coordinates" of the extremal solutions of the single equations.

These facts now are the background for the guesses we have in mind:
- for systems of fuzzy relational equations of sup-type consider the intersection of all greatest solutions of the single equations (or of good approximations for them) as a suitable approximate solution of the system;
- for systems of fuzzy relational equations of inf-type consider the union of all the smallest solutions of the single equations (or of good approximations for them) as a suitable approximate solution of the system.

And for "combined" types of equations a suitable "combined strategy" seems to provide a suitable guess for an approximate solution.

Of course, taking this approach is a kind of "brute-force" methodology. And in every case accompanying discussions of the (expected) quality of such an approximate solution, respectively of the fuzzy model constituted by it, are unavoidable. Besides this "brute-force" approach toward solving systems of fuzzy relational equations there are other, competing or supplementary, ways to treat that problem of approximately "solving" such systems.

Having in mind that in general no solution exists, it is of special interest to discuss possibilities to make any such system of fuzzy relational equations "more solvable" up to a level of approximate solvability acceptable by the intended model builder or user. In general, here two ways of thinking may be distinguished. The first one is closely related to such changes in the data set that might make a set of equations more solvable. The second one is oriented toward modofication of the structure of the equations, e.g.

by taking another form of equation, or simply by extending the structure of the equations, e.g. by adding some new variables (called explanatory ones). We list some representative methods coming from those two groups and refer to existing literature where more detailled descriptions may be found. Also we recall some underlying ideas which support those methods.

While considering the first possible stream of modification we have to pay attention to the fact that again at least two main groups might be recognized: in the first one we simply skip some elements of the data set (in fact this deletion modifies it), whereas in the second one all the elements of the data set are modified.

1. In /5/ a way of measuring the consistency of pairs of the data set, say (A_i,B_i) and (A_j,B_j), has been proposed. Roughly speaking, the procedure works as follows: A suitable consistency index has to be considered that takes into account an absolute difference between possibility of A_i versus A_j and possibility of B_i versus B_j. If for the two specified indices, say i_o and j_o, a low value is obtained (e.g. almost equal to zero) then one pair of those data is expected to be a candidate for being removed from the entire data set.

2. A different approach has been proposed in /6/. There all the fuzzy sets A_i are modified by replacing them by their following "fuzzified" version

$$A_i' = A_i \cup \underline{\alpha} \tag{26}$$

where $\underline{\alpha}$ is a fuzzy set with constant membership degree equal to α. The basic idea behind this construction (26) is that the smallest values of the membership functions of A_i are irrelevant: e.g. they may be very sensitive to existing noise and therefore they should be masked by an appropriately chosen value of α. As proven in /6/, the solvability index discussed there is an increasing function of α. Then by observing its dependence of α, we can take a suitable value of α which at one hand is low enough not to disturb the data set at a significant level, and on the other hand can led to a significant increment of the solvability index.

3. Still another approach follows again a different way. Here the fuzzy sets A_i are replaced by "sharpened" versions, more precisely: instead of the fuzzy sets A_i, $i = 1(1)N$, we take disjoint fuzzy sets A_i', i.e. such ones which fulfill the condition

$$A_i' \cap A_j' = \emptyset \qquad \text{for all} \quad 1 \le i < j \le N . \tag{27}$$

In this case, if for each pair (A_i', B_i) there exists a solution of the corresponding equation, we are sure that there exists a solution for the entire system of equations.

4. A proposal formulated by WAGENKNECHT / HARTMANN /21/,/22/ deals with modifications of the fuzzy sets B_i forming the right-hand parts of the equations. Instead of using the fuzzy sets B_i so-called fuzzy sets with tolerances have been utilized. By a fuzzy set with a tolerance they mean an interval-valued fuzzy set. At present, the tolerances are attached in a heuristic fashion to reach a situation where the entire set of equations has a solution. Nevertheless, the tolerances should be adjusted by an user and then there is no guarantee that they are not taken too broad or that their choice does not guarantee the existence of a solution.

5. Again another way has been discussed in /10/ where a certain construction has been designed to perturbate the data set and to reach a state where the system of equations has a solution.

6. In /8/ the use of the concept of probabilistic set has been discussed. In this application, by a probabilistic set there is meant a certain class of fuzzy sets that are similar each other. As a consequence, the data set is splitted into some classes, i.e. it is structured. Afterwards it is attempted to solve a reduced set of equations just taking representatives of each class, i.e. prototypes for each class.

Recapitulating this first stream of methods we can observe that they modify the data set to make the corresponding system of equations "more solvable" just by discovering a structure in it or in one of its "slightly" modified forms - modified e.g. by the elimination of "outliers" in the data set.

Discussing the second main stream of investigation it is worthwile to recall some ideas being of interest for it.

A) First, keeping in mind that now the structure of the equation is supposed to be fixed (viz. the type of composition) and moreover the fuzzy sets A_i are unchanged, one can make some parametric studies either for modifying some t-norm or φ-operator appearing in the equation, or look for a suitable value of the parameter in any parametrized class of, e.g., t-norms. This may lead to higher accordance of the data provided and the fuzzy sets generated by the equations. Unfortunately, till now any method for a systematic search in the infinite set of t-norms or t-conorms seems to be lacking.

B) Contrary to the previous approach which might be seen as a parametric adjustment, now one is interested to consider different structures of the fuzzy relational equations. Restricting ourselves to those already listed, we have four ways when fitting the basic types of equations together with the convex combination of OHSATO / SEKIGUCHI /12/ which always has a solution in case of suitable choice of the convex combinator λ, a (fuzzy) parameter controlling the influence of the "simple" types of equations combined.

C) A third way that allows to reach the main goal - to get a solvable system of equations - is to add a new variable which "separates" the fuzzy input sets A_i, $i = 1(1)N$, cf. /17/,/18/. In its spirit, this way is similar to the way mentioned in /6/ but now no one of the fuzzy sets A_i is modified. To give a straightforward explanation just take a system of equations

$$A_i \,\square\, R = B_i \,, \qquad i = 1(1)N \,. \tag{28}$$

If $\mathcal{R}_i \neq \emptyset$ for each of the solution sets of the i-th equation, then to make the system solvable it is enough to consider a new universe of discourse $\underline{C}$, $\underline{C} = \{c_1, \ldots, c_N\}$ and to define fuzzy singletons $C_1, \ldots, C_N$ as

$$C_i(c_j) =_{def} \begin{cases} 1, & \text{if } i = j \\ 0 & \text{otherwise} \,. \end{cases} \tag{29}$$

Then this new fuzzy variable defined in $\underline{C}$ allows to be sure that the system of equations

$$(A_i \times C_i) \,\square\, R = B_i \,, \qquad i = 1(1)N \tag{30}$$

has a solution. Notice, however, that now R is defined in the cartesian product of the universes $\underline{A}$, $\underline{C}$ and $\underline{B}$. In other words, the enriched structure (30) of (28) is sufficient to have the system of equations solvable. The new universe $\underline{C}$ used here is a "space of explanation" and obviously in close analogy to additional variables discussed e.g. in statistical models. A weak point of this approach perhaps is the very special choice of the separating fuzzy sets C_i as singletons - which therefore have no linguistic meaning, i.e. usually may not reflect any linguistic concept. Nevertheless, one can search for a way to fuzzify these singletons just keeping in mind not to exceed a certain overlap (measured e.g. as the height of the $C_i \cap C_j$)

between these auxiliary fuzzy sets. For a more formal discussion see /17/.

Until now we have presented ideas how a degree of solvability of a given system of equations can be enlarged. There was no discussion how the property of (partial) lack of solvability has has influence on the precision of the results obtained by means of the fuzzy model constructed in such a way. This forms the topic of the next section.

5. ON EXPRESSING QUALITY OF SOLUTIONS OF FUZZY RELATIONAL EQUATIONS

In the previous sections we have summarized how fuzzy relational equations can be solved in an exact manner or at least in approximate fashion. Now we intend to move one step forward and to discuss a way in which precision of a solution is expressed quantitatively. Immediately one concludes that this might have primordial influence on the application of constructed fuzzy relational equations. In case of broad utilization of such equations one has to have at hand a clear answer which quality of results can be expected if the fuzzy model has been chosen. A close and transparent analogy can be found in general modelling principles with the aid of statistical means. There it is a necessary step to validate a model and to express the precision attached to it. Most simply, usually, one tests the model with respect to its relevancy, and the F-test is frequently used. Moreover, the model is equipped with confidence curves associated to the equations of the model. This way of thinking has a long tradition and all its stages are commonly accepted and used in the field of statistical modelling, cf. e.g. /1/.

The same does not hold true for the fuzzy set approach. Mainly only the first of those steps is merely solved. Unfortunately, we cannot give a quantitative characterization of the quality (relevancy) of a model. Few approaches try to tackle with this problem, but in a qualitative way. The proposal formulated here forms an attempt to express and measure the relevancy of a model and in consequence leads to formation of fuzzy sets of a complex character, namely interval-valued fuzzy sets or fuzzy sets of type 2; cf. /11/,/19/.

For having a clear description of the approach we will give it in an algorithmic form. This gives a concise presentation and allows any potential users to have this idea ready for application. Additionally we include some comments to explain the consecutive steps. Having at hand a data set represented by pairs (A_i, B_i), $i = 1(1)N$, of fuzzy sets a fuzzy relation is to be constructed to - approximately - solve the corresponding fuzzy relational equations. To focus attention we restrict ourselves to one specified form of the equations, e.g. to (10),(11). Of course, the algorithm described in the following applies to any form of equations. The fuzzy relation R may be obtained via any suitable method, cf. Section 3.

Let us consider the fuzzy sets $\bar{B}_i$ resulting from the fuzzy sets A_i composed by R, i.e. $\bar{B}_i = A_i \circ R$. If the fuzzy model - here the fuzzy relation R - is perfect, which almost never occurs in practice, then for every $i = 1(1)N$ we have $\bar{B}_i = B_i$. Since this case is not a realistic one, we may present a global evaluation of the model by computing the global equality index for each pair $(\bar{B}_i, B_i)$. But, here we are more interested to take the local point of view, i.e. to look "how equal" the fuzzy sets $\bar{B}_i$ and B_i are separately at each point of the universe of discourse $\underline{B}$. By fixing some b $\underline{B}$ we get for $i = 1(1)N$ a sequence of reals indicating how closely the degrees $B_i(b)$ and $\bar{B}_i$ are equal each other. Now, instead of looking for a single number expressing en block a similarity of realized and intended outputs such as e.g. (2),(3) we build an empirical distribution function of some equality index using as a sample the values

$$\| B_i \equiv \bar{B}_i \| (b) \qquad \text{for all} \quad i = 1(1)N \ . \tag{31}$$

Let us denote this function by $F(w;b)$ with $w \in [0,1]$. By definition this distribution function is given as

$$F(w;b) =_{def} \frac{1}{N} \cdot \text{card} \{ i \in \{1,..,N\} \mid \| B_i \equiv \bar{B}_i \| (b) \leq w \} \ . \tag{32}$$

In virtue of (32), $F(w;b)$ is monotonuous in the first argument with boundary condition $F(1;b) = 1$. We interpret $F(w;b)$ as expressing the probability that the local equality index at point b does not exceed the specified value w in any case of the application of the model. Now, let us also introduce the value

$$p =_{def} 1 - F(w;b) \tag{33}$$

which will be considered as the probability that our equality index attains values greater than w at point b. Rewriting (33) as

$$\text{Prob} \{ (\text{equality index at } b) > w \} = p \tag{34}$$

we recognize a direct dependence between this value of probability p and the length of a kind of confidence interval induced by the value w. For consistency of notation denote this relationship (34) by $Q(w;b) = p$. Summarizing we have that p is a nonincreasing function of w.

Obviously, for any fuzzy set B and $w = 0$ we get in each point confidence intervals of length 1, while for $w = 1$ the length of the confidence intervals reduces to zero. From (33) we observe that with increasing values of w we have decreasing values of p. Hence this implies that if p increases the length of the confidence interval increases as well. Of course, one dislikes to accept too broad confidence intervals since this conveys no useful information. On the other hand, too narrow confidence intervals cannot be accepted due to the significantly low values of probability attached to them.

From those remarks one can deduce a rationale how to choose a suitable value of w. We want to achieve satisfaction of two contradictory goals. One is to get as high a value of probability as possible. The second is to have narrow enough confidence intervals. Denote by $l_b(w)$ the length of these intervals. To fulfill both these requirements we consider as a performance index

$$Q(w;b) \cdot (1 - l_b(w)) \ . \tag{35}$$

Since the first factor is a nonincreasing function of w and the second one a nondecreasing one, therefore it has at least one point w_o for which the index (35) achieves the maximal value:

$$Q(w_o;b) \cdot (1 - l_b(w_o)) = \max_{w \in [0,1]} Q(w;b) \cdot (1 - l_b(w)) \ . \tag{36}$$

This value w_o might be utilized for further construction of an interval-valued fuzzy set. To get it, the step to determine w_o is repeated at every point of the universe $\underline{B}$. This gives the function $w_o(b)$.

In this way the fuzzy model is characterized not only by the fuzzy relation R but also by a set of functions $Q(w;b)$ - or at least a set of numbers $w_o(b)$ is attached.

For fuzzy relational equations we proceed accordingly. For a given fuzzy set A compute $B = A \circ R$, or having B at hand solve the inverse problem. In both cases a suitable value of w is taken based on a prespecified probability p or of extremal nature as w_o. In the direct mode this strategy yields an interval-valued fuzzy set $[B_-,B_+]$ that gives an impression how precise is the fuzzy set "produced" by the fuzzy model behind the system of equations.

For the two ways just described we can give the following interpretations:
- in the first situation we say that the genuine fuzzy set, which is appearing in a system when A is given as input, has at each point b a membership value with respect to the output lying in between $B_-(b)$ and $B_+(b)$;
- for the second construction we have a fuzzy set $[B_-,B_+]$ with a highest specifity, thus the bounds $B_-(b)$ and $B_+(b)$ determine a region in the unit interval in which one is sure to get a value of the membership at a high value of probability and, simultaneously, this interval is chosen to be narrow enough for the actual purposes.

6. SELECTED METHODOLOGICAL ISSUES ARISING ON APPLICATIONAL BACKGROUND

Without going deeply into various applications where fuzzy relational equations are discussed we will focus rather on methodological aspects which arise while studying the introduced approach. For this, notice again that having an experimental data set in the form of pairs (A_i,B_i) two models can be constructed and their use may depend on some of the specifities of the problem discussed. For short we called it direct and backward use of the model.

In the direct mode of using the model, possesing the fuzzy relation R to tie A and B, viz. B is tied with A via R, we can find a responce of the model (system) for given A. Following the schema of computation of the output we get an interval-valued fuzzy set $[B_-,B_+]$ which expresses bounds in which the response of the system is contained - e.g. with a prescribed level of probability. Afterwards, if one is interested to have a state of the system which dynamics is governed by the model with R, further iterations might be performed. In other words, A is replaced by $[B_-,B_+]$. For B_- and B_+, separately, the corresponding interval-valued fuzzy sets are calculated and the resulting interval-valued set is built by taking the minimum of the two lower bounds of them and the maximum of their two upper bounds, respectively, as resulting bounds. Obviously, the width of the intervals of this interval-valued fuzzy set is broader then the two original ones on which it is based.

The resulting effect is in agreement with our intuition: the result never can be more precise than the arguments taken into account; cf. as a related paper concerning this topic e.g. /2/.

The so-called backward mode of the utilization of the fuzzy model corresponds to all the questions related to A if B is known. Of course, R has already to be computed. For instance, to answer the question which input A - if any - leads to some fixed output B we have to solve such an inverse problem. Originally, cf. /20/, assuming a nonempty set of solutions, one immediately takes a fuzzy set A fulfilling the equation, cf. Table 1. But bearing in mind the aspect of model precision it might be useful to reformulate the problem as:

> find A such that the fuzzy set resulting from A is covered by $[B_-,B_+]$, i.e. its membership values are within the bounds of this characteristic interval-valued set $[B_-,B_+]$.

However, the solution of this problem might be too tedious a task.

Therefore we can think about a different model just establishing a bit another ralation between B and A. The model just derived may answer the question which interval-valued fuzzy set $[A_-,A_+]$ is obtained for a specified B . Surely, the "new" fuzzy relation can be differ from the previous one and the precision of the new fuzzy model might also be a different one.

A schematic illustration of the presented idea, viz. to take two fuzzy models for dierct and backward mode of utilization, is displayed in Figure 1. As we mentioned, in this section we will not intend to discuss detailled applications. More information the reader may find e.g. in /7/.

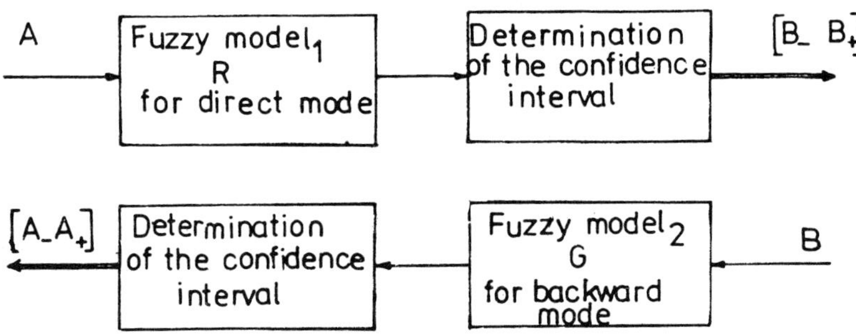

Figure 1. Direct and backward use of fuzzy models

It remains to be shown how the analysis coming from these models can be enriched by imposing imprecision attached to the interval-valued fuzzy sets. Perhaps, the forward and backward strategy can transparently being utilized for reasoning schemes in knowledge-based systems with mechanisms managing uncertainty inbuilt in them. In such a case the models correspond to modelling directions of reasoning.

7. CONCLUDING REMARKS

We have covered some issues arising from the use of fuzzy models and, in a quite general form, a fashion in which fuzzy modelling techniques are utilized. It has been pointed out how fuzzy relational equations may be viewed as a constructive tool for building fuzzy models. Special attention was paid to approximate solutions and a genuine analysis of the precision of a fuzzy model. The result has been expressed in terms of interval-valued fuzzy sets. Of course, while taking some values for the confidence interval induced by different levels of probability, finally instead of this interval-valued fuzzy set one obtains a fuzzy set of type 2. Now, contrary to some doubts that usually appear concerning this class of fuzzy sets, one has a clear method in which way they have been established and how to interpret them.

Another important goal we have achieved is a method for an effective and quantitative form to express the precision of fuzzy models, viz. by a degree of solvability of the corresponding set of equations. By introducing interval-valued fuzzy sets we reached the goal to visualize some consequences of the use of the fuzzy model for e.g. prediction tasks, and presented arguments to consider two fuzzy models for so-called direct and backward mode of use.

REFERENCES

/1/ Box, G.E.P. and Jenkins, G.M., Time Series Analysis. Forecasting and Control (Holden Day, San Francisco, 1970).

/2/ Czogała, E., Gottwald, S. and Pedrycz, W., Aspects for the evaluation of decision situations, in: Gupta, M.M. and Sanchez, E. (eds.), Fuzzy Information and Decision Processes (North-Holland, Amsterdam, 1982) pp. 41-49.

/3/ DiNola, A., Pedrycz, W. and Sessa, S., Fuzzy relation equations with equality and difference composition operators, Fuzzy Sets and Systems (to appear).

/4/ Gottwald, S., Characterizations of the solvability of fuzzy equations, Elektron. Informationsverarbeitung Kybernetik 22 (1986) 67-91.

/5/ Gottwald, S. and Pedrycz, W., Analysis and synthesis of fuzzy controller, Problems Control Information Theory 14 (1985), no.1, 33-45.

/6/ Gottwald, S. and Pedrycz, W., Solvability of fuzzy relational equations and manipulation of fuzzy data, Fuzzy Sets and Systems 18 (1986) 45-65.

/7/ Gottwald, S. and Pedrycz, W., On the suitability of fuzzy models: an evaluation through fuzzy integrals, Intern. J. Man - Machine Studies 24 (1986) 141-151.

/8/ Hirota, K. and Pedrycz, W., Analysis and synthesis of fuzzy systems by the use of probabilistic sets, Fuzzy Sets and Systems 10 (1983) 1-13.

/9/ Izumi, K., Tanaka, H. and Asai, K., Adjointness of fuzzy systems, Fuzzy Sets and Systems 20 (1986) 211-222.

/10/ Li H.-X., Fuzzy perturbation analysis. Part I: Directional perturbation, Fuzzy Sets and Systems 17 (1985) 189-197; Part II: Undirectional perturbation, ibid. 19 (1986) 165-175.

/11/ Mizumoto, M. and Tanaka, K., Some properties of fuzzy sets of type 2, Inform. and Control 31 (1976) 312-340.

/12/ Ohsato, A. and Sekiguchi, T., Convexly combined form of fuzzy relational equations and its application to knowledge representation, in: 1983 Proc. Intern. Conf. Systems, Man and Cybernet., Bombay (Dec. 29, 1983 - Jan. 7, 1984), vol. 1 (IEEE - India Council, Bombay - New Delhi, 1983) pp. 294-299.

/13/ Ohsato, A. and Sekiguchi, T., Maximin solution of the convexly combined form of composite fuzzy relation equations (in Japanese), Trans. Soc. Instrument and Control Engineers (Japan) 21 (1985) 423-428.

/14/ Pedrycz, W., On generalized fuzzy relational equations and their applications, J. Math. Anal. Appl. 107 (1985) 520-536.

/15/ Pedrycz, W., Applications of fuzzy relational equations for methods of reasoning in presence of fuzzy data, Fuzzy Sets and Systems 16 (1985) 163-175.

/16/ Pedrycz, W., Equality indices based on triangular norms and inverse problem (submitted to Fuzzy Sets and Systems).

/17/ Pedrycz, W., Approximate solutions of fuzzy relational equations (submitted to Fuzzy Sets and Systems).

/18/ Pedrycz, W., Relevancy of fuzzy models (submitted to Intern. J. General Systems).

/19/ Sambuc, R., Fonctions Φ - floues. Application a l'aide au diagnostic en pathologie thyroidienne, Thèse, Univ. Marseille, 1975.

/20/ Sanchez, E.,Resolution of composite fuzzy relation equations, Inform. and Control 30 (1976) 38-48.

/21/ Wagenknecht, M. and Hartmann, K., On the solution of direct and inverse problems for fuzzy equation systems with tolerances, in: Bocklisch, S. et al., (eds.), Fuzzy Sets Applications, Methodological Approaches, and Results; Mathematical Research 30 (Akademie-Verlag, Berlin, 1986) pp. 37-44.

/22/ Wagenknecht, M. and Hartmann, K., Fuzzy modelling with tolerances, Fuzzy Sets and Systems 20 (1986) 325-332.

Fuzzy Logic in Knowledge-Based Systems,
Decision and Control
M.M. Gupta and T. Yamakawa (Editors)

MULTIFACTORIAL FUNCTIONS IN FUZZY SETS THEORY

Li Hongxing

Section of Mathematics, Tianjin Institute of Textile Engineering, Tianjin, China

Multifactorial function is very important new concept which can be used to many aspects in fuzzy sets theory and appeared firstly in the paper [1] where it was used to define fuzzy perturbation function and studied the stability for the solutions of fuzzy relation equations by using fuzzy perturbation. In another paper [2], by means of multifactorial function, multifactorial fuzzy sets and multifactorial degree of nearness, that are two new concepts too, were given, and they used to multifactorial pattern recognition and clustering analysis with fuzzy characteristics.
In fact, generally speaking, the most of the mathematics models depended on several factors should use multifactorial function. Especially, fuzzy decision-making, fuzzy game, fuzzy programming and fuzzy linear programming with several objective function should.
Here we study the properties for multifactorial function, how to generate some new multifactorial function, and introduce in detail its applications in fuzzy pattern recognition, fuzzy decision-making, fuzzy clustering, fuzzy game and fuzzy programming, etc.

Keywards: Multifactorial functions, pattern recognition, fuzzy decision, fuzzy game, fuzzy clustering, fuzzy programming.

1. DEFINITION OF MULTIFACTORIAL FUNCTION

In $[0,1]^m$ a partial ordering "$\leqslant$" is defined by

$$X \leqslant Y \quad \text{iff} \quad (x_j \leqslant y_j,\ j=1,\dots,m),$$

where $X=(x_1,\dots,x_m)$, $Y=(y_1,\dots,y_m)\in[0,1]^m$.
It is easy to know that $([0,1]^m, \leqslant)$ is a complete distributive lattice with the grestest element $I=(1,\dots,1)$ and the least element $\theta=(0,\dots,0)$.
Stipulate five transformations p_j, q_j, s_{ij}, r_j, k_j: $[0,1]^m \longrightarrow [0,1]$, where

$$p_j(X)=(x_1,\dots,x_{j-1},0,x_{j+1},\dots,x_m), \tag{1.1}$$

$$q_j(X)=(1,\dots,1,x_j,1,\dots,1), \tag{1.2}$$

$$s_{ij}(X)=(x_1,\dots,x_i,\dots,x_j,\dots,x_m)$$
$$=(x_1,\dots,x_j,\dots,x_i,\dots,x_m), \tag{1.3}$$

$$r_j(X)=(0,\dots,0,x_j,0,\dots,0), \tag{1.4}$$

$$k_j(X)=(x_1,\dots,x_{j-1},1,x_{j+1},\dots,x_m). \tag{1.5}$$

DEFINITION 1.1. A mapping M_m: $[o,1]^m \longrightarrow [0,1]^m$ is called an m-ary multifactorial function if it satisfies:

(m.1) $X \leq Y$ implies $M_m(X) \leq M_m(Y)$;

(m.2) $\min_j[x_j] \leq M_m(X) \leq \max_j[x_j]$,

which is simply denoted by M_m-func.
Evidently M_1-func is an identity mapping from [0,1] to [0,1], when m=1.
For M_m-func we may also give the following conditions:

(m.3) $M_m(p_j(X))=0$;

(m.4) $M_m(q_j(X))=x_j$;

(m.5) $M_m(s_{ij}(X))=M_m(X)$;

(m.6) $M_m(r_j(X))=x_j$;

(m.7) $M_m(k_j(X))=1$;

(m.8) There exist M_r-func, M_{m-r}-func and M_2-func such that

$$M_m(X)=M_2(M_r(x_1,\ldots,x_r),\ M_{m-r}(x_{r+1},\ldots,x_m));$$

(m.9) There exists M_2-func such that

$$M_m(X)=M_2(M_2(\ldots\ M_2(M_2(x_1,x_2),\ x_3)\ \ldots),\ x_m);$$

(m.10) For any $X=(x_1,\ldots,x_m)$, $Y=(x_m,\ldots,x_{2m-1})$, have

$$M_m(M_m(X),\ x_{m+1},\ldots,x_{2m-1})=M_m(x_1,\ldots,x_{m-1},\ M_m(Y)).$$

EXAMPLE 1. The following functions $[0,1]^m \longrightarrow [0,1]$ are multifactorial functions:

$$L(X)=\min_j[x_j], \tag{1.6}$$

$$V(X)=\max_j[x_j], \tag{1.7}$$

$$\Sigma(X)=\sum_{j=1}^{m} a_j x_j, \tag{1.8}$$

where $a_j \in [0,1]$ and $\sum_{j=1}^{m} a_j=1$;

$$M_m(X)=\bigvee_{j=1}^{m}(a_j x_j), \tag{1.9}$$

where $a_j \in [0,1]$ and $\bigvee_{j=1}^{m} a_j=1$;

$$M_m(X)=\bigvee_{j=1}^{m}(a_j \wedge x_j), \tag{1.10}$$

where $a_j \in [0,1]$ and $\bigvee_{j=1}^{m} a_j=1$;

$$M_m(X)=\Big(\prod_{j=1}^{m} x_j\Big)^{1/m}, \tag{1.11}$$

$$M_m(X)=(\frac{1}{m}\sum_{j=1}^{m} x_j^p)^{1/p}, \tag{1.12}$$

where $p>0$;

$$M_m(X)=(\sum_{j=1}^{m} a_j x_j^p)^{1/p}, \tag{1.13}$$

where $p>0$, $a_j \in [0,1]$ and $\sum_{j=1}^{m} a_j=1$.

It is easy to know the following cases:

(1.6) satisfies (m.i), i=3,4,5,8,9, 10;
(1.7) does (m.i), i=5,...,10;
(1.8) does (m.8), when $a_j=1$ does (m.6);
(1.9) does (m.8), when $a_j=1$ does (m.6);
(1.10) does (m.8), when $a_j \geq x_j$ does (m.6);
(1.11) does (m.i), i=3,5,10, and when $x_j=1$ does (m.4);
(1.12) does (m.5);
(1.13) does (m.6) when $a_j=1$.

2. PROPERTIES OF M_m-func

Write $@_m=(M_m |\ M_m$ is an M_m-func), and in $@_m$ a partial ordering "$\dashv$" is defined by

$$M_m \dashv M'_m \text{ iff } (\forall X)(M_m(X) \leq M'_m(X)).$$

It is easy to prove:

PROPOSITION 2.1. $(@_m, \dashv)$ is a complete dirtributive lattice, and L is the least element and V is the greatest element in $@_m$.

PROPOSITION 2.2. If $M_m \in @_m$, then

(i) $(x_j=a,\ j=1,\ldots,m)$ implies $M_m(X)=a$;

(ii) $M_m(I)=1$, $M_m(\theta)=0$.

Proof. (m.2) implies (i) implies (ii). Q.E.D.

PROPOSITION 2.3. Write $I-X=(1-x_1,\ldots,1-x_m)$ and let $M_m \in @_m$. If

$$M_m^*(X)=1-M_m(I-X),$$

then $M_m^* \in @_m$.

The proof is omitted since it is simple.

Clear $L^*(X)=V(X)$, or $V^*(X)=L(X)$.

EXAMPLE 2. From the proposition 2.3 and the example 1 the following functions are M_m-func:

$$\Sigma^*(X)=1-\sum_{j=1}^{m} a_j(1-x_j), \tag{2.1}$$

where $a_j \in [0,1]$ and $\sum_{j=1}^{m} a_j=1$;

$$M_m(X)=1-\bigvee_{j=1}^{m}(a_j(1-x_j)), \tag{2.2}$$

where $a_j\in[0,1]$ and $\bigvee_{j=1}^{m} a_j=1$;

$$M_m(X)=1-\bigvee_{j=1}^{m}(a_j\wedge(1-x_j)), \tag{2.3}$$

where $a_j\in[0,1]$ and $\bigvee_{j=1}^{m} a_j=1$;

$$M_m(X)=1-(\prod_{j=1}^{m}(1-x_j))^{1/m}, \tag{2.4}$$

$$M_m(X)=1-(\frac{1}{m}\sum_{j=1}^{m}(1-x_j)^p)^{1/p}, \tag{2.5}$$

where $p>0$;

$$M_m(X)=1-(\sum_{j=1}^{m}a_j(1-x_j^p))^{1/p}, \tag{2.6}$$

where $p>0$, $a_j\in[0,1]$ and $\sum_{j=1}^{m} a_j=1$.

PROPOSITION 2.4. If $M_m\in@_m$, then

(i) M_m satisfies (m.3) iff M_m^* does (m.7);

(ii) M_m does (m.4) iff M_m^* does (m.6);

(iii) M_m does (m.i) iff M_m^* does (m.i), i=5,8,9,10.

Proof. (i) necessity: $\forall$ X, we have

$$M_m^*(k_j(X))=1-M_m(I-k_j(X))=1-M_m(p_j(I-X))=1-0=1$$

sufficiency: It is similar.

(ii) necessity: $\forall$ X, we have

$$M_m^*(r_j(X))=1-M_m(I-r_j(X))=1-M_m(q_j(I-X))=1-(1-x_j)=x_j$$

sufficiency: That is similar.

(iii) When i=5, it is clear. We just prove that i=8 as the following:

necessity: $\forall$ X, we have

$$\begin{aligned}M_m^*(X)&=1-M_m(I-X)\\&=1-M_2(M_r(1-x_1,\ldots,1-x_r),\ M_{m-r}(1-x_{r+1},\ldots,1-x_m))\\&=M_2^*(1-M_r(1-x_1,\ldots,1-x_r),\ 1-M_{m-r}(1-x_{r+1},\ldots,1-x_m))\\&=M_2^*(M_r^*(x_1,\ldots,x_r),\ M_{m-r}^*(x_{r+1},\ldots,x_m))\end{aligned}$$

sufficiency: That is similar.

The proof that i=9 is just like that i=8. Now we prove that i=10:

n-ecessity: $\forall$ $X=(x_1,\ldots,x_m)$, $Y=(x_m,\ldots,x_{2m-1})$, we have

$$M_m^*(M_m^*(X),\ x_{m+1},\dots,x_{2m-1})$$
$$=1-M_m(1-M_m^*(X),\ 1-x_{m+1},\dots,1-x_{2m-1})$$
$$=1-M_m(M_m(I-X),\ 1-x_{m+1},\dots,1-x_{2m-1})$$
$$=1-M_m(1-x_1,\dots,1-x_m,\ M_m(I-Y))$$
$$=1-M_m(1-x_1,\dots,1-x_m,\ 1-M_m^*(Y))$$
$$=M_m^*(x_1,\dots,x_m,\ M_m^*(Y))$$

sufficiency: That is similar. Q.E.D.

PROPOSITION 2.5. Let $M_2 \in @_2$, and put

$$M_m(X)=M_2(M_2(\ \dots\ M_2(M_2(x_1,x_2),x_3)\ \dots,x_m), \tag{2.7}$$

then $M_m \in @_m$, and if M_2 satisfies (m.i), then M_m does (m.i), i=3,...,7.
The proof is omitted since it is simple.

PROPOSITION 2.6. Let $M_m^o \in @_m$, $t \in [0,1]$.

(i) If we put

$$M_m(X)=\begin{cases} M_m^o(X), & M_m^o(X)\in[t,1],\\ \min_j(x_j), & M_m^o(X)\in[0,t),\end{cases}$$

then $M_m \in @_m$, and when M_m^o satisfies (m.i), M_m does (m.i), too, i=3,4,5,8,9,10;

(ii) If we put

$$M_m(X)=\begin{cases} \bigvee_{j=1}^{m} x_j, & M_m^o(X)\in[t,1],\\ M_m^o(X), & M_m^o(X)\in[0,t),\end{cases}$$

then $M_m \in @_m$, and when M_m^o satisfies (m.i), M_m does (m.i), too, i=i,...,10.
The proof is omitted because it is clear.

PROPOSITION 2.7. Let $M_n \in @_n$, $M_m^{(k)} \in @_m$, k=1,...,n, and put

$$M_m(X)=M_n(M_m^{(1)}(X),\dots,\ M_m^{(n)}(X)),$$

then $M_m \in @_m$. When $M_m^{(k)}$, k=1,...,n, satisfy (m.i), M_m does, too, i=1,...,7,10; when a $M_m^{(k)}$ and a M_n satisfy (m.i), M_m does (m.i), too, i=3,7.

The proof is omitted since it is clear.

EXAMPLE 3. Take $M_n=L$, $M_m^{(k)}(X)=\sum_{j=1}^{m} a_{kj}x_j$, k=1,...,n, where $a_{kj}\in[0,1]$, $\sum_{j=1}^{m} a_{kj}=1$, k=1,...,n, and put

$$M_m(X)=\min_k\left(\sum_{j=1}^{m} a_{kj}x_j\right),$$

then $M_m \in @_m$.

NOTE: Let $M_n \in @_n$, M_r^1, $M_r^2,\dots,M_r^n \in @_r$, M_m^{11}, $M_m^{12},\dots,M_m^{1r}$, M_m^{21}, $M_m^{22},\dots,M_m^{2r},\dots,$

M_m^{n1}, M_m^{nr},...,$M_m^{nr} \in @_m$, and put

$$M_m(X)=M_n(M_r^1(M_m^{11}(X),\ldots,M_m^{1r}(X)),\ldots, M_r^n(M_m^{n1}(X),\ldots, M_m^{nr}(X))),$$

then $M_m \in @_m$ clearly.

EXAMPLE 4. Take $M_n=V$, $M_r^k=L$, $M_m^{ki}(X)=\sum_{j=1}^{m} a_{kij}x_j$, and put

$$M_m(X)=\bigvee_{k=1}^{n}(\min_i(\sum_{j=1}^{m} a_{kij}x_j)),$$

then $M_m \in @_m$, where $a_{kij} \in [0,1]$ and $\sum_{j=1}^{m} a_{kij}=1$.

3. THE GENERATION OF M_m-func

Let $g:[0,1] \longrightarrow [0,1]$ be a continuous and strictly monotone increasing function with $g(0)=0$ and $G(1)=1$. Evidently its inverse function $G:[0,1] \longrightarrow [0,1]$ exists, which is also a continuous and strictly monotone increasing function with $G(0)=0$ and $G(1)=1$.

M_m-func can be generated by using the function g, here has

THEOREM 3.1. Take $M_m^0 \in @_m$, and put

$$M_m(X)=G(M_m^0(g(x_1),\ldots,g(x_m))), \tag{3.1}$$

then $M_m \in @_m$, and when M_m^0 satisfies (m.i), M_m does (m.i), too, i=3,...,10.

Proof. (m.1),...,(m.7) are true clearly, so we only prove (m.8),(m.9) and (m.10).

(m.8): Because M_m^0 satisfies (m.8), there exist M_r^0, M_{m-r}^0 and M_2^0 such that

$$M_m^0(X)=M_2^0(M_r^0(x_1,\ldots,x_r), M_{m-r}^0(x_{r+1},\ldots,x_m))$$

Take

$$M_r(x_1,\ldots,x_r)=G(M_r^0(g(x_1),\ldots,g(x_r))),$$

$$M_{m-r}(y_1,\ldots,y_{m-r})=G(M_{m-r}^0(g(y_1),\ldots,g(y_{m-r}))),$$

$$M_2(u_1,u_2)=G(M_2^0(g(u_1), g(u_2))),$$

we have

$$\begin{aligned}&M_2(M_r(x_1,\ldots,x_r), M_{m-r}(x_{r+1},\ldots,x_m))\\&=G(M_2^0(g(M_r(x_1,\ldots,x_r)), g(M_{m-r}(x_{r+1},\ldots,x_m))))\\&=G(M_2^0(M_r^0(g(x_1),\ldots,g(x_r)), M_{m-r}^0(g(x_{r+1}),\ldots,g(x_m))))\\&=G(M_m^0(g(x_1),\ldots,g(x_m)))\\&=M_m(X)\end{aligned}$$

(m.9): Because M_m^0 satisfies (m.9), there exists M_2^0 such that

$$M_m^0(X)=M_2^0(M_2^0(\ \ldots\ M_2^0(M_2^0(x_1,x_2),x_3)\ \ldots), x_m),$$

and take $M_2(u_1,u_2)=G(M_2^0(g(u_1), g(u_2)))$. We have

$$M_2(M_2(\ldots M_2(M_2(x_1,x_2),x_3)\ldots),x_m)$$

$$=G(M_2^0(g(M_2(\ldots M_2(M_2(x_1,x_2),x_3)\ldots)),\ g(x_m)))$$

....

$$=G(M_2^0(M_2^0(\ldots\ M_2^0(M_2^0(g(x_1),\ g(x_2)),\ g(x_3))\ldots),\ g(x_m)))$$

$$=G(M_m^0(g(x_1),\ldots,g(x_m)))$$

$$=M_m(X)$$

(m.10): $\forall\ X=(x_1,\ldots,x_m),\ Y=(x_m,\ldots,x_{2m-1})$, we have

$$M_m(M_m(X),\ x_{m+1},\ldots,x_{2m-1})$$

$$=G(M_m^0(g(M_m(X)),\ g(x_{m+1}),\ldots,\ g(x_{2m-1})))$$

$$=G(M_m^0(M_m^0(g(x_1),\ldots,g(x_m)),\ g(x_{m+1}),\ldots,g(x_{2m-1}))$$

$$=G(M_m^0(g(x_1),\ldots,g(x_{m-1}),\ M_m^0(g(x_m),\ldots,g(x_{2m-1}))))$$

$$=G(M_m^0(g(x_1),\ldots,g(x_{m-1}),\ g(G(M_m^0(g(x_m),\ldots,g(x_{2m-1}))))))$$

$$=M_m(x_1,\ldots,x_{m-1},\ M_m(Y))$$ Q.E.D.

EXAMPLE 5. Take $M_m^0=L$, $g(x)=x$, clearly $G(x)=x$, so $M_m(X)=\min_j(x_j)$; if we take $M_m^0=\sum$, $g(x)=x^p$, $p>0$, then $G(x)=x^{1/p}$, so

$$M_m(X)=(\sum_{j=1}^{m}a_jx_j^p)^{1/p}.$$

EXAMPLE 6. Take $g(x)=\sin\frac{\pi}{2}x$, then $G(x)=\frac{2}{\pi}\arcsin x$, so the following functions are M_m-func:

$$M_m(x)=\frac{2}{\pi}\arcsin[\min_j(\sin\frac{\pi}{2}x_j)],$$

$$M_m(X)=\frac{2}{\pi}\arcsin(\bigvee_{j=1}^{m}\sin\frac{\pi}{2}x_j),$$

$$M_m(X)=\frac{2}{\pi}\arcsin(\sum_{j=1}^{m}a_j\sin\frac{\pi}{2}x_j),$$

where $a_j\in[0,1]$ and $\sum_{j=1}^{m}a_j=1$;

$$M_m(X)=\frac{2}{\pi}\arcsin[\min_j(1-a_j(1-\sin\frac{\pi}{2}x_j))],$$

where $a_j\in[0,1]$ and $\bigvee_{j=1}^{m}a_j=1$;

$$M_m(X)=\frac{2}{\pi}\arcsin[\min_j((1-a_j)\vee\sin\frac{\pi}{2}x_j)],$$

where $a_j\in[0,1]$ and $\bigvee_{j=1}^{m}a_j=1$;

$$M_m(X)=\frac{2}{\pi}\arcsin[1-[\prod_{j=1}^{m}(1-\sin\frac{\pi}{2}x_j)]^{1/m}],$$

$$M_m(X)=\frac{2}{\pi}\arcsin[1-[\frac{1}{m}\sum_{j=1}^{m}(1-\sin\frac{\pi}{2}x_j)^p]^{1/p}],$$

where $p>0$;

$$M_m(X)=\frac{2}{\pi}\arcsin[1-[\sum_{j=1}^{m}a_j[1-(\sin\frac{\pi}{2}x_j)^p]]^{1/p}],$$

where $p>0$, $a_j\in[0,1]$ and $\sum_{j=1}^{m}a_j=1$.

Let $h:[0,+\infty]\longrightarrow[0,+\infty]$ be a continuous and strictly monotone decreasing function with $h(0)=+\infty$, $h(1)=1$ and $h(+\infty)=0$. evidently, its inverse function $H:[0,+\infty]\longrightarrow[0,+\infty]$ exists, which is also a continuous and strictly monotine decreasing function with $H(0)=+\infty$, $H(1)=1$ and $H(+\infty)=0$.
M_m-func ca n be generated by using the function h, here has

THEOREM 3.2. Take $M_m^0\in@_m$, and put

$$M_m(X)=H([M_m^0(1/h(x_1),\ldots\ldots,1/h(x_m))]^{-1}),\qquad(3.2)$$

then $M_m\in@_m$, and when M_m^0 satisfies (m.i), M_m does (m.i), too, $i=3,\ldots,10$.

Proof. Put $g(x)=1/h(x)$, then its inverse function $G(x)=H(1/x)$, and clearly g satisfies the conditions of the theorem 1. And since $H(x)=G(1/x)$, too, so we have

$$M_m(X)=H([M_m^0(1/h(x_1),\ldots\ldots,1/h(x_m)]^{-1})=G(M_m^0(g(x_1),\ldots,g(x_m))),$$

that means the theorem is true by the theorem1. Q.E.D.

EXAMPLE 7. Take $h(x)=\frac{1}{2}(e-e^{-1})\operatorname{csch}x$, then

$$H(x)=\frac{1}{2}\ln\frac{[1+[4x^2/(e-e^{-1})^2]]^{\frac{1}{2}}+1}{[1+[4x^2/(e-e^{-1})^2]]^{\frac{1}{2}}-1},$$

thus the following functions are M_m-func:

$$M_m(X)=\frac{1}{2}\ln\frac{[1+[4/[\min_j(e^{x_j}-e^{-x_j})]^2]]^{\frac{1}{2}}+1}{[1+[4/[\min_j(e^{x_j}-e^{-x_j})]^2]]^{\frac{1}{2}}-1};$$

$$M_m(X)=\frac{1}{2}\ln\frac{[1+[4/[\bigvee_{j=1}^{m}(e^{x_j}-e^{-x_j})]^2]]^{\frac{1}{2}}+1}{[1+[4/[\bigvee_{j=1}^{m}(e^{x_j}-e^{-x_j})]^2]]^{\frac{1}{2}}-1};$$

$$M_m(X)=\frac{1}{2}\ln\frac{[1+[4/[\sum_{j=1}^{m}a_j(e^{x_j}-e^{-x_j})]^2]]^{\frac{1}{2}}+1}{[1+[4/[\sum_{j=1}^{m}a_j(e^{x_j}-e^{-x_j})]^2]]^{\frac{1}{2}}-1},$$

where $a_j\in[0,1]$ and $\sum_{j=1}^{m}a_j=1$.

4. APPLICATIONS TO PATTERN RECOGNITION

4.1. Putting Forward the Problems

Suppose n standard patterns with m fuzzy characteristics are known, where the m fuzzy characteristics of the i-th pattern are A_{ij}, $j=1,\dots,m$, which are fuzzy sets on universe U_j, $j=1,\dots,m$, i.e. $A_{ij}\in F(U_j)$, $j=1,\dots,m$, respectively. Put

$$U^{*}=\prod_{j=1}^{m} U_j \tag{4.1}$$

If $u_0\in U^{*}$ is an object to be recognized, then we have

Problem 1. How to recognize to which of the n patterns u_0 relatively belongs?

If an object to be recognized has also m fuzzy characteristics $B_j\in F(U_j)$, $j=1,\dots,m$, then we have

Problem 2. How to recognize to which of n patterns the pattern to be recognized is relatively closest?

Put $F_m=[\ \overline{A}=(A_1,\dots,A_m)\mid A_j\in F(U_j), j=1,\dots,m]$, i.e.

$$F_m=\prod_{j=1}^{m} F(U_j) \tag{4.2}$$

Then any element $\overline{A}$ is a generalized fuzzy vector. If we put $\overline{A}_i=(A_{i1},\dots,A_{im})$, $i=1,\dots,n$, and $\overline{B}=(B_1,\dots,B_m)$, then

Problem 1 means how to recognize to which one of $\overline{A}_1,\dots,\overline{A}_n$ u_0 relatively belongs.

Problem 2 means how to recognize to which one of $\overline{A}_1,\dots,\overline{A}_n$ $\overline{B}$ is relatively closest.

4.2. Multifactorial Fuzzy Sets

In F_m the inclusion relation " [" and the equality relation " = " are defined as follows:

$$\overline{A}\ [\ \overline{B} \quad \text{iff} \quad (A_j \subset B_j,\ j=1,\dots,m),$$

$$\overline{A}=\overline{B} \quad \text{iff} \quad \overline{A}\ [\ \overline{B} \text{ and } \overline{B}\ [\ \overline{A}.$$

If " [" is regarded as a partially ordered relation, the it is easy to prove:

PROPOSITION 4.1. $(F_m,[)$ is a complete distributive lattice, and $\overline{\theta}=(\emptyset,\dots,\emptyset)$ is the least element and $\overline{I}=(U_1,\dots,U_m)$ is the greatest element in F_m.

DEFINITION 4.1. For any $\overline{A}\in F_m$, a fuzzy set on U^{*} is constructed by $\overline{A}$ and denoted by

$$A=:[\overline{A}]=:[A_1,\dots,A_m]. \tag{4.3}$$

A is called a multifactorial fuzzy set if

$$\mu_A(u)=M_m(\ \mu_{A_1}(u_1),\dots,\ \mu_{A_m}(u_m)), \tag{4.4}$$

where $M_m\in\mathcal{M}_m$ and $u=(u_1,\dots,u_m)\in U^{*}$. Multifactorial fuzzy set is simply named M_m-FS. The set of all M_m-FS is denoted by $F(U^{*},M_m)$.

It is easy to see:

PROPOSITION 4.2. (i) $\emptyset$, $U^* \in F(U^*, M_m) \subset F(U^*)$;
(ii) $f: F_m \longrightarrow F(U^*, M_m)$, $\overline{A} \longmapsto [\overline{A}]$, is an ordered-preserving mapping from $(F_m, [\ )$ to $(F(U^*, M_m), \subset)$.

THEOREM 4.1. $(F(U^*, M_m), \cup, \cap)$ is a distributive lattice with the greatest element U^*, and the least element $\emptyset$.

Proof. The distributivity is inherited to $F(U^*, M_m)$ by $F(U^*)$. Now we prove that $F(U^*, M_m)$ is a lattice. $\forall\ A, B \in F(U^*, M_m)$, there exist $\overline{A}, \overline{B}$, such that $A = [\overline{A}]$, $B = [\overline{B}]$. Write $C = A \cup B$, we only need to prove that there exists $\overline{C} \in F_m$ such that $C = [\overline{C}]$. In fact, we take $\overline{C} = (C_1, \ldots, C_m) \in F_m$, which satisfies:

$$(\mu_{C_1}(u_1), \ldots, \mu_{C_m}(u_m)) = \begin{cases} (\mu_{A_1}(u_1), \ldots, \mu_{A_m}(u_m)), & \mu_A(u) \in [\mu_B(u), 1], \\ (\mu_{B_1}(u_1), \ldots, \mu_{B_m}(u_m)), & \text{otherwise.} \end{cases}$$

It is easy to check $C = [\overline{C}]$, thus $C \in F(U^*, M_m)$. That $A \cap B \in F(U^*, M_m)$ can be proved in the same way, so $F(U^*, M_m)$ is a lattice. Q.E.D.

The theorem describes an algeraic structure for M_m-FS.

An equivalence relation " # " on F_m may be determined by the following:

$$\overline{A} \# \overline{B} \quad \text{iff} \quad [\overline{A}] = [\overline{B}],$$

so we obtain a quotient set of F_m: $F_m' =: F_m / \# =: [\ (\overline{A}) \mid \overline{A} \in F_M\ ]$, where $(\overline{A})$ is an equivalence class for $\overline{A}$.

In F_m' two operations $\cup'$ and $\cap'$ are defined by

$$(\overline{C}) = (\overline{A}) \cup' (\overline{B}) \quad \text{iff} \quad [\overline{C}] = [\overline{A}] \cup [\overline{B}],$$

$$(\overline{C}) = (\overline{A}) \cap' (\overline{B}) \quad \text{iff} \quad [\overline{C}] = [\overline{A}] \cap [\overline{B}].$$

It is easy to prove:

PROPOSITION 4.3. $(F_m', \cup', \cap')$ and $(F(U^*, M_m), \cup, \cap)$ are two algebraic systems of isomorphism.

PROPOSITION 4.4. Let $A = [\overline{A}]$, $B = [\overline{B}]$, $C = [\overline{A}_1 \cup \overline{B}_1, \ldots, \overline{A}_m \cup \overline{B}_m]$, $D = [\overline{A}_1 \cap \overline{B}_1, \ldots, \overline{A}_m \cap \overline{B}_m]$, then
(i) $A \cup B \subset C$, $D \subset A \cap B$;
(ii) $\min_j (\mu_{A_j^c}(u_j)) \leq \mu_{A^c}(u) \leq \bigvee_{j=1}^{m} \mu_{A_j^c}(u_j)$.

The proof is omitted since it is simple.

4.3. Method For Solving Problem 1

Let $\overline{A}_1, \ldots, \overline{A}_n \in F_m$ be n known patterns, $u_0 \in U^*$ be an object to be recognozed. Since the patterns differ from the object on levels of concepts (the former are gereralized fuzzy vectors, but latter is an element in U^*), this problem of pattern recognition is called multifactorial pattern recognition of distinct levels.

Based on the principle of the highest grade of membership (see [4]) we have Direct Method For Multifactorial Pattern Recognition on Distinct levels:

Take $A_i=[\overline{A}_i]$, i=1,...,n. If there is i such that

$$\mu_{A_i}(u_o)=\max(\mu_{A_1}(u_o),\ldots,\mu_{A_m}(u_o)),$$

then it is decided that u_o belongs to $\overline{A}_i$.

4.4. Multifactorial Degree of Nearness

For solving Problem 2, first we solve the problem of nearness for two generalized fuzzy vectors.

DEFINITION 4.2. Let N be degree of nearness which is defined by axiomatization, $M_m \in @_m$. The mapping

$$N^*: F_m X F_m \longrightarrow [0,1]$$

$$(\overline{A},\overline{B}) \longmapsto N^*(\overline{A},\overline{B})=:M_m(N(\overline{A}_1,\overline{B}_1),\ldots,N(\overline{A}_m,\overline{B}_m))$$

is called multifactorial degree of nearness, and denoted simply by M_m-DN.

It is easy to prove:

THEOREM 4.2. N^* satisfies the axioms of degree of nearness [4], i.e.

(i) $N^*(\overline{A},\overline{A})=1$, $N^*(\overline{I},\overline{\theta})=0$;

(ii) $N^*(\overline{A},\overline{B})=N^*(\overline{B},\overline{A})$;

(iii) $\overline{A}\subset\overline{B}\subset\overline{C}$ implies $N^*(\overline{A},\overline{C})\leqslant \min(N^*(\overline{A},\overline{B}),N^*(\overline{B},\overline{C}))$.

4.5. Methods For Solving Problem 2

Let $\overline{A}_1,\ldots,\overline{A}_n \in F_m$ be n known patterns and $\overline{B}\in F_m$ be a pattern to be recognized. Since the known pattern and the pattern to recognized are the level in the concept (they are all generalized fuzzy vectors), this problem of pattern recognition is called multifactorial pattern recognition of the same level. Based on the principle for choosing degree of bearness (see [4]) we have

Direct Method 1 Multifactorial Pattern Recognition of the Same Level:

Let N be a degree of nearness, take $A_i=[\overline{A}_i]$, i=1,...,n, $B=[\overline{B}]$. If there exists i such that

$$N(A_i,B)=\max(N(A_1,B),\ldots,N(A_n,B)),$$

then it is decided that $\overline{B}$ is closest to $\overline{A}_i$.

Direct Method 2 Multifactorial Pattern Recognition of the Same Level:

If there exists i such that

$$N^*(\overline{A}_i,\overline{B})=\max(N^*(\overline{A}_1,\overline{B}),\ldots,N^*(\overline{A}_n,\overline{B})),$$

then it is decided that $\overline{B}$ is closest to $\overline{A}_i$.

5. APPLICATION TO CLUSTERING ANALYSIS

5.1. Putting Forward the Problem

Let $X=[x_1,\ldots,x_n]$ be a set of objects to be clustered, every object x_i has m fuzzy characteristics $A_{ij} \in F(U_j)$, j=1,...,m. Then we have

Problem 3. How to establish a fuzzy similarity matrix for X?

The problem means how to establish a fuzzy similarity matrix on X in accordance with the fuzzy characteristics $\overline{A}_i$ of $x_i \in X$, i=1,....,n ?

5.2. The Methods For Solving Problem 3

Based on M_m-FS we have

Method 1. Let N be a degree of nearness, and take $A_i=[\overline{A}_i]$, i=1,...,n. The similarity coefficient of A_i and A_j may be calculated by

$$r_{ij}=N(A_i,A_j), \quad i,j=1,\dots,n. \tag{5.1}$$

Based on M_m-DN we have

Method 2. The similarity coefficient of A_i and A_j may be calculated by

$$r_{ij}=N^*(\overline{A}_i,\overline{A}_j), \quad i,j=1,\dots,n.$$

6. APPLICATIONS TO EXTREMUM PROBLEM

6.1. Case of One Object and One Restriction

Let f be a bounded real function on a universe U ($U\neq\emptyset$), $A\in F(U)$. Consider the conditional extremum of f on fuzzy restricted set A. Write

$$\sup(f)=\sup[f(u)\mid u\in U],$$

$$\inf(f)=\inf[f(u)\mid u\in U].$$

According to Zadeh' definition (see [5]), the maximal set M_f of f on U is stipulated as the following:

$$\mu_{M_f}(u)=\frac{f(u)-\inf(f)}{\sup(f)-\inf(f)} \tag{6.1}$$

Note 1: If $\inf(f)=-\infty$, then may choose an appropriate lower bound $m>\inf(f)$, ($m<\sup(f)$), which is defined by

$$\mu_{M_f}(u)=\begin{cases}\dfrac{f(u)-m}{\sup(f)-m}, & f(u)\geq m,\\ 0, & f(u)<m.\end{cases}$$

Note 2: The minimal set of f on U m_f is defined by

$$m_f=M_f^c \,. \tag{6.2}$$

Regarding M_f as a fuzzy object set, take the decision fuzzy set $B\in F(U)$:

$$\mu_B(u)=M_2(\mu_A(u),\mu_{M_f}(u)), \tag{6.3}$$

where $M_2\in\text{@}_2$.

If there exists $u_0\in U$, such that

$$\mu_B(u_0)=hgt(B), \tag{6.4}$$

where $hgt(B)=:\sup[\mu_B(u)|u\in U]$, then u_0 is the maxmal point and $f(u_0)$ is the maximal value, of f on the restriction A.

6.2. Case with Several Object and Several Restrictions

Let $A_1,\ldots,A_n \in F(U)$ be n fuzzy restricted sets. Take $M_n \in @_n$, and form a new fuzzy restricted set $A\in F(U)$:

$$\mu_A(u)=M_n(\mu_{A_1}(u),\ldots,\mu_{A_n}(u)). \tag{6.5}$$

Let $f_1,\ldots,f_m$ be m objective functions. Their fuzzy objective sets $M_{f_1},\ldots,M_{f_m} \in F(u)$ can be respectively found, and by them form a new fuzzy objective set $M_f \in F(U)$:

$$\mu_{M_f}(u)=M_m(\mu_{M_{f_1}}(u),\ldots,\mu_{M_{f_m}}(u)), \tag{6.6}$$

where $M_m \in @_m$.

For A and M_f, the extreme value and extreme point may be found by above model.

7. APPLICATIONS TO FUZZY LINEAR PROGRAMMING

7.1. Linear Programming with Fuzzy Restriction

Consider the problem of linear programming with fuzzy restriction:

$$Z=c_1x_1+c_2x_2+\ldots\ldots+c_nx_n \tag{7.1}$$

$$\left.\begin{array}{l} a_{11}x_1+a_{12}x_2+\ldots\ldots+a_{1n}x_n \lesssim b_1 \\ a_{21}x_1+a_{22}x_2+\ldots\ldots+a_{2n}x_n \lesssim b_2 \\ \quad\ldots\ldots\ldots\ldots \\ a_{m1}x_1+a_{m2}x_2+\ldots\ldots a_{mn}x_n \lesssim b_m \\ x_1,x_2,\ldots\ldots,x_n \geqslant 0 \end{array}\right\} \tag{7.2}$$

where (7.1) is a linear objective function, (7.2) is the fuzzy restriction and "$a \lesssim b$" means 'a approximately less than or equal to b'.

Let $\overline{X}=[X\in R^n|X\geqslant\theta=:(0,\ldots,0)]$, where R is the set of all real numbers. For every

$$\sum_{j=1}^{n} a_{ij}x_j \lesssim b_i,$$

there is the fuzzy set $D_i \in F(\overline{X})$:

$$\mu_{D_i}(X)=f_i\left(\sum_{j=1}^{n} a_{ij}x_j\right)=\begin{cases} 1, & \sum_{j=1}^{n} a_{ij}x_j \leqslant b_i, \\ 1-\frac{1}{d_i}\left(\sum_{j=1}^{n} a_{ij}x_j-b_i\right), & \sum_{j=1}^{n} a_{ij}x_j \in (b_i,\ b_i+d_i], \\ 0, & \sum_{j=1}^{n} a_{ij}x_j > b_i+d_i, \end{cases}$$

where d_i is a elastic parameter chosen appropriately with $d_i \geqslant 0$. Write:

$$A=(a_{ij})_{m\times n},\ X=(x_1,\dots,x_n)^T,\ B=(b_1,\dots,b_m)^T,\ C=(c_1,\dots,c_m)^T,$$

then (7.1) and (7.2) may b-e shown as the following:

$$Z=CX, \tag{7.3}$$

$$\left.\begin{array}{l} AX \lessapprox B \\ X \geqslant \theta \end{array}\right\} \tag{7.4}$$

By $D_1,\dots,D_m$ we take the fuzzy restricted set $D\in F(\overline{X})$, of (7.3):

$$\mu_D(X)=M_m(\ \mu_{D_1}(X),\ \dots,\ \mu_{D_m}(X)\ ), \tag{7.5}$$

where $M_m \in @_m$.

Let Z_0 be the optimal solution of the following linear programming:

$$Z=CX$$

$$\left.\begin{array}{l} AX \leqslant B \\ X \geqslant \theta \end{array}\right\} \tag{7.6}$$

In order to find the optimal solution of the objective function (7.3) on the fuzzy restriction (7.4),first change (7.3) into the following restriction:

$$CX \geqslant Z_0 \tag{7.7}$$

and second (7.7) is changed to:

$$CX \gtrapprox Z_0 \tag{7.8}$$

So we have the fuzzy objective set $E\in F(\overline{X})$:

$$\mu_E(X)=g(\sum_{i=1}^{n} c_i x_i)$$

$$=\begin{cases} 0, & \sum_{i=1}^{n} c_i x_i \leqslant Z_0, \\ \frac{1}{d_0}(\sum_{i=1}^{n} c_i x_i - Z_0), & \sum_{i=1}^{n} c_i x_i \in (Z_0,\ Z_0+d_0], \\ 1, & \sum_{i=1}^{n} c_i x_i > Z_0+d_0, \end{cases}$$

where Z_0+d_0 is the optimal value of (7.6) in where b_i is changed into b_i+d_i.

Take $M_2 \in @_2$ an-d the decision set $B\in F(\overline{X})$:

$$\mu_B(X)=M_2(\ \mu_D(X),\ \mu_E(X)). \tag{7.9}$$

If there exists $X_0 \in \overline{X}$, such that $\mu_B(X_0)=hgt(B)$, then X_0 is just the optimal solution of (7.3) on (7.4).

7.2. Linear Programming with Several Objective Functions

Let us consider the linear programming:

$$\left.\begin{aligned} Z_1&=c_{11}x_1+c_{12}x_2+\cdots\cdots+c_{1n}x_n \\ Z_2&=c_{21}x_1+c_{22}x_2+\cdots\cdots+c_{2n}x_n \\ &\cdots\cdots\cdots\cdot \\ Z_r&=c_{r1}x_1+c_{r2}x_2+\cdots\cdots+c_{rn}x_n \end{aligned}\right\} \qquad (7.10)$$

with the restriction:

$$\left.\begin{aligned} AX&\leq B \\ X&\geq\theta \end{aligned}\right\} \qquad (7.11)$$

Write $C=(c_{ij})_{r\times n}$, $Z=(Z_1,\ldots,Z_r)^T$, then above problem is changed into:

$$\left.\begin{aligned} Z&=CX \\ AX&\leq B \\ X&\geq\theta \end{aligned}\right\} \qquad (7.12)$$

Write $\overline{X}^*=[X\in\overline{X}|AX\leq B]$. For every objective function $Z_i=\sum_{j=1}^{n}c_{ij}x_j$ in (7.10), on $\overline{X}^*$ we can find the optimal value:

$$Z_i^0=\max[Z_i|\ Z_i=\sum_{j=1}^{n}c_{ij}x_j,\ X\in\overline{X}^*].$$

For every Z_i^0, appropriately give a elastic parameter $d_i>0$, and the smaller d_i, the more importance, that it means. Thus we have the fuzzy objective set $E_i\in F(\overline{X}^*)$:

$$\mu_{E_i}(X)=g_i(\sum_{j=1}^{n}c_{ij}x_j)=\begin{cases} 0, & \sum_{j=1}^{n}c_{ij}x_j<Z_i^0-d_i, \\ 1-\frac{1}{d_i}(Z_i^0-\sum_{j=1}^{n}c_{ij}x_j), & Z_i^0-d_i\leq\sum_{j=1}^{n}c_{ij}x_j<Z_i^0, \\ 1, & Z_i^0\leq\sum_{j=1}^{n}c_{ij}x_j. \end{cases}$$

By $E_1,\ldots,E_r$, form a fuzzy objective set $E\in F(\overline{X}^*)$:

$$\mu_E(X)=M_r(\mu_{E_1}(X),\ldots,\mu_{E_r}(X)), \qquad (7.13)$$

where $M_r\in\mathcal{M}_r$.

If there exists $X_0\in\overline{X}^*$, such that $\mu_E(X_0)=hgt(E)$, then $Z^*=CX_0$ is the optimal solution of (7.12) which its degree of membership is $t^*=hgt(E)$.

8. APPLICATION TO FUZZY GAME

Let U_1,U_2 be two tactical sets, and $D_1 \in F(U_1)$, $D_2 \in F(U_2)$ are fuzzy tactical sets. $f_1(d_1,d_2)$ and $f_2(d_1,d_2)$ respectively is the win function on both sides. Suppose the both sides make the game according to the adventurous principle. We find the solution of balance for the game.
Let $M_{f_1}(d_2) \in F(U_1)$ be the maximal set on fuzzy restriction D_1, which $f_1(.,d_2)$ is regarded as a one-variable function when d_2 is fixed arbitrarily. Symmetrically we have $M_{f_2}(d_1) \in F(U_2)$. By $M_{f_1}(d_2)$ and $M_{f_2}(d_1)$ we have the balance solution $E \in F(U_1 X U_2)$:

$$\mu_E(d_1,d_2)=M_2(\mu_{M_{f_1}(d_2)}(d_1), \mu_{M_{f_2}(d_1)}(d_2)), \tag{8.1}$$

where $M_2 \in @_2$.
If there exists $(d_1^0,d_2^0) \in U_1 X U_2$, such that

$$\mu_E(d_1^0,d_2^0)=hgt(E), \tag{8.2}$$

then (d_1^0,d_2^0) is the balance solution determined.

9. APPLICATION TO FUZZY DECISION-MAKING

9.1. Fuzzy Decision-making with Several Criterions

Let $A=[a_1,...,a_n]$ b-e the set of some events, $B=[b_1,...,b_m]$ be the game set, and $C=[c_1,...,c_p]$ be the set of p criterions.
For every criterion c_k, there is a game matrix:

$$R^{(k)}=(r_{ij}^{(k)})_{nxm}, \tag{9.1}$$

where $r_{ij}^{(k)}$ is the win degree of the situation when b_j is used to deal with a_i on c_k.

Take $M_p \in @_p$, and put

$$r_{ij}^*=M_p(r_{ij}^{(1)}, r_{ij}^{(2)},...,r_{ij}^{(p)}), \tag{9.2}$$

then have the multifactorial decision matrix:

$$R^*=(r_{ij}^*)_{nxm} \tag{9.3}$$

If there is j_0 such that

$$r_{ij}^*=max[r_{i1}^*,...,r_{im}^*],$$

then b_{j_0} is the optimal decision with respect to a_i.

9.2. Multifactorial Fuzzy Decision-making

Let $U=[u_1,\ldots,u_n]$ be the factor set, $V=[v_1,\ldots,v_m]$ be the set of m decisions. First we make one-factor decision, i.e. making the mapping:

$$f: U \longrightarrow F(V)$$

$$u_i \longmapsto f(u_i)=:R_i=:(r_{i1},\ldots,r_{im}). \quad (9.4)$$

Second taking $M_n \in @_n$ make the multifactorial decision $B=(b_1,\ldots,b_m)\in F(V)$:

$$b_j=M_n(r_{1j},\ldots,r_{nj}).$$

If there is j_0 such that

$$b_{j_0}=\max[b_1,\ldots,b_m],$$

then we take the decision $v_{j_0}\in V$.

9.3. Estimation of degree

Let $U=[u_1,\ldots,u_n]$ be the set of the factors which influence the good and bad degree of some thing T. First make the estimation of degree of one-factor:

$$f: U \longrightarrow [0,1]$$

$$u_i \longmapsto f(u_i)=:a_i.$$

Take $M_n \in @_n$. Second make the multifactorial degree estimation:

$$a=M_n(a_1,\ldots,a_n),$$

then a is just the point estimation of degree of T with respect to U, which is the good and bad degree.

REFERENCES

[1] Hongxing Li, Fuzzy Perturbation Analysis, Part 2, Fuzzy Sets and Systems, Vol.19, 2(1986) pp.165-175.

[2] Hongxing Li, Multifactorial Fuzzy Sets and Multifactorial Degree of Nearness, Fuzzy Sets and Systems, Vol.19, 3(1986) pp. 291-297.

[3] Hongxing Li, Fuzzy Perturbation Analysis, Part 1, Fuzzy Sets and Systems, Vol.17, 2(1985) pp. 189-197.

[4] Peizhuang Wang, Fuzzy Sets Theory and Its Applications (Shanghai Technical Publishers, 1983).

[5] Zadeh, L.A., On Fuzzy Algorithms, Memo UCB/ERL M-325 (Univ. of California, Berkeley, 1972).

[6] Zimmermann,H.J., Fuzzy Programming and Linear Programming with Several objective Functions, Fuzzy Sets and Systems, 1(1978) pp. 45-55.

[7] Zimmermann, H.J., Description and Optimization of Fuzzy Systems, Int.j.Gen. Syst., 2(1976) pp.209-215.

[8] Hongxing Li, Inverse Problem of Multifactorial Decision, Fuzzy Mathematics. Vol.5, 1(1985) pp.41-48.

[9] Bellman, R.E. and Zadeh, L.A., Decision-making in A Fuzzy Environment, Mange. Sci. 17(1970) pp.141-164.

[10] Dubois,D. and Prade,H., Fuzzy Sets and Systems (Acad. Press, New York, 1980).

Fuzzy Logic in Knowledge-Based Systems,
Decision and Control
M.M. Gupta and T. Yamakawa (Editors)
Elsevier Science Publishers B.V. (North-Holland), 1988

AUTONOMOUS EVOLUTIVE SYSTEMS WITH AMBIGUOUS STATES *

Dan BUTNARIU

Department of Mathematics, University of Haifa,
31999 Haifa, Israel.

In this paper a model of autonomous evolutive system whose components may achieve specific states in various degrees is studied. It is shown that, in given conditions, the variations of the "state parameters" of such a system while the system moves from its intial state into another state can be approximated by an iterative method. Some applications are also discussed.

KEYWORDS: Autonomous evolutive system, Fuzzy measure, Radon-Nikodym derivative, Fixed point.

1. INTRODUCTION

A problem arising frequently in practice when one has to deal with complex systems is that the actual states of the system at different stages of its evolution can not be precisely determined and, even in the case that the actual state of the system (at a specific stage of its evolution) is known to be A, one can merely estimate the plausibility degree that the system will "evoluate" from A to another state B than to determine which of all the states will be "the next" in the evolution of the system. Scientists interested in modelling biological, economical or social systems are well aware of this difficulty since they are frequently unable to determine precisely the actual state of the systems which they are dealing with without modifying the nature

---

* This work was made possible by the financial support of the Israeli Ministry of Absorption.

of the system in itself. During the last twenty years, people involved in hardware realization are themself encountering similar difficulties since the systems which they are trying to immagine are more complex and, by consequence, it is more and more difficult to ellaborate adequate mathematical tools to study them in the standard analytical way. For these reasons, the literature of the last two decades is aboundant in scientific studies suggesting alternative methods for modelling the behaviour of systems which are such that only "vague" or "unprecise" knowledge can be used in order to analyse their evolution. For instance, S. Eilenberg [9] approaches the so called "ambiguous automata" which are essentially discrete systems with ambiguous states. In many other works Zadeh's concept of *fuzzy set* [17] is considered to be the basis for the desired "new approach". The reader may refer to the results and commented references contained in [8] and [13] in order to have a better up-to-date image of what is happening in this field.

In this paper we intend to describe a mathematical tool of modelling the behaviour of what we are calling *autonomous evolutive system with ambiguous states*. Intuitively, the systems which we have in mind may be seen as large sets of interconnected "atomic components", called generically *points*, whose states at different stages of the *autonomous* evolution of the system can not be determined without essential modifications of the system in itself, but we assume that there are objective means of evaluating the probability that, at a given stage, a component of the system is in a specific state A and one has the possibility to make reasonable accurate estimations of the extent in which modifications of the state of a point x may lead to modifications of the state of the other components of the system. Our purpose is to try to find a method of *determining (a priori) what variations of the "state parameters" of the system may be expected while the system "moves" in a specific state (which may be "ambiguous" in the sense that different points may achieve this state in various degrees)*. The formal definition of the concept of *autonomous evolutionary system* is given in Section 2. Also in Section 2, we give sufficient conditions in which the equations describing the behaviour of an autonomous evolutionary system can be "solved" and their solutions can be effectively computed. In Section 3, technics suggested in [11] and [10] are employed to prove our statements. In Section 4, potential applications of the concepts and results presented in Sections 2 and 3, to solving specific problems arising in Reliability Theory, Hardware Realization and in Economics are described.■

2. MAIN CONCEPTS AND RESULTS

From a strictly formal point of view, the concept of *autonomous evolutive system* studied in this paper is defined by the following set of data:

(a) A probability space (X,C,μ); the elements of X are called *points*; the elements of C are called *states* and, for a state $A \in C$, $\mu(A)$ is the probability that the system will "move" (will evoluate) in the state A . A [0,1]-valued (X,C)-measurable function S is called *ambiguous state*; in this case, when x is a point, then S(x) is seen as the extent in which x is in the state S and

$$\bar{\mu}(S) := \int S(x)d\mu(x)$$

is seen as the probability that the systems will evoluate in such a way that each point x of it will be in the state named S in the degree S(x).

(b) A [0,1]-valued $(X,C)\times(X,C)$-measurable function E called *evolution function*; for any point x it induces the ambiguous state E_x defined by $E_x(y) := E(x,y)$ and the ambiguous state E^x defined by $E^x(y) := E(y,x)$. The ambiguous state E_x is seen as the state of the "interconnections" of x with the other points of the system and $E_x(y)$ is the degree (the intensity) of the connection of y with x inside the system.

(c) There are finitely many (significant) "state parameters" denoted μ_1, μ_2, ..., μ_n whose (expected real) values $\mu_i(A)$ $(1\le i\le n)$ while the system "moves" in the state A have to be (a priori) determined. It is known that for the system to move in the state A the values of its state parameters have to be "initially" $\mu_i^o(A)$ $(1\le i\le n)$. One assume that:

(i) the functions $A \longmapsto \mu_i^o(A)$ are *given* μ -singular measures (over (X,C)), and $\mu_i^o(A)$ is called *initial value* μ_i in

the state A $(1 \le i \le n)$ and

(ii) the state parameters μ_i are (finite) measures over (X,C) and they are of the form

$$\mu_i = \mu_i^o + \mu_i^\alpha$$

where $\mu_i^\alpha(A)$ denotes the expected variation (deviation from the initial value) of the state parameter μ_i when the system evoluates in the state A. The measures μ_i^α are assumed to be μ-absolutely continuous and this means that the expected variations $\mu_i^\alpha(A)$ of the values of the state parameters are zero whether it is unprobable that the system will move in the state A.[2]

(d) Variations of the state of the points forming the system are reflected by variations of the values of its state parameters. We assume that there are given n real $(X,C)\times(\mathbb{R}^n,\mathbf{B})$-random variables $F_i(x,\lambda_1, \ldots, \lambda_n)$ $(1 \le i \le n)$ (with $\mathbf{B}$ = the family of the Borel subsets of $\mathbb{R}^n$) such that the expected value of μ_i when the system is expected to evoluate in the state A is

$$\mu_i(A)=\mu_i^o(A) + \int_A F_i(x,\bar{\mu}_1^\alpha(E_x),\ldots, \bar{\mu}_n^\alpha(E_x))d\mu(x), \qquad (1 \le i \le n). \qquad (1)$$

where, for x in A and $1 \le i \le n$,

$$\bar{\mu}_i^\alpha(E_x) := \int E_x(y)d\mu_i^\alpha(y)$$

"measures" the variation of the state parameter μ_i due to the aggregated effect of the modification of the state of x over the state of the other points of the system through their connections with x.

The question is *if (and in what conditions concerning the given data involved in the definition of the autonomous evolutive*

---

[2] One may recall that any finite measure on (X,C) can be uniquelly written as a sum of a μ-singular and of a μ-absolutely continuous measure over (X,C) -- see [12, pp. 125-126].

system there are measures $\mu_1, \ldots, \mu_n$ *satisfying the equations* (1) *and how can these measures be computed provided that they exist.* It is our purpose in that follows to give an answer to this question. To this end, for any finite measure ν over (X,C), we denote $\frac{d\nu}{d\mu}$ the Radon-Nikodym derivative of the μ-absolutely continuous part of ν with respect to the probability measure μ (see [11]). With this notation, our problem of "solving" the system of equations (1) *(which is a system of equations with measures as unknown)* may be equivalently restated as follows: *in what conditions (concerning the given data) there are measures* μ_1, $\mu_2, \ldots, \mu_n$ *whose* μ*-singular parts are* $\overset{o}{\mu}_1, \ldots, \overset{o}{\mu}_n$ *respectively and such that*

$$\frac{d\mu_i}{d\mu}(x) = F_i(x, \bar{\mu}_1^{-o}(E_x), \ldots, \bar{\mu}_n^{-o}(E_x)), \quad \mu\text{-a.e. on } X, \ (1 \le i \le n). \quad (2)$$

One may observe that the equations (2) can be restated as

$$\frac{d\mu_i}{d\mu}(x) = F_i(x, \bar{\mu}_1(E_x), \ldots, \bar{\mu}_n(E_x)), \quad \mu\text{-a.e.} \ (1 \le i \le n) \quad (3)$$

where

$$\bar{\mu}_i(E_x) := \int E_x d\mu_i$$

whether one replaces $F_i(x, \lambda_1, \ldots, \lambda_n)$ by $F_i(x, \lambda_1 - \bar{\mu}_1^{o}(E_x), \ldots, \lambda_n - \bar{\mu}_n^{o}(E_x))$ for $i=1,2,\ldots,n$. That is why, we may deal with the problem of solving the system of equations (3) in order to give an answer to our initial problem concerning the existence of solutions for the equations (1) (or, equivalently, (2)). Problems similar with (3) are studied in a different context in [10] and [11] and it is shown there that they are natural generalizations of important classes of differential equations with initial or boundary conditions. The following results are extending in some respects the Theorem II.2.1 in [10], but we are paying a price for these extensions to be possible, namely we restrict μ to be a probability measure (or at least a finite measure).

Theorem: *If the function* $F = (F_1, \ldots, F_n)$ *from* $X \times \mathbb{R}^n$ *to* $\mathbb{R}^n$ *is measurable, bounded and "uniformly continuous" in the sense that* *for any* $\varepsilon > 0$ *there exists* $\delta(\varepsilon) > 0$ *such that*

$$\| F(x,t) - F(x,s) \| < \varepsilon \tag{4}$$

whenever $x \in X$, $s, t \in \mathbb{R}^n$ *and* $\| s - t \| < \delta(\varepsilon)$, *then the system* (3) *has at least one solution, i.e. there exists at least one vector measure* $(\mu_1, \ldots, \mu_n)$ *over* (X, C) *whose coordinates are satisfying the equations* (3).

A natural consequence of this result is the following:

Corollary: *If in the previously stated Theorem the uniform continuity condition is replaced by the following "uniform Lipschitz condition":*

$$\exists\, L \in \mathbb{R}_+ \text{ such that } L.\sup_{x\in X}\{\mu(E_x)\} < 1 \quad \text{and}$$
$$\| F(x,s) - F(x,t) \| < L \| s - t \| \quad (\forall x \in X,\ \forall s,t \in \mathbb{R}^n); \tag{5}$$

then the solution of (3) *exists and it is unique. Moreover, in this case, the unique solution of* (3) *can be uniformly approximated by the sequence of vector measures* $\{(\varphi_1^k, \ldots, \varphi_n^k)\}_{k=0}^{\infty}$ *defined by*

$$\varphi_i^o := \mu_i^o \quad \text{and}$$
$$\varphi_i^{k+1}(A) := \mu_i^o(A) + \int_A F_i(x, \bar{\varphi}_1^k(E_x), \ldots, \bar{\varphi}_n^k(E_x))\,d\mu(x) \quad (k \geq 0) \tag{6}$$

for any $i=1,\ldots,n$ and for any A in C.[3] ∎

3. PROOFS OF THE MAIN RESULT AND OF ITS COROLLARY

In this Section we prove the results stated in Section 2. In order to avoid notational complication we restrict ourself to the proofs in the case n=1 since the further extension to $n > 1$ is obvious.

---

[3] Here, as everywhere in that follows, we denote $\bar{\nu}$ the extension the measure ν over C to the set of all ambiguous states; that the function $\bar{\nu}(S) := \int S(x)\,d\nu(x)$ for every S in C^*.

In other words, we intend to show that *if* $(X,\mathcal{C},\mu)$ *is a probability space,* E *is an* $(X,\mathcal{C}) \times (X,\mathcal{C})$*-measurable* $[0,1]$*-valued function and* F *is a bounded* $(X,\mathcal{C}) \times (\mathbb{R},\mathcal{B})$*-random variable subjected to the "uniform continuity" condition (4), then there exists a finite measure* φ *whose* μ*-singular part is exactly the given* μ*-singular finite measure* μ^o $(:=\mu_1^o)$ *and such that the following equation is satisfied*

$$\frac{d\varphi}{d\mu}(x) = F(x,\bar{\varphi}(E_x)) \qquad \mu\text{-a.e.} \tag{7}$$

(or, equivalently, the following equation is satisfied

$$\varphi(A) = \mu^o(A) + \int_A F(x,\bar{\varphi}(E_x))d\mu(x) \qquad (\forall\ A \in \mathcal{C})) \tag{8}$$

In the proof we use some notions, notations and results of the Theory of Fuzzy Measures as developed in [17], [15], [14] and in author's works [2]-[6]. That is why, we start by recalling some of these facts.

(i) Let $\mathcal{C}^*$ be the family of all the ambiguous states (i.e., of all measurable $[0,1]$-valued functions on X). This is a σ-additive class of fuzzy sets (in the sense of [2]) and it is closed under the following operations:

$$A \oplus B := \min(1;\ A + B),\ A \ominus B := \max(0;\ A + B - 1),$$

$$A \triangle B := \| A - B \| .$$

(ii) A a function $\bar{\nu}$ from $\mathcal{C}^*$ to $\mathbb{R}$ is a *fuzzy measure* (on $\mathcal{C}^*$) iff there exists a real measure ν_* on $\mathcal{C}$ such that

$$\bar{\nu}(S) = \int S(x)d\nu_*(x)$$

for every S in $\mathcal{C}^*$ (and in this case, ν_* is defined by $\nu_*(A) := \bar{\nu}(1_A)$ $(\forall\ A \in \mathcal{C})$) - cf. [15], [14] or [6, Theorem 2.6]. We denote **M** the set of all fuzzy measures over $\mathcal{C}^*$. This is a Banach space provided with the algebraic operations induced from $\mathbb{R}$ and with the usual variation norm $\|\bar{\nu}\|$. It is shown in [4] that, if $\bar{\nu}$ is a fuzzy measure, then the function

$$\bar{\nu}^*(S) = \sup\{ \sum_{i=1}^{n} \|\bar{\nu}(S_i)\|;\ S_1,\ldots,S_n \in C^*,\ \bigoplus_{i=1}^{n} S_i = S \} \tag{9}$$

is non-negative fuzzy measure and $\|\bar{\nu}\| = \bar{\nu}^*(\mathbf{1})$ (with $\mathbf{1}$:= the constant 1 ambiguous state).

(iii) Let $\bar{\nu}$ be a fuzzy measure. One say that it is *μ-absolutely continuous (μ-singular)* iff $\bar{\nu}^*_*$ is μ-absolutely continuous (μ-singular) (see [3]). We denote **AC** the set of all μ-absolutely continuous fuzzy measures in **M** and one may easily see that this is a Banach subspace of **M.** Also, an immediate consequence of the theorem of Radon-Nikodym is that any fuzzy measure can be uniquelly written as a sum of a μ-singular and of a μ-absolutely continuous fuzzy measure.

With these in mind we start the proof of the theorem stated above.

Claim I: If φ is a fuzzy measure, then the real function $u[\varphi](x) := \varphi(E_x)$ *is* (X,C)*-measurable and*

$$\|\varphi(A) - \varphi(B)\| \le \varphi^*(A \,\Delta\, B), \tag{10}$$

for every A *and* B *in* C^*.

The measurability of $u[\varphi]$ follows from Fubini's theorem since our function is an integral of the measurable and bounded function E. In order to prove (10), we first consider that φ is non-negative. Then,

$$\varphi(A) - \varphi(B \ominus A) = \varphi(A \oplus (B \ominus A)) = \varphi(\max(A;B)) \ge \varphi(B)$$

and

$$\varphi(B) - \varphi(A \ominus B) = \varphi(B \oplus (A \ominus B)) = \varphi(\max(A;B)) \ge \varphi(A)$$

since φ is monotonic. Hence,

$$\|\varphi(A) - \varphi(B)\| \le \varphi(A \ominus B) + \varphi(B \ominus A) = \varphi(A \,\Delta\, B) = \varphi^*(A \,\Delta\, B)$$

and (10) holds in this case. If φ is not non-negative, then it can be represented as $\varphi = \varphi^{+} - \varphi^{-}$ where φ^{+} and φ^{-} are the non-negative fuzzy measures

$$\varphi^{+} := \frac{1}{2}(\varphi^{*} + \varphi) \quad \text{and} \quad \varphi^{-} := \frac{1}{2}(\varphi^{*} - \varphi).$$

Thus, we have

$$|\varphi(A) - \varphi(B)| \le |\varphi^{+}(A) - \varphi^{+}(B)| + |\varphi^{-}(A) - \varphi^{-}(B)|$$

$$\le \varphi^{+}(A \,\Delta\, B) + \varphi^{-}(A \,\Delta\, B) = \varphi^{*}(A \,\Delta\, B)$$

and our claim is proved.

Claim II: Consider X *provided with the topology induced on it by the semi-metric* $\rho(x,y) := \bar{\mu}(E_x \,\Delta\, E_y)$. *If* φ *is in* **AC**, *then the function* $u[\varphi]$ *defined above is* ρ*-continuous.*

Indeed, for ever $\varepsilon > 0$, there exists a number $\delta(\varepsilon) > 0$ such that $\varphi^{*}(C) < \varepsilon$ whenever $\bar{\mu}(C) < \delta(\varepsilon)$ (that because φ is into **AC**). By consequence, we have

$$|u[\varphi](x) - u[\varphi](y)| = |\varphi(E_x) - \varphi(E_y)| \le \varphi^{*}(E_x \,\Delta\, E_y) < \varepsilon$$

whenever $\rho(x,y) < \delta(\varepsilon)$, since the first claim is proved to be true.

Let $\mathcal{H}$ be the family of all the functions $u[\varphi]$ with φ into **AC**. Since

$$a.u[\varphi] + b.u[\psi] = u[a.\varphi + b.\psi]$$

whenever $a, b \in \mathbb{R}$ and $\varphi, \psi \in$ **AC**, it follows that $\mathcal{H}$ is a linear space of ρ-continuous functions. This space is provided with the supremum norm

$$\|u\| := \sup_{x \in X} \{|u(x)|\}.$$

The norm $\|u[\varphi]\|$ can not exceed $\|\varphi\|$ by vertue of the already proved Claim II and of the properties of the fuzzy measures envisaged in [6]. Hence, the supremum norm is finite over $\mathscr{H}$. It is clear that for every φ in **AC** the function $u[\bar{\mu}^{o} + \varphi]$ is measurable. Thus, the function which assigns to x in X the real value $F(x,u[\bar{\mu}^{o}+\varphi](x))$ is integrable (being bounded and measurable over X) and the fuzzy measure

$$w_{\varphi}(A) := \int_X A(x).F(x,\ u[\bar{\mu}^{o} + \varphi](x))d\mu(x)$$

is well-defined and contained into **AC**. Hence the operator $\mathbb{A}$

$$(\mathbb{A}u[\varphi])(x) := u[w_{\varphi}](x) \quad (x \in X,\ \varphi \in \mathbf{AC})$$

is well defined from $\mathscr{H}$ into $\mathscr{H}$. let $\mathscr{A}$ be the range of $\mathbb{A}$.

Claim III: $\mathscr{A}$ is an uniformly bounded family of functions in the normed linear space $\mathscr{H}$.

Indeed, one can easily see that for every u in $\mathscr{H}$, $\|\mathbb{A}u\|$ can not be greather than $K.\mu(X) < \infty$, where

$$K := \sup_{(x,t)\in X\times\mathbb{R}} \{\|F(x,t)\|\}.$$

Claim IV: The family of real functions $\mathscr{A}$ is equicontinuous.

Indeed, by the definition of $\mathbb{A}$ and the proved Claim I, we deduce that

$$\|\mathbb{A}u(x) - \mathbb{A}u(y)\| \le \|w_{\varphi}(E_x) - w_{\varphi}(E_y)\| \le w^{*}_{\varphi}(E_x \wedge E_y) \qquad (11)$$

whenever x, y are contained in X, and $u=u[\varphi]$ for some φ in **AC** . Now, if one consider the non-negative and μ-absolutely continuous fuzzy measure ω_{φ} defined by

$$\omega_{\varphi}(A) := \int_X A(x).\|F(x,\ u[\bar{\mu}^{o}+\varphi](x))\|d\mu(x) \qquad (A \in \mathbf{C}^{*})$$

one may observe that

$$\sum_{i=1}^{n} \| w_{\varphi}(A_i) \| \leq \sum_{i=1}^{n} \omega_{\varphi}(A_i)$$

for any finite family $A_1, \ldots, A_k$ of ambiguous states and this implies that

$$w^*_{\varphi}(A) \leq \omega_{\varphi}(A)$$

for any $A \in \mathbf{C}^*$ (cf. (9)). Hence, taking into account (11), we get

$$\| \mathbb{A}u(x) - \mathbb{A}u(y) \| \leq w^*_{\varphi}(E_x \Delta E_y) \leq \omega_{\varphi}(E_x \Delta E_y)$$

where, by the definition of ω_{φ}, it is clear that

$$\omega_{\varphi}(E_x \Delta E_y) \leq K.\rho(x,y)$$

for every x and y in X. Combining the last two relations, one obtain that

$$\| \mathbb{A}u(x) - \mathbb{A}u(y) \| \leq K.\rho(x,y)$$

for every x and y in X and and for anu u in $\mathcal{K}$ and this implies that $\mathcal{A}$ is equicontinuous.

Since $\mathcal{A}$ is uniformly bounded and equicontinuous, one may use the results in [15, Theorem 4.1 combined with Lemma 2.2] in order to deduce that $\mathcal{A}$ is a relative compact subset of $\mathcal{K}$ and, by consequence, cl($\mathcal{A}$) is compact in $\mathcal{K}$.

Let $\mathbb{V}$ be the operator from $\mathcal{A}$ to **AC** defined by

$$(\mathbb{V}u)(A) = \int_X A(x).F(x, \bar{\mu}^{O}(E_x) + u(x))d\mu(x) \quad , \quad (\forall A \in \mathbf{C}^*) \qquad (12)$$

Using the uniform continuity property of F, it is not difficult to see that the operator $\mathbb{V}$ is continuous over $\mathcal{K}$. Let us denote $\mathcal{V}$ the image of cl($\mathcal{A}$) by $\mathbb{V}$. Since cl($\mathcal{A}$) is compact and $\mathbb{V}$ is continuous, it follows that $\mathcal{V}$ is a compact subset of **AC**. Then, by Mazur's theorem, we have that conv($\mathcal{V}$), the convex envelope of $\mathcal{A}$ into **AC** is also compact into **AC**. Now, if we consider the operator $\mathbb{P}\varphi := \mathbb{V}u[\varphi]$ from **AC** into itself, this is continuous (since $\mathbb{V}$ is continuous and the function $\varphi \longmapsto u[\varphi]$ is also continuous because $\|u[\varphi] - u[\psi]\| \le \|\varphi - \psi\|$ for any two fuzzy measures φ and ψ). By the definition of $\mathbb{P}$ it is clear that $\mathbb{P}$ transforms conv($\mathscr{A}$) into a subset of $\mathscr{V}$. Hence, using Shauder's fixed point theorem for the restriction of $\mathbb{P}$ to the compact set conv($\mathscr{A}$), we deduce that there is a fixed point of $\mathbb{P}$ contained in conv($\mathscr{A}$). Let ξ be this fixed point and let us denote

$$\bar{\varphi} := \bar{\mu}^{o} + \xi .$$

Then

$$\bar{\varphi}(A) = \bar{\mu}^{o}(A) + \xi(A) = \bar{\mu}^{o}(A) + \mathbb{P}\xi(A)$$

$$= \bar{\mu}^{o}(A) + \int_X A(x).F(x,u[\bar{\mu}^{o}+\xi](x))d\mu(x) , \qquad (\forall A \in C^*)$$

and this implies that the measure φ over C defined by $\varphi := \bar{\varphi}_*$ satisfies the equation (8). Thus, the proof of the Theorem is complete. ∎

Now, we are in position to prove the already stated Corollary. To this end, one may observe that the uniform Lipschitz condition (5) implies the uniform continuity condition (4) and thus, in the hypothesis of the Corollary there exists a solution for (7) (and (8)). It remains to show that this solution is unique and it can be uniformly approximated by the sequence of measures $\{\varphi^k\}_{k=0}^{\infty}$ where

$$\varphi^{o} := \mu^{o} \quad \text{and}$$

$$\varphi^{k+1}(A) := \mu^{o}(A) + \int_A F(x, \bar{\varphi}^{k}(E_x))d\mu(x) , \qquad (\forall A \in C) \tag{13}$$

for every $k \ge 0$. In order to prove that, observe that if φ is a solution of (8), then the fuzzy measure $\bar{\varphi}^{\alpha}$ is necesarilly a fixed point of the operator $\mathbb{P}$ defined above. Or, this implies that the function $u := u[\bar{\varphi}^{\alpha}]$ is necesarilly a fixed point of $\mathbb{A}$ whether φ^{α} denotes the μ-absolutely continuous part of the measure φ.

Claim VI: Under the hypothesis of the Corollary the operator $\mathbb{A}$ *from* **AC** *into itself has an unique fixed point.*

Indeed, the operator $\mathbb{A}$ has the fixed point $u' := u[\bar{\varphi}]$ where φ is the solution of (8) (and of (7)). If $u'' := u[\psi]$ is another fixed point of $\mathbb{A}$ into **AC**, then we have

$$\|u'(x) - u''(x)\| = \|\mathbb{A}u'(x) - \mathbb{A}u''(x)\|$$

$$\leq \int_X E_x(y).\|F(y,\bar{\mu}^o(E_y)+u'(y)) - F(y,\bar{\mu}^o(E_y)+u''(y))\|d\mu(y)$$

$$\leq L.\bar{\mu}(E_x).\|u'(x) - u''(x)\| \leq L.\sup_{x\in X}\{\bar{\mu}(E_x)\}.\|u' - u''\|.$$

Thus,

$$\|u' - u''\| \leq H.\|u' - u''\|$$

where $H:=L.\sup_{x\in X}\{\bar{\mu}(E_x)\} < 1$. This inequality holds iff $u''=u'$. Hence, the fixed point of $\mathbb{A}$ is unique. In this case, let us assume that φ and ψ are solutions of (8). Then, their μ-singular parts coincide with μ^o and their μ-absolutely continuous parts are

$$\varphi^{\alpha}(A) = \int_A F(x,u[\bar{\varphi}](x))d\mu(x) = \int_A F(x,u[\bar{\psi}](x))d\mu(x) = \psi^{\alpha}(A)$$

for every A in **C**. Hence,

$$\varphi = \mu^o + \varphi^{\alpha} = \mu^o + \psi^{\alpha} = \psi$$

since the Radon-Nikodym decomposition of any measure (in a μ-singular and a μ-absolutely continuous part) is unique. Hence, there is at most one solution of (8) in this case.

At this stage, we want to show that the measures defined at (13) are converging to the solution φ of (8) (in the norm of **M**). To this end, we consider the functions

$$u = u[\bar{\varphi}^{\alpha}] \text{ and } u_k = u[\overline{(\varphi^k)^{\alpha}}], \quad (k \geq 0),$$

where, as usual, for a measure ψ, ψ^{α} denotes its μ-absolutely continuous part and $\bar{\psi}^{\alpha}$ denotes the corresponding fuzzy measure of ψ^{α}. It is clear that $u_k = \mathbb{A}u_{k-1}$ for every $k \geq 1$. Thus,

$$\|u_k - u^*\| \leq \frac{H^k}{1-H}\|u_1 - u_o\|$$

(with H defined above) where $u^* := u[\varphi]$ since $\mathbb{A}$ is contractive. Tacking into account that $H < 1$, this implies that $\{u^k\}_{k=o}^{\infty}$ converges to u^* in the norm of $\mathcal{H}$. Since the operator $\mathbb{V}$ is continuous, it follows that $\mathbb{V}u_k \longrightarrow \mathbb{V}u^*$ as $k \longrightarrow \infty$ and this implies that

$$\overline{\mathbb{P}(\varphi^k)^{\alpha}} \longrightarrow \mathbb{P}\bar{\varphi}^{\alpha} = \bar{\varphi}^{\alpha} \quad \text{as} \quad k \longrightarrow \infty$$

and, by the definition of $\mathbb{P}$, this means that $(\varphi^k)^{\alpha} \longrightarrow \varphi^{\alpha}$. Since all the measures involved have the same μ-singular component μ^{σ}, this implies that $\varphi^k \longrightarrow \varphi$ and the Corollary is completely proved.

4. APPLICATIONS

In this Section several potential applications of the concepts and results intruduced above are presented.

(I) A question arising in the context of many reliability theoretical problems is the following: *What variations are to be expected for the values of the "outputs" of a system when there is an a priori known risk that specific sets of its components may go out of order ?* A similar question arises when one tries to realize complex hardware and one has to find methods of identifying "failing subsystems" by the specific variations of the outputs of the system when the failing components are malfunctioning.

In order to answer this question one may consider that the given system is represented as a finite set X of interconnected "points" (devices) whose functioning is interdependent. Let $\mu(\{x\})$ be the (usualy known) probability that the point x will fail working. The values $\mu(\{x\})$ $(x \in X)$ are inducing a probability measure μ on C $:= \mathcal{P}(X)$ (= the set of all subsets of X) defined by

$$\mu(A) := \sum_{\alpha \in A} \mu(\{\alpha\})$$

whenever A is a subset of X. It is clear that $\mu(A)$ is the probability that one of the points in A will fail working. Usually, when a point y of the system is going out of order the functioning of other components of the system is affected in less or greater extents. Let us denote by $E(x,y)$ the degree in which the point x of the system will be affected when the point y goes out of order. In other words, $E(x,y)$ is the extent in which the functioning of x is reduced whether y goes out of order. Let μ_1, ..., μ_n be the output parameters (state parameters) whose variations determined by the going out of order of the components of the system are of interest. One may assume that their values in "normal" conditions are tacken as basic levels of the variations, that is $\mu_i^o \equiv 0$ $(1 \le i \le n)$. It happens currently in practice that there are "statistic methods" of determining the "effects" of going out of order of individual points x on the state parameters of the system. Precisely, one may assume that there are n known functions $F_i(x,\lambda_1, \ldots, \lambda_n)$ $(1 \le i \le n)$ such that $F_i(x, \bar{\mu}_1(E_x), \ldots, \bar{\mu}_n(E_x))$ is the variation of the state parameter μ_i when the point x is going out of order.Now, it is clear that the expected variation of the values of the state parameters μ_i in the event that one of the points in the set A fails to work is

$$\mu_i(A) = \sum_{x \in A} F_i(x,\bar{\mu}_1(E_x), \ldots, \bar{\mu}_n(E_x)).\mu(\{x\}) \tag{14}$$

and this is equivalent with the system (3) in this specific case. The Theorem and its Corollary proved above give sufficient conditions in which the (usually large) system (14) has solution and point out a method of solving it. (Note that it happens frequently that the system (14) is not linear).■

(II) A question arising in Economics is that of *how to evaluate the expected revenues from a "production process" (productive economy) in which various commodities may be involved in the production of other commodities ?* The difficulty of the problem is generated by the fact that, even if one knows the way in which the expected revenues from producing specific commodities are depending on the demand, one has to take into account that the demand itself depend on the extent in which the specific commodity will be involved in the production of other commodities inside of the given "production process" and thus, the revenue which may be expected from producing a given set of commodities is essentially dependent on the revenue which is expected from producing the commodities in the production of which the given set of commodities is involved[4]. In that follows we intend to suggest a method of solving this problem.

Consider a "productive economy" which can "generate" a known set X of commodities (in the sense that it can produce any commodity in X and to make it available to the production of the other commodities in X if needed). At each stage of the production process a set $A \subseteq X$ of commodities will be produced. The set of all subsets A of commodities which can be produced at a given stage of the production generates a σ-algebra C of subsets of X. It is known that each commodity x in X may be used in order to produce a set $E_x \in C$ of other commodities. Let us denote by $\mu(A)$ the probability that a commodity x contained into the set $A \in C$ of commodities will be demanded (in order to generate the commodities in E_x). We denote by $\varphi(A)$ the expected revenue from the production of the set of commodities $A \in C$ and we assume that there are no revenues which can be obtained from producing sets of commodities which are not demanded inside the productive economy; that is, $\varphi(A) = 0$ whether $\mu(A) = 0$. Also, we assume that φ is a measure over C, i.e. the production of disjoint subsets of commodities will lead to cummulated revenues. With these notations, $\varphi(E_x)$ represents the revenue expected from the production of the set of commodities E_x into which the commodity x is involved. The "price" which one may expect to be payed for the commodity x when it is produced is essentially dependent on the nature of the commodity x and on the way in which this commodity can be used for the production of other commodities. In other words, the revenue which one may expect to obtain for a commodity x is dependent not only on the nature of x, but also on the revenues which are further expected from using x in the production process. Formally, we

---

[4] For economic details, see P. Sraffa, "Production of Commodities by Mean of Commodities", Cambridge University Press, 1960.

assume that there is a known function $F(x, \lambda)$ which assigns to every x the revenue which may be expected for it whether it can be used in order to generate other commodities with an expected revenue λ. Hence, the expected revenue from the production of a set of commodities A is naturally expressed by the formula

$$\varphi(A) = \int_A F(x, \varphi(E_x)).d\mu(x) \tag{15}$$

and this is an equation in which the measure φ over C is the unknown. If the function F is sufficiently continuous (in the sense that there are no "jumps" of the expected revenues when the expected revenues from the involvement in the production of other commodities of a given commodity are varying in specific limits), then the equation (15) has a solution which can be effectivelly computed .■

ACKNOWLEDGEMENTS

The author is grateful to Professor Adolf Haimovici for encouraging him to write this paper and for helpful comments concerning its content.

REFERENCES

[1] Aumann, R.J. and Shapley, L.S.: "Values of Non-Atomic Games", *Princeton University Press, 1974.*

[2] Butnariu, D.: "Additive Fuzzy Measures and Integrals", *J. of Math. Anal. Appl., 93, (1983), 436-452.*

[3] Butnariu, D.: "Decompositions and Range of Additive Fuzzy Measures", *Fuzzy Sets and Systems, 10, (1983), 135-155.*

[4] Butnariu, D.: "Fuzzy Measurability and Integrability", *J. of Math. Anal. Appl., 117, (1986), 385-410.*

[5] Butnariu, D.: "Non-Atomic Fuzzy Measures and Games", *Fuzzy Sets and Systems, 17, (1985), 39-52.*

[6] Butnariu, D.: "Values and Cores of Fuzzy Games With infinitely Many Players", *Int. J. of Game Theory, 16, 1, (1987), 43-68.*

[7] Butnariu, D.: "A Note On The measurability Of The Fuzzy Mappings", *in: "Devlopments of the Theory of Fuzzy Sets" edited by A. Kandel, New Orleans, 1986.*

[8] Dubois, D. and Prade, H.: "Fuzzy Sets and Systems - Theory and Applications", *Academic Press, 1980.*

[9] Eilenberg, S.: "Theory of Automata Languages and Machines", *Academic Press, 1974.*

[10] Haimovici, A.: "Une Généralisation de la Notion d'Équation Diffèrentielle", *An. St. ale Univ. "Al. I. Cuza"-Iasi, Serie I, (1963), 377-382.*

[11] Haimovici, A.: "Differential Equations With Set-Functions As Unknowns", *The Publishing House of the Romanian Academy of*

Sciences, 1976.
[12] Halmos, P.R.: "Measure Theory", *Springer Verlag, 1974.*
[13] Kandel, A. and Lee, S.C.: "Fuzzy Switching and Automata Theory and Applications", *Crane Russak, New York, 1985.*
[14] Klement, E.P.: "Characterisation of Fuzzy Measures Constructed by Mean of Triangular Norms", *J. Math. Anal. Appl., 86, (1982), 345-358.*
[15] Klement, E.P., Lowen, R. and Schwyhla, W.: "Fuzzy Probability Measures", *Fuzzy Sets and Systems, 5, (1981), 21-30.*
[16] Myers, B.S.: "Equicontinuous Sets of Mappings", *Annals of Mathematics, 47, (1946), 496-502.*
[17] Zadeh, L.A.: "Fuzzy Sets", *Information and Control, 8, (1965), 338-352.*
[18] Zadeh, L.A.: "Probability Measures of Fuzzy Events", *J. Math Anal. Appl., 23, (1968), 421-427.*■

Fuzzy Logic in Knowledge-Based Systems,
Decision and Control
M.M. Gupta and T. Yamakawa (Editors)

USE OF FUZZY SET THEORY IN INTRODUCING GRADED CONSEQUENCE IN MULTIPLE VALUED LOGIC

M. K. CHAKRABORTY*

Key Words : Fuzzy Relation, Multi-valued logic,consequence relation, Conclusion Relation.

I. INTRODUCTION

The purpose of this paper is to explore the possibility of introducing the notion 'graded consequence' in a multiple valued logic.

If in two-valued logic, logical consequence is a relation between a set of statements called premisses and a statement called conclusion there is no theoretical barrier in extending this relation to a fuzzy relation (4) and in calling the grade of relatedness 'the strength of derivation'. There has been occassional mentioning of the necessity of this 'strength' in recent literature (see 3,5,6,9,10). Buehler in (3) has dealt with this notion to a good detail but keeping in mind the need of relating logical conclusions to inference procedure in biological experimentations.

Our aim is to search for some theoretical foundations to such consequences. The most natural starting point is the axiomatization of the corresponding notion in two-valued situation. This started from Tarski (12) and has taken a very concrete shape in Shoesmith and Smiley (11) who have extened the notion further to multiple conclusion logic. In section II we shall build up the general mathematical framework and in the next section shall state our main proposal and provide justifications for it.

We have used some basic ideas of fuzzy set theory obtainable in neumerous literature on this subject e.g. in Kaufmann (8).

II. MATHEMATICAL FRAMEWORK

Let V be the universe which is taken as non empty. Subsets of V will be denoted by letters X,Y,Z... etc. and the elements of V by A,B,C,... etc.

DEFINITION 1 Conclusion relation :

A fuzzy relation from the power set of V to V is called a conclusion relation on V.

A conclusion relation will be denoted by $\mid\!\sim$ and the grade in which a subset X of V is related to an element B of V will be denoted by $gr(X \mid\!\sim B)$.

*Department of Pure Mathematics, Calcutta University, 35 Ballygunge Circular Road, Calcutta - 700 019, India.

The ideas in this paper are the outcome of the author's research when he was a Fellow of the Indian Institute of Advanced Study, Shimla, India, from April to June, 1986. It was complete and finalised during the author's stay at RWTH, Aachen, W. Germany as a DAAD Fellow in the same year.

DEFINITION 2 A conclusion relation $\vdash\!\sim$ is said to be an α-consequence relation if and only if it satisfies the following conditions :

I. α-overlap : If $B \in X$ then $gr(X \vdash\!\sim B) \geqslant \alpha$, $\alpha > o$,

II. Dilution : If $X' \subseteq X$ then $gr(X' \vdash\!\sim B) \leqslant gr(X \vdash\!\sim B)$,

III. Cut : $gr(X \vdash\!\sim B) \geqslant gr(X \cup Z \vdash\!\sim B) \wedge \inf_{A \in Z} gr(X \vdash\!\sim A)$,

for any subsets X,Z of V and any element B.

NOTE Conditions I,II and III are generalizations of the similar conditions for consequence relations in ordinary two valued logic (See Shoesmith and Smiley).

We define below two more cut-conditions and transitivity.

DEFINITION 3 Cut for finite set :

$$gr(X \vdash\!\sim B) \geqslant gr(X \cup Z \vdash\!\sim B) \wedge \min_{A \in Z} gr(X \vdash\!\sim A)$$

where Z is a finite set.

DEFINITION 4 Cut for an element :

$$gr(X \vdash\!\sim B) \geqslant gr(X \cup A \vdash\!\sim B) \wedge gr(X \vdash\!\sim A)$$

where A is an element of V.

DEFINITION 5 Transitivity :

$$gr(X \vdash\!\sim B) \geqslant gr(X \vdash\!\sim A) \wedge gr(A \vdash\!\sim B).$$

PROPOSITION 1 $gr(B \vdash\!\sim B) \geqslant \alpha$ for an α-consequence relation $\vdash\!\sim$.

PROPOSITION 2 Cut for an element and dilution imply transitivity.

PROPOSITION 3 If $\vdash\!\sim$ satisfies dilution then cut for an element implies cut for finite set and conversely.

Proof is by applying induction on the number of elements in Z.

DEFINITION 6 A conclusion relation $\vdash\!\sim$ is compact if and only if

$$gr(X \vdash\!\sim A) \leqslant \sup_{\substack{X' \subseteq X \\ X' \text{ is finite}}} gr(X' \vdash\!\sim A).$$

NOTE If in the definition 6, the supremum is attained then

$$gr(X \vdash\!\sim A) \leqslant gr(X' \vdash\!\sim A) \text{ for some finite } X' \subseteq X.$$

We can define a notion of 'strong compactness' as follows :

$$gr(X \vdash\!\sim A) = gr(X' \vdash\!\sim A) \text{ for some finite } X' \subseteq X.$$

When $\vdash\!\sim$ satisfies dilution and is compact,

$$gr(X \vdash\!\sim A) = \sup_{\substack{X' \subseteq X \\ X' \text{ finite}}} gr(X' \vdash\!\sim A).$$

If we further assume that this sup is attained, the notions of compactness and strong compactness coincide.

PROPOSITION 4 Let $\mid\!\sim$ be a compact conclusion relation and satisfies dilution. Then

Cut for finite set implies cut.

PROOF. By compactness,

$$(1) \ldots\ldots gr(X \cup Z \mid\!\sim B) = \sup_{\substack{X' \subseteq X, X' \text{ finite} \\ Z' \subseteq Z, Z' \text{ finite}}} gr(X' \cup Z' \mid\!\sim B).$$

By dilution,

$$gr(X' \cup Z' \mid\!\sim B) \leqslant gr(X \cup Z' \mid\!\sim B) \text{ for each } X' \subseteq Z.$$

Hence

$$(2) \ldots\ldots \sup_{\substack{X' \subseteq X, X' \text{ finite} \\ Z' \subseteq Z, Z' \text{ finite}}} gr(X' \cup Z' \mid\!\sim B) \leqslant \sup_{Z' \subseteq Z, Z' \text{ finite}} gr(X \cup Z' \mid\!\sim B).$$

By hypothesis, for each finite $Z' \subseteq Z$

$$gr(X \mid\!\sim B) \geqslant gr(X \cup Z' \mid\!\sim B) \wedge \min_{A \in Z'} gr(X \mid\!\sim A)$$

$$\geqslant gr(X \cup Z' \mid\!\sim B) \wedge \inf_{A \in Z} gr(X \mid\!\sim A) \text{ because } Z' \subseteq Z.$$

So,

$$gr(X \mid\!\sim B) \geqslant \sup_{Z' \subseteq Z} gr(X \cup Z' \mid\!\sim B) \wedge \inf_{A \in Z} gr(X \mid\!\sim A)$$

$$\geqslant \sup_{\substack{X' \subseteq X, X' \text{ finite} \\ Z' \subseteq Z, Z' \text{ finite}}} gr(X' \cup Z' \mid\!\sim B) \wedge \inf_{A \in Z} gr(X \mid\!\sim A), \text{ by (2)}$$

$$= gr(X \cup Z \mid\!\sim B) \wedge \inf_{A \in Z} gr(X \mid\!\sim A), \quad \text{by (1).}$$

Hence the proposition is proved.

COROLLARY Granted dilution, cut for an element is equivalent to Cut.

DEFINITION 7 For any conclusion relation $\mid\!\sim$, let a relation $\mid\!\sim'$ be defined as follows

$$gr(X \mid\!\sim' A) = gr(X \mid\!\sim A) \wedge \sup_{\substack{X' \subseteq X \\ X' \text{ finite}}} gr(X' \mid\!\sim A).$$

In the following propositions we get some properties of $\mid\!\sim'$.

PROPOSITION 5 If $\mid\!\sim$ satisfies dilution, then

$$gr(X \mid\!\sim' A) = \sup_{X' \subseteq X, X' \text{ finite}} gr(X' \mid\!\sim A).$$

PROPOSITION 6 (i) If $\mid\!\sim$ satisfies α-overlap then $\mid\!\sim'$ satisfies the same.

(ii) If $\mid\!\sim$ satisfies dilution then $\mid\!\sim'$ satisfies the same.

(iii) If $\mid\!\sim$ satisfies Cut then $\mid\!\sim'$ satisfies the same.

PROOF (i) Let $A \in X$. Then $gr(X \mid\!\sim A) \geqslant \alpha$.

Also, $gr(A \mid\!\sim A) \geqslant \alpha$ and hence $\sup_{\substack{X' \subseteq X \\ X' \text{ finite}}} gr(X' \mid\!\sim A) \geqslant \alpha$.

So, $$gr(X \mid\!\sim' A) = gr(X \mid\!\sim A) \wedge \sup_{\substack{X' \subseteq X \\ X' \text{ finite}}} gr(X' \mid\!\sim A) \geqslant \alpha .$$

(ii) Let $\mid\!\sim$ satisfies dilution. Let $Y \subseteq X$.

Then $$gr(X \mid\!\sim' A) = \sup_{\substack{X' \subseteq X, \\ X' \text{ finite}}} gr(X' \mid\!\sim A) \qquad \text{(prop.5)}$$

$$\geqslant \sup_{\substack{Y' \subseteq Y \\ Y' \text{ finite}}} gr(Y' \mid\!\sim A)$$ because the finite subsets of Y are also finite subsets of X.

So, $$gr(X \mid\!\sim' A) \geqslant gr(Y \mid\!\sim' A) . \qquad \text{(prop.5)}$$

(iii) We first prove that

$$\sup_{\substack{X' \subseteq X, \\ X' \text{ finite}}} gr(X' \mid\!\sim B) \geqslant \sup_{\substack{Y' \subseteq X \cup Z \\ Y' \text{ finite}}} gr(Y' \mid\!\sim B) \wedge \inf_{A \in Z} \sup_{\substack{X' \subseteq X \\ X' \text{ finite}}} gr(X' \mid\!\sim A) \quad \ldots(1)$$

If (1) is proved then assertion (iii) can be established as follows.

$$gr(X \mid\!\sim' B) = gr(X \mid\!\sim B) \wedge \sup_{\substack{X' \subseteq X, \\ X' \text{ finite}}} gr(X' \mid\!\sim B)$$

$$\geqslant [gr(X \cup Z \mid\!\sim B) \wedge \inf_{A \in Z} gr(X \mid\!\sim A)] \wedge \sup_{\substack{X' \subseteq X, \\ X' \text{ finite}}} gr(X' \mid\!\sim B)$$

(by the assumption that $\mid\!\sim$ satisfies cut).

Then by (1)

$$gr(X \mid\!\sim' B) \geqslant [gr(X \cup Z \mid\!\sim B) \wedge \inf_{A \in Z} gr(X \mid\!\sim A)] \wedge [\sup_{\substack{Y' \subseteq X \cup Z \\ Y' \text{ finite}}} gr(Y' \mid\!\sim B) \wedge \inf_{A \in Z} \sup_{\substack{X' \subseteq X \\ X' \text{ finite}}} gr(X' \mid\!\sim A)]$$

$$= [gr(X \cup Z \mid\!\sim B) \wedge \sup_{\substack{Y' \subseteq X \cup Z \\ Y' \text{ finite}}} gr(Y' \mid\!\sim B)] \wedge \inf_{A \in Z} [gr(X \mid\!\sim A) \wedge \sup_{\substack{X' \subseteq X \\ X' \text{ finite}}} gr(X' \mid\!\sim A)]$$

$$= gr(X \cup Z \mid\!\sim' B) \wedge \inf_{A \in Z} gr(X \mid\!\sim' A).$$

Now, to prove (1) we proceed as follows.

$$\sup_{\substack{Y' \subseteq X \cup Z \\ Y' \text{ finite}}} gr(Y' \mid\!\sim B) \wedge \inf_{A \in Z} \sup_{\substack{X' \subseteq X \\ X' \text{ finite}}} gr(X' \mid\!\sim A)$$

$$= [\sup_{X' \subseteq X} gr(X' \mid\!\sim B) \vee \sup_{\substack{X' \subseteq X \\ Z' \subseteq Z}\text{ finite}} gr(X' \cup Z' \mid\!\sim B)] \wedge \inf_{A \in Z} \sup_{\substack{X' \subseteq X \\ X' \text{ finite}}} gr(X' \mid\!\sim A)$$

$$= [\sup_{X' \subseteq X} gr(X' \mid\!\sim B) \wedge \inf_{A \in Z} \sup_{X' \subseteq X} gr(X' \mid\!\sim A)] \vee \sup_{X' \subseteq X} [\sup_{Z' \subseteq Z} gr(X' \cup Z' \mid\!\sim B) \wedge \inf_{A \in Z} gr(X' \mid\!\sim A)]$$

$$\ldots\ldots (2)$$

Again

$$\inf_{A\in Z} gr(X' \mid\!\sim A) \leqslant \inf_{A\in Z'} gr(X' \mid\!\sim A), \text{ for each } X' \text{ and } Z'.$$

So,

$$\inf_{A\in Z} gr(X' \mid\!\sim A) \leqslant \sup_{Z'\subseteq Z} \inf_{A\in Z'} gr(X' \mid\!\sim A) \quad \text{for each } X'.$$

So,

$$\sup_{Z'\subseteq Z} gr(X'\cup Z' \mid\!\sim B) \wedge \inf_{A\in Z} gr(X' \mid\!\sim A) \leqslant \sup_{Z'\subseteq Z} gr(X'\cup Z' \mid\!\sim B) \wedge \sup_{Z'\subseteq Z} \inf_{A\in Z'} gr(X' \mid\!\sim A)$$

for each X'.

So,

$$\sup_{X'\subseteq X} [\sup_{Z'\subseteq Z} gr(X'\cup Z' \mid\!\sim B) \wedge \inf_{A\in Z} gr(X' \mid\!\sim A)] \leqslant$$

$$\sup_{X'\subseteq X} [\sup_{Z'\subseteq Z} gr(X'\cup Z' \mid\!\sim B) \wedge \sup_{Z'\subseteq Z} \inf_{A\in Z'} gr(X' \mid\!\sim A)]$$

$$= \sup_{Z'\subseteq Z} [\sup_{X'\subseteq X} (gr(X'\cup Z' \mid\!\sim B) \wedge \inf_{A\in Z'} gr(X' \mid\!\sim A))]$$

$$\leqslant \sup_{X'\subseteq X} gr(X' \mid\!\sim B) \quad \text{because}$$

$$gr(X'\cup Z' \mid\!\sim B) \wedge \inf_{A\in Z'} gr(X' \mid\!\sim A) \leqslant gr(X' \mid\!\sim B) \quad \text{by assumption and hence}$$

$$\sup_{X'\subseteq X} [gr(X'\cup Z' \mid\!\sim B) \wedge \inf_{A\in Z'} gr(X' \mid\!\sim A)] \leqslant \sup_{X'\subseteq X} gr(X' \mid\!\sim B) \text{ for any } Z' \subseteq Z.$$

Hence from (2) we get

$$\sup_{Y'\subseteq X\cup Z} gr(Y' \mid\!\sim B) \wedge \inf_{A\in Z} \sup_{X'\subseteq X} gr(X' \mid\!\sim A) \leqslant$$

$$[\sup_{X'\subseteq X} gr(X' \mid\!\sim B) \wedge \inf_{A\in Z} \sup_{X'\subseteq X} gr(X' \mid\!\sim A)] \vee \sup_{X'\subseteq X} gr(X' \mid\!\sim B)$$

$$= \sup_{X'\subseteq X} gr(X' \mid\!\sim B). \qquad \text{This is (1).}$$

(In the above lines X', Y', Z' are finite subsets of X, Y, Z respectively.)

Hence proposition 6 (iii) is proved.

PROPOSITION 7 $gr(X \mid\!\sim' B) = gr(X \mid\!\sim B)$ when X is finite.

PROPOSITION 8 For any conclusion relation $\mid\!\sim$, $\mid\!\sim'$ defined as above is a compact subrelation.

PROOF. Let X be any set and let X' be any finite subset of X.

Then

$$gr(X' \mid\!\sim' B) = gr(X' \mid\!\sim B) \text{ by prop. 7.}$$

So,

$$\sup_{\substack{X'\subseteq X \\ X' \text{ finite}}} gr(X' \mid\!\sim' B) = \sup_{\substack{X'\subseteq X \\ X' \text{ finite}}} gr(X' \mid\!\sim B) \quad .$$

So,

$$\sup_{\substack{X'\subseteq X\\ X' \text{ finite}}} gr(X' \mathrel{|\!\!\sim'} B) \geqslant gr(X \mathrel{|\!\!\sim} B) \wedge \sup_{\substack{X'\subseteq X\\ X' \text{ finite}}} gr(X' \mathrel{|\!\!\sim} B) = gr(X \mathrel{|\!\!\sim'} B).$$

This completes the proof.

PROPOSITION 9 $\mathrel{|\!\!\sim'}$ is the maximal compact subrelation contained in $\mathrel{|\!\!\sim}$.

PROOF. Let $\mathrel{|\!\!\sim''}$ be any compact subrelation of $\mathrel{|\!\!\sim}$.
We have the following,

(i) $gr(X \mathrel{|\!\!\sim''} B) \leqslant gr(X \mathrel{|\!\!\sim} B)$.

(ii) $gr(X \mathrel{|\!\!\sim''} B) \leqslant \sup_{\substack{X'\subseteq X\\ X' \text{ finite}}} gr(X' \mathrel{|\!\!\sim''} B)$.

(iii) $\sup_{\substack{X'\subseteq X\\ X' \text{ finite}}} gr(X' \mathrel{|\!\!\sim''} B) \leqslant \sup_{\substack{X'\subseteq X\\ X' \text{ finite}}} gr(X' \mathrel{|\!\!\sim} B)$.

Hence from (i) and (iii)

$$gr(X \mathrel{|\!\!\sim''} B) \wedge \sup_{\substack{X'\subseteq X\\ X' \text{ finite}}} gr(X' \mathrel{|\!\!\sim''} B) \leqslant gr(X \mathrel{|\!\!\sim} B) \wedge \sup_{\substack{X'\subseteq X\\ X' \text{ finite}}} gr(X' \mathrel{|\!\!\sim} B) = gr(X \mathrel{|\!\!\sim'} B).$$

So, by (ii)

$$gr(X \mathrel{|\!\!\sim''} B) \leqslant gr(X \mathrel{|\!\!\sim'} B).$$

This completes the proof.

Combining propositions 6,8 and 9, we get

PROPOSITION 10. For any conclusion relation $\mathrel{|\!\!\sim}$, there exists a maximal compact subrelation $\mathrel{|\!\!\sim'}$. If $\mathrel{|\!\!\sim}$ is a consequence relation then $\mathrel{|\!\!\sim'}$ is also a consequence relation.

PROPOSITION 11. Let $\{\mathrel{|\!\!\sim}_i\}$ be a collection of α -consequence relations. Then $\mathrel{|\!\!\sim} = \bigcap_i \mathrel{|\!\!\sim}_i$ is an α -consequence relation.

PROOF. α -overlap and dilution are proved in a straight forward way.

For cut we give the following proof.

$$gr(X \mathrel{|\!\!\sim}_i B) \geqslant gr(X \cup Z \mathrel{|\!\!\sim}_i B) \wedge \inf_{A \in Z} gr(X \mathrel{|\!\!\sim}_i A).$$

Hence,

$$\inf_i gr(X \mid\!\sim_i B) \geqslant \inf_i [gr(XUZ \mid\!\sim_i B) \wedge \inf_{A \in Z} gr(X \mid\!\sim_i A)]$$

$$= \inf_i gr(XUZ \mid\!\sim_i B) \wedge \inf_i \inf_{A \in Z} gr(X \mid\!\sim_i A)$$

$$= \inf_i gr(XUZ \mid\!\sim_i B) \wedge \inf_{A \in Z} \inf_i gr(X \mid\!\sim_i A).$$

So,

$$gr(X \mid\!\sim B) \geqslant gr(XUZ \mid\!\sim B) \wedge \inf_{A \in Z} gr(X \mid\!\sim A).$$

PROPOSITION 12. Any conclusion relation $\mid\!\sim$ can be extended to a minimal 1-consequence relation.

LEMMA. Any ordinary relation R between the subsets X and the elements B of V has an extension $\vdash_R$ that satisfies I,II and III.

PROOF OF THE LEMMA.

Let < X,B > be any general element of R.

Let I be the set of all subsets T of V that satisfy the following condition :

T $\in$ I if and only if $\forall$ < X,B > $\in$ R, X $\subseteq$ T implies B $\in$ T.

It can be easily seen that I is nonempty.

We now define a relation $\vdash_R$ as follows.

"$\vdash_R$ excludes a pair <Y,D> where Y $\subseteq$ V, D $\in$ V, only if for some T $\in$ I, Y $\subseteq$ T and B $\notin$ T".

$\vdash_R$ is obviously an extension of R and satisfies the three conditions (See Smiley).

PROOF OF THE PROPOSITION. We define an ordinary relation R as follows :

<X,B> $\in$ R if and only if $gr(X \mid\!\sim B) > 0$.

By the lemma R can be extended to $\vdash_R$. So, $\mid\!\sim \subseteq R \subseteq \vdash_R$ and $\vdash_R$ satisfies I,II and III. In fact, $\vdash_R$ is a crisp relation which is an extention of $\mid\!\sim$ and which satisfies I,II and III.

Hence by proposition 11, there exists a smallest consequence relation that contains $\mid\!\sim$.

This completes the proof.

NOTATION. The above smallest consequence relation is the intersection of all single consequence extensions of $\mid\!\sim$ and shall be denoted by $\mid\!\sim^c$ and called the closure of $\mid\!\sim$ under 1-overlap, dilution and cut.

Let now $\mid\!\sim$ be any conclusion relation and $\mid\!\sim^c$ be its closure under I,II and III. By Proposition 9, there exists a maximal compact subrelation $\mid\!\sim'$ of $\mid\!\sim^c$ defined according to definition 7. By proposition 6, $\mid\!\sim'$ satisfies I,II and III.

In the general situation $\mid\!\sim'$ can be smaller than $\mid\!\sim$.

But when $\mid\!\sim$ is compact obviously, $\mid\!\sim \subseteq \mid\!\sim' \subseteq \mid\!\sim^{c}$. As $\mid\!\sim^{c}$ is the smallest extension of $\mid\!\sim$ satisfying I,II and III and $\mid\!\sim'$ is also one such extension **of** $\mid\!\sim$, we have $\mid\!\sim' = \mid\!\sim^{c}$. Thus we get the following proposition.

PROPOSITION 13. If $\mid\!\sim$ is compact then its closure $\mid\!\sim^{c}$ is also compact.

III. THE DEFINITION OF CONSEQUENCE IN MULTI-VALUED LOGIC

The results and discussions in section II were general, in the sense that V, the universe, could be any set and $\mid\!\sim$ could be any fuzzy relation, not necessarily the consequence relation of logic.

We now take V as the universe of sentences or formulas supposed to have truth values in [0,1] and propose the following.

"A fuzzy relation $\mid\!\sim$ defined between the subsets of V and the elements of V will be called a consequence relation (in the sense of logical consequence) if and only if it is a 1-consequence relation according to definition 2".

The above is, in fact, the same as the demand of Shoesmith and Smiley (11) in the special case of two valued logic.

In further support to the above claim we shall see in the following passages that the above definition conforms to the following interpretation of consequence in two valued logic :

(B is a consequence of X) iff (for any state of affair, every member of X is true implies B is also true).

In other words

(Ω) (B is a consequence of X) iff ($\forall$ T(X $\subseteq$ T implies B $\in$ T) where T is the ordinary subset of the universe of true propositions determined by a particular state of affair.

In our case as the logic is multiple valued, any state of affair assigns values (Truth values) ranging from 0 to 1 to each element of V. So a particular state of affair can be identified (and represented) by a fuzzy subset T of V. Depending on the argumentative pre-suppositions one may find all possible states of affair (11, p.14). So by all possible states of affair, we shall understand a set I of fuzzy subsets of V - the underlying pre-suppositions are immeterial for our purpose.

Now, in (1,7) a notion of 'grade of inclusion' of fuzzy subsets is defined as follows :

$$gr(A \subseteq B) = \inf_{x \in U} (\mu_A(x) \rightarrow \mu_B(x))$$

where U is the universe, A,B, two fuzzy subsets of U, μ_A, μ_B there respective membership functions and $\rightarrow$ is any fuzzy implication operator.

In our case X is an ordinary finite subset of the universe V of propositions. Let T_i be any particuar state of affair which is a fuzzy subset of V andlet us choose an implication operator $\rightarrow$ for which $0 \rightarrow a = 1$. (Of all the nine implication operators discussed by Bandler and Kohout in (2), only one lacks this property).

So we have,

$$gr(X \subseteq T_i) = \inf_{x \in V} (\mu_X(x) \rightarrow T_i(x))$$

$$= \bigwedge_{x \in X} (\mu_X(x) \rightarrow T_i(x)) \wedge \inf_{x \in \overline{X}} (\mu_X(x) \rightarrow T_i(x))$$

$$= \bigwedge_{x \in X} (\mu_X(x) \rightarrow T_i(x))$$

where $\overline{X}$ is the complement of X in V.

Similarly,

$$gr(B \in T_i) = (\mu_B(B) \rightarrow T_i(B))$$

$$= 1 \rightarrow T_i(B).$$

Let us now choose for the conditional in the second part of (Ω) a fuzzy implication operator => that satisfies the following three conditions viz.

I' : $a \Rightarrow a = 1$

II' : $a \leqslant b$ implies $a \Rightarrow c \geqslant b \Rightarrow c$.

III' : $a \Rightarrow c \geqslant a \Rightarrow b \wedge b \Rightarrow c$.

We have at our disposal at least two such operators viz.

Standard Strict

$$a \Rightarrow b = \begin{cases} 1 & \text{iff } a \leqslant b \\ 0 & \text{otherwise} \end{cases}$$

and

Standard Star

$$a \Rightarrow b = \begin{cases} 1 & \text{iff } a \leqslant b \\ b & \text{otherwise.} \end{cases}$$

The reproduction of (Ω) in the multiple valued situation is the following :

$$gr(X \mid\!\sim B) = \inf_{T_i} \left(\bigwedge_{x \in X} (\mu_X(x) \rightarrow T_i(x)) \Rightarrow (1 \rightarrow T_i(B)) \right).$$

We shall now show that the fuzzy relation $\mid\!\sim$ defined as above satisfies

1-overlap,

dilution

and Cut for an element.

As we restrict ourselves to define $\mid\!\sim$ for finite X only, $\mid\!\sim$ is automatically compact and by proposition 4, $\mid\!\sim$ satisfies cut if cut for an element holds.

That $\mid\!\sim$ satisfies 1-overlap and dilution could be checked easily. For cut for an element we state below a lemma.

<u>LEMMA</u>. Any binary operator =>, satisfying II' satisfies the condition $p \Rightarrow q \geqslant ((p \wedge r) \Rightarrow q) \wedge (p \Rightarrow r)$ if and only if it satisfies III'.

Now,

$$gr(X \mid\!\sim B) = \inf_{T_i}[\bigwedge_{x \in X} (\mu_X(x) \rightarrow T_i(x)) \Rightarrow (1 \rightarrow T_i(B))],$$

$$gr(XUA \mid\!\sim B) = \inf_{T_i}[\bigwedge_{x \in XUA} (\mu_{XUA}(x) \rightarrow T_i(x)) \Rightarrow (1 \rightarrow T_i(B))]$$

and $$gr(X \mid\!\sim A) = \inf_{T_i}[\bigwedge_{x \in X} (\mu_X(x) \rightarrow T_i(x)) \Rightarrow (1 \rightarrow T_i(A))].$$

By putting p for $\bigwedge_{x \in X} (\mu_X(x) \rightarrow T_i(x))$

q for $(1 \rightarrow T_i(B))$

and r for $(1 \rightarrow T_i(A))$

we can easily see using the lemma that

$$p \Rightarrow q \geqslant (p \wedge r \Rightarrow q) \wedge (p \Rightarrow r).$$

And taking the infimum over T_i

$$gr(X \mid\!\sim B) \geqslant gr(XUA \mid\!\sim B) \wedge gr(X \mid\!\sim A).$$

Thus if a consequence relation $\mid\!\sim$ is defined by Ω in the multivalued case for finite premisses, $\mid\!\sim$ must satisfy I, II and III.

<u>CONCLUSION</u>. It may be noted that the cut condition could be defined as follows :

III(a) :

$$gr(X \mid\!\sim B) \geqslant gr(XUZ \mid\!\sim B) \wedge \prod_{A \in Z} gr(X \mid\!\sim A).$$

Keeping all other definitions in tact, we can observe that all the results remain valid if one replaces III by III(a).

In section III accordingly, we have to replace III' by III'(a) which is as follows

$$(p \Rightarrow q) \geqslant (p \Rightarrow r).(r \Rightarrow q).$$

In fact, in the metalinguistic assertion (Ω) the "iff" is a two-valued bi-implication, and the "implies" on the right hand side of iff can be taken as any implication operator => defined by Goguen in a Closg. If * is the binary operator in terms of which => is defined (see Goguen 6), the cut condition should also have to be defined in terms of * . This is not unnatural, because the assertions I,II and III are also metalinguistic.

We have not taken any particular language in the whole discussion. The only assumption is that the sentences have truth values in [0,1]. It will be interesting to investigate the connection between the relation $\mid\!\sim$ and the implication operator, if any, within the object language.

REFERENCES :

1. Bandler W. and Kohout L.J., Fuzzy power sets andfuzzy implication operator, Fuzzy Sets and Systems, 4, 1980, 13-30.

2. ________ Semantics of implication operators and fuzzy relational products, in Fuzzy Reasoning and its Applications, eds, Mamdani, Gaines, Academic Press, 1981, 210-246.

3. Büehler G. Albrecht, Numerical Evaluation of the validity of experimental proofs in Biology, Synthese 33, 1976, 283-312.

4. Chakraborty M.K., Sarkar, S. and Das, M. Some aspects of [0,1] fuzzy relations and a few suggestions towards its use, in Approximate reasoning in Expert Systems, eds, Gupta, Kandel, Bandler, Kiszka, North Holland, 1985, 139-156.

5. Driankov Dimiter, Uncertainty Calculus with verbally defined belief intervals, Technical Report, University of Linkoping, Sweden.

6. Goguen J.A., The logic of inexact concepts, Synthese, 19, 1969, 325-373.

7. Gottwald S., Set theory for fuzzy sets of higher level, Fuzzy Sets and Systems 2, 1979, 125-151.

8. Kaufmann, A. Introduction to the theory of Fuzzy Subsets, Vol.1,Academic Press, New York, 1975.

9. Lakoff, G., Hedges : A study in meaning criteria and the logic of fuzzy concepts, Jour. of Phil. Logic, 2, 1973, 458-508.

10. Machina, K.F., Vague predicates, Amer. Phil. Quarterly, 9, 1972,225-233.

11. Shoesmith, D.J and Smiley, T.J., Multiple Conclusion Logic, Cambridge University Press, 1978.

12. Tarski, A., Methodology of deductive sciences, in Logic, Semantics,Metamathematics, Oxford, 1956.

13. Zadeh, L.A., Fuzzy logic and approximate reasoning (in memory of Grigore Moisil), Synthese, 30, 1975, 407-428.

Fuzzy Logic in Knowledge-Based Systems,
Decision and Control
M.M. Gupta and T. Yamakawa (Editors)

AN AXIOMATIC APPROACH TO FUZZY MULTICRITERIA ANALYSIS

Javier MONTERO

Departamento de Estadística e I.O., Facultad de Ciencias Matemáticas
Universidad Complutense, 28040 Madrid, Spain

In this paper, an axiomatic approach to rational decision making is proposed. In particular, the successive aggregation into a global criterion of the judgements (about each alternative) with respect to all criteria is analyzed for a given multicriteria decision-making problem with weighted criteria. Some aggregation operations are axiomatically justified when such judgements are defined by a desirability function, and an analogous treatment can be developed when judgements are defined by a fuzzy outranking relation.

Keywords: Aggregation, Multicriteria Decision Making, Probabilistic Set.

1. INTRODUCTION

Most formal models of decision-making proceed from simple evaluation of the desirability of each feasible alternative. Such an evaluation summarizes the decision-maker's goal. Classical Decision Theory proposes to find an utility function which assigns an objective value for each alternative, and then the resulting optimization problem must be solved. But though this mathematical problem is well defined, it is not always representative of reality: on the one hand, comparision of alternatives is rarely made according on a single point of view, and even preferences on a single point of view are in many cases modelled with difficulty by an utility function (Roy-Vincke [1]); on the other hand, if we admit a complex problem with multiplicity of objectives, the idea of being the best alternative is not clear, and since we often use imprecise concepts in our analysis, criteria themselves may be of fuzzy nature (Blin [2]).

As pointed out by Saaty [3], the link between fuzziness and multicriteria problems is clear due to the fact that the complexity of experience acquired by our mind is fuzzy (fuzziness in perception, fuzziness in meaning). Moreover, the existence of a natural isomorphism between the class of fuzzy sets and the class of multicriteria decision-making procedures, as proved by Blin [2], shows the deepness of such a link.

In this paper, multiattribute decision-making problems are considered when criteria (or goals) are to be aggregated successively into a single global criterion, in such a way that at each moment we can obtain an aggregated criterion which summarizes the approval of each alternative under any family of criteria, by applying an adecuate operation aggregation.

Several authors have dealt with aggregation operations in different contexts (cf. [4,5,6,7,8]), though the point of departure is usually Bellman-Zadeh's paper [9]. In this article, logical connectives applied to probabilistic sets (Hirota [10]) are to be studied in a more general context than that investigated in Czogala et al. [4] and Czogala-Zimmerman [5].

In any case, aggregation operations will allow us to consider successive new criteria, by amalgamating them one by one. When a particular aggregation operation has been selected as rational for a given decision-making problem, classical methods in decision-making can be applied in order to choose optimal alternatives, by considering the distribution function of a probabilistic set (cf. [4,5,10,11]).

2. GENERALIZED AGGREGATION OPERATIONS

Let X be the set of feasible alternatives in a given decision-making problem. Each alternative can be analyzed under an infinite countable set of criteria D. If we denote by W the space of observable states, we shall suppose that the degree of approval of each alternative for any possible state is defined under each criterion. In other words, an approval function

$$\mu : X \times D \times W \longrightarrow [0,1]$$

assigns the degree of approval $\mu(x,c,w)$ of alternative x under criterion c when w is the true state. If a probability space (W,Δ,P) is defined on W, and the mapping $\mu(x,c,\cdot)$ is supposed to be measurable for any fixed alternative x and any fixed criterion c (that is, $\{w\in W/\mu(x,c,w)\in B\}\in\Delta$ for any Borel set B on the unit interval), then it is clear that a probabilistic set in the sense of Hirota [10]

$$\mu_c : X \times W \longrightarrow [0,1]$$

has been associated to each criterion c, with $\mu_c(x,w)=\mu(x,c,w)$. It is well known that measurability of $\mu(x,c,\cdot)$ holds if and only if

$$\{w\in W/\mu(x,c,w)\leq r\}\in\Delta \quad \forall r\in[0,1]$$

and it must be pointed out that when the probability measure P is defined on the family of all subsets of W, $\Delta=\wp(D)$ (the usual situation in case of discrete space of states), then any mapping $\mu(x,c,\cdot)$ is assured to be measurable.

For our purpose, the following problem should be solved: from the given approval function, we must be able to build a mapping

$$\mu^{+}: \ X \times @(D) \times W \ \text{-----}\!> \ [0,1]$$

satisfying some logical conditions (as above, @(D) represents the family of all subsets of D). In particular,

$$\mu^{+}(x,\{c\},w)=\mu(x,c,w) \quad \forall c \in D$$

and if we denote $\mu^{+}(x,A,w)=\mu^{+}_{A}(x,w)$, an aggregation operation ought to be defined in such a way that for any pair of disjoint subsets of criteria $A,B\in @(D)$, the probabilistic set μ^{+}_{AUB} is obtained from the probabilistic sets μ^{+}_{A} and μ^{+}_{B}.

Following the sheme of [12], we can denote

$$F_{W}(X)=\{\mu:X\times W\text{--->}[0,1]/\mu(x,\cdot) \text{ is measurable } \forall x\in X\}$$

and $M=F_{W}(X)\times @(D)$. Then the generalized approval aggregation operation will be defined as a correspondence

$$o : \ M \times M \ \text{-----}\!> \ M$$

$$(\mu_{A},A)o(\mu_{B},B)=(\mu_{AUB},AUB) \quad \forall A,B\neq\phi, \ A\cap B=\phi$$

which assigns the aggregated approval μ_{AUB} to each pair of approvals μ_{A} and μ_{B} from two disjoint and non-empty sets of criteria. Associativity and commutativity are supposed to be verified by definition:

$$((\mu_{A},A)o(\mu_{B},B))o(\mu_{C},C)=(\mu_{A},A)o((\mu_{B},B)o(\mu_{C},C))$$

$$(\mu_{A},A)o(\mu_{B},B)=(\mu_{B},B)o(\mu_{A},A)$$

and other ethical conditions can be required in adition:

i) $\mu_{A}(x,w)=\mu_{B}(x,w)=\mu(x,w) \implies \mu_{AUB}(x,w)=\mu(x,w)$

ii) $\mu_{A}(x,w)\geq\mu'_{A}(x,w), \ \mu_{B}(x,w)\geq\mu'_{B}(x,w) \implies \mu_{AUB}(x,w)\geq\mu'_{AUB}(x,w)$

In section 4 the problem with unequal criteria will be considered. But previously we shall study how their relative importance can be aggregated.

3. AMALGAMATING CRITERIA WEIGHTS

Let us suppose that the set of criteria is defined by a membership function

$$\beta : \ D \ \text{-----}\!> \ [0,1]$$

$\beta(c)$ meaning the weight of criterion c. Following [13], we must be able to define the weight of any subset of criteria, in such a way that we can expect that approval function only depends on criteria through their weights.

If we denote $N=[0,1]\times @(D)$, each pair $(b,A)\in N$ represents a set of criteria A with relative weight b. The weight aggregation operation

$$\S : \ N \times N \ \text{-----}\!> \ N$$

$$(b(A),A)\S(b(B),B)=(b(AUB),AUB) \quad \forall A,B\neq\phi, \ A\cap B=\phi$$

assigns the aggregated weight $b(A\cup B)$ to each pair of weights $b(A)$ and $b(B)$ from two disjoint and non-empty sets of criteria. Analogously to the approval aggregation operation, commutativity and associativity are supposed by definition, and both aggregation operations must be connected in a rational way. In this section, some weight aggregation operations are characterized.

We shall suppose that $b(\{c\})=\beta(c)$, and under the following conditions:

i') $b(A)=b(B)=b \implies b(A\cup B)=b$

ii') $b(A)\geq b'(A),\ b(B)\geq b'(B) \implies b(A\cup B)\geq b'(A\cup B)$

it makes sense to say that a set of criteria A with weight $b(A)$ is d-decisive over a set of criteria B ($A\cap B=\emptyset$) with weight $b(B)$ ($b(A)\neq b(B)$) if and only if $b(A\cup B)=d.b(A)+(1-d).b(B)$.

THEOREM 1.- Let us consider weight aggregation operations verifying conditions i'), ii') and

$$\text{iii')}\quad \left.\begin{array}{l} card(A)=card(A'),\ b(A)=b(A') \\ card(B)=card(B'),\ b(B)=b(B') \end{array}\right\} \implies b(A\cup B)=b(A'\cup B')$$

Then the aggregation rule defined as

$$\bar{b}(A\cup B)=(card(A).\bar{b}(A)+card(B).\bar{b}(B))/card(A\cup B) \qquad \forall A\cap B=\emptyset$$

makes maximum the minimum decisiveness d of criteria.

Proof: Let us consider any aggregation rule b verifying such conditions, and let us denote $d=d_{1,n}(p,q)$ the decisiveness of a criterion with weight p over a set of n criteria with weight q. On the one hand, since $d_{1,1}(p,q)=1-d_{1,1}(q,p)$ due to commutativity, it is clear that $d_{1,1}(p,q)\leq\frac{1}{2}$ for some $p\neq q$. On the other hand, assuming $d_{1,n-1}(p,q)\leq 1/n$ for some $p\neq q$, let us consider a set of n-1 criteria B and $A=\{a\}$, $C=\{c\}$, with a and c two distinct criteria not belonging to B. Taking $b(A)=p$, $b(B)=q$ and $b(C)=q$, then

$$b(A\cup(B\cup C))=d_{1,n}(p,q).p+(1-d_{1,n}(p,q)).q=$$
$$=b((A\cup B)\cup C)=(1-d_{1,n}(q,r)).r+d_{1,n}(q,r).q$$

where $r=b(A\cup B)$. Since

$$b(A\cup B)=d_{1,n-1}(p,q).p+(1-d_{1,n-1}(p,q)).q$$

it must be

$$d_{1,n}(p,q)=(1-d_{1,n}(q,r)).d_{1,n-1}(p,q)$$

and therefore

$$d_{1,n}(p,q)\leq(1-d_{1,n}(q,r))/n$$

due to induction hypothesis. If $d_{1,n}(q,r)\leq 1/(n+1)$, the result holds; and if $d_{1,n}(q,r)>1/(n+1)$, then $d_{1,n}(p,q)\leq 1/(n+1)$, and in any case

$$\inf_{p\neq q} d_{1,n}(p,q) \leq 1/(n+1)$$

Since the proposed rule is characterized by the fact that its decisiveness takes the value

$$d_{1,n}(p,q)=1/(n+1) \quad \forall p\neq q$$

then the theorem is proved. Moreover,

$$\bar{b}(A)= \Sigma_{c\in A}\ \beta(c)/\mathrm{card}(A)$$

It is clear that if we modify condition iii') in order to get independence of the sizes of criteria sets, then the weight aggregation operation can be analyzed through a classical fuzzy connective: any aggregation operation such that

iv') $b(A)=b(A')$, $b(B)=b(B')$ $\Longrightarrow$ $b(A\cup B)=b(A'\cup B')$

is characterized by a mapping

$$h : [0,1] \times [0,1] \longrightarrow [0,1]$$

verifying commutativity and associativity, in such a way that

$$b(A\cup B)=h(b(A),b(B)) \quad \forall A,B\neq\emptyset,\ A\cap B=\emptyset$$

Therefore, under condition iv'), classical results on fuzzy connectives can be translated here. For example:

THEOREM 2.- Let h be a fuzzy connective verifying continuity, commutativity, associativity, idempotent law and non-decreasingness. Then we can define an associated weight aggregation rule, and it must exist an $a\in[0,1]$ such that

$$h(p,q)=\max(p,q) \quad \text{if} \quad p,q\leq a$$

$$h(p,q)=\min(p,q) \quad \text{if} \quad p,q\geq a$$

and $h(p,q)=a$ otherwise.

Proof: see Fung-Fu [7], taking into account that condition iv') is equivalent to Fung-Fu's law of independent components. When a=1 or a=0, they talk about an "optimistic" rule and a "pessimistic" rule, respectively. When $a\neq 0,1$ they say that the rule is a "mixed" rule. Moreover, $a=h(0,1)$.

Analogously to Fung-Fu's approach [7], in Dubois-Prade's article [6] the condition of independent components is supposed to be verified. Under some conditions ($h(1,1)=1$, $h(0,0)=0$, commutativity, associativity, continuity and a compatibility condition with respect to the Pareto approach), they define the "conjunction operators" (if $h(p,q)\leq\min(p,q)$ $\forall p,q$), the "disjunction operators" (if $h(p,q)\geq\max(p,q)$ $\forall p,q$) and the "averaging operators" (if $\min(p,q)\leq h(p,q)\leq$ $\leq\max(p,q)$ $\forall p,q$). As expected, similar results to those of Fung-Fu [7] are easily found. In particular, they prove that the only associative averaging operators are the pessimistic, optimistic and mixed rules.

It must be pointed out that many interesting fuzzy connectives are special cases of t-norms or t-conorms (cf. Klement [14]): any commutative and associative fuzzy connective h is called t-norm if non-decreasingness holds and

$h(p,1)=p$; when $h(p,0)=p$ holds instead of this last condition, such a fuzzy connective is called t-conorm. Examples of important t-norms and t-conorms can be seen in Czogala-Zimmermann [5] and Weber [15].

4. AMALGAMATING UNEQUAL CRITERIA

The concept of generalized aggregation operation has been introduced in section 2. Let us consider now a fixed weight aggregation operation, in such a way that the weights of all subsets of criteria are defined. We shall suppose $b(A)\neq 0$ for each non-empty set of criteria A.

It is clear that the above considerations on fuzzy connectives can be applied here under some conditions: any aggregation operation such that

$$\text{iv)}\quad \mu_A(x,w)=\mu_{A'}(x,w),\ \mu_B(x,w)=\mu_{B'}(x,w) \implies \mu_{A\cup B}(x,w)=\mu_{A'\cup B'}(x,w)$$

is characterized by a family of mappings

$$h_{x,w}:\ [0,1]\times[0,1] \longrightarrow [0,1] \quad \forall x\in X,\ \forall w\in W$$

verifying commutativity and associativity, in such a way that

$$\mu_{A\cup B}(x,w)=h_{x,w}(\mu_A(x,w),\mu_B(x,w)) \quad \forall A,B\neq\emptyset,\ A\cap B=\emptyset$$

and therefore the following result holds:

THEOREM 3.- Let us suppose a family of fuzzy connectives $\{h_{x,w}\}$ verifying each one continuity, associativity, idempotent law and non-decreasingness. Then we can define an associated generalized aggregation rule, and it must be such that

$$h_{x,w}(p,q)=\max(p,q) \quad \text{if } p,q\leq a_{x,w}$$
$$h_{x,w}(p,q)=\min(p,q) \quad \text{if } p,q\geq a_{x,w}$$

and $h_{x,w}(p,q)=a_{x,w}$ otherwise, for some $a_{x,w}\in[0,1]$.
Proof: analogous to theorem 2.

In many cases we can suppose that there exists a mapping h such that $h_{x,w}=h$ $\forall x\in X$, $\forall w\in W$. But in general the rule of aggregation for each alternative can depend upon the possible states (the rule of evaluation may change with the state of environment).

Dubois-Prade [6] discuss the amalgamating of more than two criteria by direct recursivity, and therefore the necessity of associative law. If associativity no longer holds, they define, not in an axiomatic way, symmetrical aggregations through simultaneous combination of consistent values, when criteria have equal importance. When dealing with unequal criteria, they propose a weighted mean of criteria, but they themselves emphasize the lack of generality of such an aggregation. The following result will characterize a linear combination, by considering that under conditions i) and ii) it makes sense to say that a set of criteria A with approval μ_A is π-decisive over a set of criteria B ($A\cap B=\emptyset$) with approval μ_B relative to a pair (x,w) such that $\mu_A(x,w)\neq\mu_B(x,w)$ if and only if $\mu_{A\cup B}(x,w)=\pi.\mu_A(x,w)+(1-\pi).\mu_B(x,w)$.

THEOREM 4.- Let us consider criteria aggregations verifying conditions i) and ii). Then the aggregation rule such that

$$\bar{\mu}_{A\cup B}(x,w)=(\text{card}(A).\bar{b}(A).\bar{\mu}_A(x,w)+\text{card}(B).\bar{b}(B).\bar{\mu}_B(x,w))/(\text{card}(A\cup B).\bar{b}(A\cup B))$$

makes maximum the minimum ratio decisiveness $\pi/\beta(c)$ of criteria.

Proof: let us denote $\pi=\pi^{x,w}_{A,B}(p,q)$ the decisiveness of a set of criteria A with $\mu_A(x,w)=p$ over a disjoint set of criteria B with $\mu_B(x,w)=q$ ($p\neq q$) for an arbitrary aggregation rule verifying such conditions. We shall prove that

$$\inf_{p\neq q} \pi^{x,w}_{\{c\},A-\{c\}}(p,q) \leq \beta(c)/\Sigma_{e\in A}\beta(e) \qquad \forall A,\ \forall c\in A$$

and since the proposed rule is characterized by the fact that

$$\pi^{x,w}_{\{c\},A-\{c\}}(p,q) = \beta(c)/\Sigma_{e\in A}\beta(e) \qquad \forall c\in A,\ \forall p\neq q$$

the theorem will be proved. On the one hand, if we suppose

$$\pi^{x,w}_{\{c\},\{e\}}(p,q) > \beta(c)/(\beta(c)+\beta(e))$$

for any $c\neq e$, then

$$\pi^{x,w}_{\{e\},\{c\}}(q,p) < \beta(e)/(\beta(c)+\beta(e))$$

due to

$$\pi^{x,w}_{A,B}(p,q) = 1-\pi^{x,w}_{B,A}(q,p) \qquad \forall p\neq q,\ \forall A,B$$

On the other hand, if we suppose that for a set A of n criteria and $c\notin A$

$$\pi^{x,w}_{\{c\},A}(p,q) \leq \beta(c)/(\beta(c)+\Sigma_{e\in A}\beta(e))$$

for some $p\neq q$, then we can define

$$\mu_{\{c\}}(x,w)=p\ ,\ \mu_A(x,w)=\mu_{\{e\}}(x,w)=q$$

for a criterion $e\notin A$ ($e\neq c$). Analogously to theorem 1, it is easy to prove that

$$\pi^{x,w}_{\{c\},A\cup\{e\}}(p,q) = (1-\pi^{x,w}_{\{e\},A\cup\{c\}}(q,r)).\pi^{x,w}_{\{c\},A}(p,q)$$

where $r=\mu_{A\cup\{c\}}(x,w)$, and in any case there exists some criterion verifying the required relation against a set of n+1 criteria.

Two questions arise at this point: 1) fixed a weight aggregation, when is it consistent, in a sense to be specified, with a criteria aggregation?, and 2) fixed a weight aggregation, can a criteria aggregation be defined depending on the aggregated criteria only through their weights and sizes? The second question has positive answer when the following condition is verified:

$$\text{iii)}\quad \left.\begin{array}{l}\text{card}(A)=\text{card}(A'),\ b(A)=b(A'),\ \mu_A=\mu_{A'}\\ \text{card}(B)=\text{card}(B'),\ b(B)=b(B'),\ \mu_B=\mu_{B'}\end{array}\right\} \Longrightarrow \mu_{A\cup B}=\mu_{A'\cup B'}$$

and it is clear, for example, that criteria aggregation $\bar{\mu}$ of theorem 4 verifies such a condition when the weight aggregation $\bar{b}$ of theorem 1 is given. But not any criteria aggregation under iii) can be defined through any weight aggregation. For example, $\bar{\mu}$ is not compatible with neither pessimistic, optimistic or mixed rules. Moreover, it seems reasonable to impose that criteria decisiveness increases with weight and size:

$$\text{v) } card(A)\geq card(B),\ b(A)\geq b(B) \implies \pi^{X,W}_{A,B}(p,q)\geq \pi^{X,W}_{B,A}(q,p)$$

The following results are related to the first question:

THEOREM 5.- Given a mixed weight aggregation, there is no criteria aggregation verifying condition i) and

$$\text{v') } b(A\cup B)=b(A)\neq b(B) \implies \mu_{A\cup B}=\mu_A \quad \forall A\cap B=\emptyset$$

Proof: Let us consider $a\in(0,1)$ and A, A', B, B' disjoint sets of criteria such that $\mu_A=\mu_{B'}\neq\mu_B=\mu_{A'}$, and $b(A')\leq b(A)<a<b(B)\leq b(B')$. Since $b(A'\cup B)=b(A\cup B\cup A')=a$, it must be $\mu_{A\cup B\cup A'}=\mu_{B\cup A'}=\mu_B$ due to unanimity of condition i). Analogously it is proved that $\mu_{B\cup A\cup B'}=\mu_A$ in contradiction with the previous consideration.

THEOREM 6.- Pessimistic weight aggregation is not consistent with a criteria aggregation verifying conditions i), ii), v) and v').

Proof: Trivial, since $\pi^{X,W}_{A,B}(p,q)=1$ should hold for any pair of disjoint sets of criteria A and B such that $b(A)<B(B)$.

It must be pointed out that condition v'), introduced in [13], means that approval remains the same when importance does not change. It is equivalent to impose

$$d_{A,B}(a,b)=1 \implies \pi^{X,W}_{A,B}(p,q)=1$$

when conditions i) and ii) are supposed to be verified. Moreover, it is easy to see that condition v') can be relaxed in theorem 6, for example by assuming

$$d_{A,B}(a,b)=1 \implies \pi^{X,W}_{A,B}(p,q)>\tfrac{1}{2}$$

Optimistic weight aggregation can be consistent with criteria aggregations verifying conditions of theorem 6, and under these conditions the set

$$A = \{c\in D \ / \ \beta(c)= \max_{d\in D} \beta(d)\}$$

verifies, in case of being non-empty, that $\mu_{A\cup B}=\mu_A$ for any disjoint set B of criteria (A is said to be absolutely decisive). Since A contains criteria with the same relative importance, aggregation operations for equal criteria can be applied in order to get the aggregated approval.

Other interesting aggregation rules for unequal criteria, verifying conditions i) and ii), are the following:

$$\hat{\mu}_{AUB}(x,w)=[\hat{\mu}_A(x,w)]^{card(A).\bar{b}(A)/card(AUB).\bar{b}(AUB)}\cdot [\hat{\mu}_B(x,w)]^{card(B).\bar{b}(B)/card(AUB).\bar{b}(AUB)}$$

and

$$\hat{\mu}'_{AUB}(x,w)=1-[1-\hat{\mu}'_A(x,w)]^{card(A).\bar{b}(A)/card(AUB).\bar{b}(AUB)}\cdot [1-\hat{\mu}'_B(x,w)]^{card(B).\bar{b}(B)/card(AUB).\bar{b}(AUB)}$$

which are types of weighted pessimistic rules and weighted optimistic rules, respectively. But they do not hold condition v): aggregated approval can be fixed by criteria with low relative weight. Their axiomatic justification can be obtained by considering a multiplicative model, instead of the aditive model developed here.

5. CONCLUDING REMARKS

In this paper, criteria have been supposed to be represented by probabilistic sets, to be amalgamated by using generalized aggregation operations, which must take into account that those criteria relevant for the evaluation are of different importance for the final judgement. The set of criteria has been defined as a fuzzy set which allows us to build some valid system of relative weights for each subset of criteria, and the amalgamated approval represents how well each alternative satisfies a given family of criteria. But two things must be pointed out: on the one hand, a purely mathematical justification of an aggregation operation is nor sufficient, and it must be tested empirically in order to stablish the adequacy of the model to human behavior (a first analysis in this sense can be seen in [16]). On the other hand, though decision analysis depends on the state of the environment, weights of each criterion has been supposed to remain constant, whatever the state or degree of approval, and non-interaction of criteria (something that may be difficult to justify) has been assumed (see [17] for a discussion). Another kind of models must be developed in order to avoid this problem.

In any case, in this context it makes sense to look for alternatives with highest degree of expected approval, for a fixed probability distribution on the space of states (see [5]): given a set A of criteria we must solve

$$\max_{x\in X} E_P[\mu_A(x,\cdot)]$$

where $\mu_A(x,w)=\mu(x,A,w)$ is the aggregated approval of alternative x when w is the state of the environment, and E_P means the mathematical expectation with respect to the probability measure P.

Following Hirota [10], other classical methods in Decision Theory, like moment analysis, can be applied.

Finally, it must be pointed out that here we have considered approval function taking values on the unit interval. Analogous treatment must be tried in a more general topological framework. Moreover, here we have developed an aggregation approach instead of an order-focussed approach (see Zimmermann [18] for a study of these two distinct families of approaches). But results obtained in this paper are easily translated to the aggregation of Roy's outranking relations [19], by considering approval functions

$$\mu : (X \times X) \times D \times W \longrightarrow [0,1]$$

in such a way that each value $\mu(x,y,c,w)$ represents the degree to which alternative x outranks alternative y under criteria c and state w. In this context, Saaty's procedure (see [3]) for obtaining a ratio scale of importance can be useful. In any case, general considerations about decision-making in management (see [20], for example) must be taken into account under both approaches.

REFERENCES

[1] Roy, B. and Vincke, P., Multicriteria analysis: survey and new directions, European Journal of Operational Research 8 (1981) pp. 207-218.

[2] Blin, J.M., Fuzzy sets in multiple criteria decision-making, TIMS Studies in the Management Science 6 (1977) pp. 129-146.

[3] Saaty, T.L., Exploring the interface between hierarchies, multiple objectives and fuzzy sets, Fuzzy Sets and Systems 1 (1978) pp. 57-68.

[4] Czogala, E., Gottwald, S. and Pedrycz, W., Logical connectives of probabilistic sets, Fuzzy Sets and Systems 10 (1983) pp. 299-308.

[5] Czogala, E. and Zimmermann, H.J., The aggregation operations for decision making in probabilistic fuzzy environment, Fuzzy Sets and Systems 13 (1984) pp. 223-239.

[6] Dubois, D. and Prade, H., Criteria aggregation and ranking of alternatives in the framework of fuzzy set theory, TIMS Studies in the Management Science 20 (1984) pp. 209-240.

[7] Fung, L.W. and Fu, K.S., An axiomatic approach to rational decision making in a fuzzy environment, in: Zadeh, L.A., Fu, K.S., Tanaka, K. and Shimura, M. (eds.), Fuzzy Sets and their Applications to Cognitive and Decision Processes (Academic Press, New York, 1975) pp. 227-256.

[8] Yager, R.R., Fuzzy decision making including unequal objectives, Fuzzy Sets and Systems 1 (1978) pp. 87-95.

[9] Bellman, R.E. and Zadeh, L.A., Decision-making in a fuzzy environment, Management Science 17 (1970) pp. 141-164.

[10] Hirota, K., Concepts of probabilistic sets, Fuzzy Sets and Systems 5 (1981) pp. 31-46.

[11] Czogala, E., On distribution function description of probabilistic sets and its applications in decision-making, Fuzzy Sets and Systems 10 (1983) pp. 21-29.

[12] Montero, F.J., A note on Fung-Fu's theorem, Fuzzy Sets and Systems 13 (1985) pp. 259-269.

[13] Montero, F.J., Aggregation of fuzzy opinions in a non-homogeneous group, in print.

[14] Klement, E.P., Construction of fuzzy algebras using triangular norms,

J. Math. Anal. Appl. 8 (1982) pp. 543-565.
[15] Weber, S., A general concept of fuzzy connectives negations and implications based on t-norms and t-conorms, Fuzzy Sets and Systems 11 (1983) pp. 115--134.
[16] Thole, U., Zimmermann, H.J. and Zysno, P., On the suitability of minimum and product operators for the intersection of fuzzy sets, Fuzzy Sets and Systems 2 (1979) pp. 167-180.
[17] Efstathiou, J., Practical multi-attribute decision-making and fuzzy set theory, TIMS Studies in the Management Science 20 (1984) pp. 307-320.
[18] Zimmermann, H.J., Multi criteria decision making in crisp and fuzzy environments, in: Jones, A., Kaufmann, A. and Zimmermann, H.J. (eds.), Fuzzy Set Theory and Applications (Reidel, Dordrecht, 1986) pp. 234-256.
[19] Roy, B., Partial preference analysis and decision-aid. The fuzzy outranking relation concept, in: Bell, D., Keeney, R. and Raiffa, H. (eds.), Conflicting Objectives in Decision (Wiley, New York, 1977) pp. 40-78.
[20] Montero, F.J. and Tejada, J., Choices under fuzzy preferences, Proceedings FISAL'86 (Universitat de les Illes Balears, Portocolom, 1986).

Fuzzy Logic in Knowledge-Based Systems,
Decision and Control
M.M. Gupta and T. Yamakawa (Editors)

FORMALIZATION OF QUANTIFIER PHRASES

Walenty Ostasiewicz
Academy of Economics in Wroclaw, Poland.

Generalized quantifiers are considered from the point of view of class calculus in such a way that they form the generalization of classical Aristotelian square.
The possible applications in operation research are indicated, esspecially for constructing the fuzzy algorithms.

Keywords: quantifiers, generalized, fuzzy algorithms

1. INTRODUCTION

In much of our reasoning me make use of the expressions containing the follows phrases: *for all, at least one, one and only one, none*, etc..
Let us refer to such words or phrases as *quantifiers*. These quantifiers are conceived to be some operators which from a propositional function form a proposition.
For example the expression "x is polite" is a propositional function which is true or false depending on x. Functions of this type are called *predicates* and symbolically are denoted as P(x). The variable x is called free variable, it can be replaced by any element of the given set X named the domain of P or the range of x. Applying operator "every" to the expression "x is polite" gives us the proposition (but not predicate) "every x is polite". Variable x is now bounded by the operator "every", denoting it by symbol $\forall$ the proposition under consideration takes the following symbolic form $\forall x \in X : P(x)$ or the form $\forall x P(x)$, if there is no need to specify the domain of P.
Generally the expression of the form

$$Qx \in X : P(x),$$

where Q stands for some quantifier, and P denotes predicate on X, is called a *quantifier phrase*, quantified proposition or quantified statement.

The rules and operations holding for such expressions are comprised in logical theory known as claculus of function or in recent literature as a quantification theory.
There is no doubt that "there are all sorts of subtle problems about natural-language quantifiers which have no counter part in logicians' so called quantification theory" (see [4]). Avoiding any philosophical and methodological discussion we confine ourselves to consider only those aspects of natural-language quantifiers which can be formalized within the framework of Boolean algebra of classes. The particular aim is to generalize the classical square of opposition considered within the traditional syllogistical logic.

2. CLASSICAL SQUARE OF OPPOSITION

In traditional logic the following four quantifier phrases were considered:

all S are P	symbolized as	S A P
no S is P	symbolized as	S E P
some S are P	symbolized as	S I P
some S are not P	symbolized as	S O P

These phrases were being classified twofold (see [5]): according to qualitative criterion as affirmatives and negatives, and according to quantitative criterion as general and particular. To remember this classification the following *versus memoriales* were cointed:

Asserit A negat E, sed universaliter ambo
Asserit I negat O, sed particulariter ambo.

Letters A and I came as first two vocals from the latin word affirmo, and letters E and O from the word nego.

We can represent this classification graphically in a two-coordinate system as shown in Fig. 1.

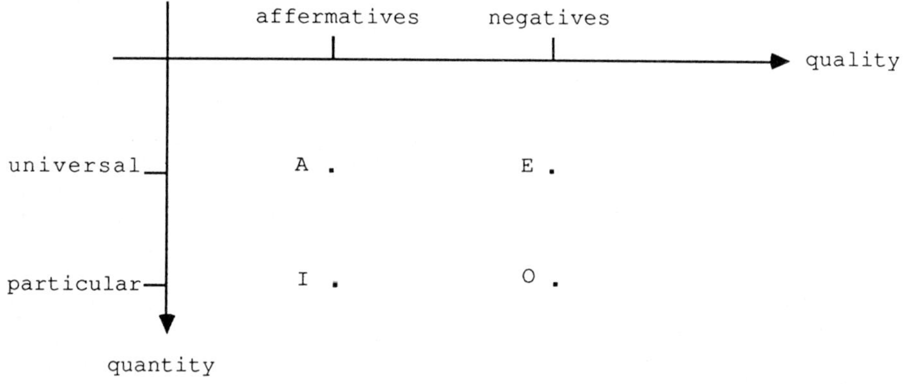

Fig. 1

The four quantifiers under consideration are encoded here us "all" (A), "some" (I), "none" (E) and "anyone" but all" (O) and we will use the different symbols for denoting them, namely the following symbols: ∀,∃, И , ᗺ respectively. Symbols ∀ and ∃ as a reverse letters A and E (first letters of the words: all and exists) are frequently used in the literature. Symbols И and ᗺ as a reverse letters N (none) and B (but all) are introduced here only for unified presentation.

Among all the possible quantified statements of the form QxP(x) where Q is one of the four quantifiers there are some relationships.

Usually the following four basic relations are considered:

1) contrary: which means that one statement excludes the other;
2) sub-contrary: one statement does not exclude the other but both statements cannot be false;
3) contradictory: one statement is the negation of the other;
4) entailment: one statement is implied by the other.

These four relations are usually presented in the form of the square known as square of opposition or as Aristotelian square, shown in Fig. 2 (see [5,6]).

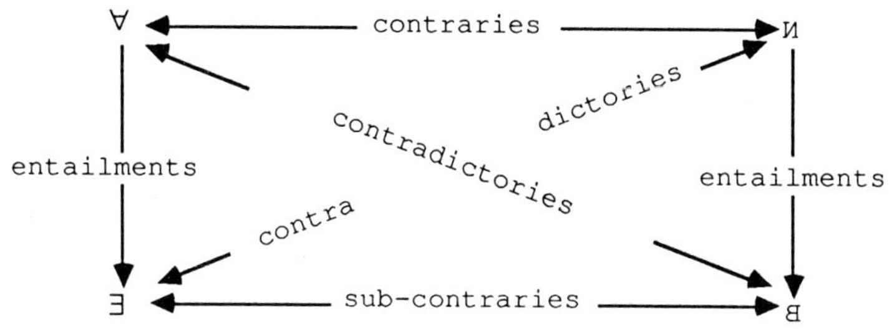

Fig. 2

Since

$$\text{И}\ xP(x) \Leftrightarrow \overline{\exists xP(x)},$$

$$\text{ꓭ}\ xP(x) \Leftrightarrow \overline{\forall xP(x)},$$

only two quantifiers are usually considered: exists and all.
On the other hand, one of these two quantifiers can be expressed by the other:

$$\forall xP(x) \Leftrightarrow \exists x\overline{P(x)},$$

$$\exists xP(x) \Leftrightarrow \forall x\overline{P(x)},$$

it means that only *one* quantifier is to be considered as a primitive terms.
Two quantifiers are however in use only for practical reasons: compactness of expressions, intuitive appealing, clarity etc.
For the aim of sequel generalization we consider now the algebric structure of the set formed from the four quantifiers extended by two new "quantifiers" denoted by **1** and θ and defined below.
The sentence

$$\mathbf{1}x\in X:P(x)$$

is true for any predicate P, and the sentence

$$\theta x\in X:P(x)$$

is false independently of predicate P.
It is clear that for any $Q \in \{\forall, \exists, \text{И}, \text{ꓭ}\}$ holds:

$$Qx\in X:P(x) \Rightarrow \mathbf{1}x\in X:P(x),$$

$$\theta x\in X:P(x) \Rightarrow Qx\in X:P(x).$$

So that the set of six quantifiers forms the lattice with respect to the partial relation →. This lattice is shown in Fig. 3.

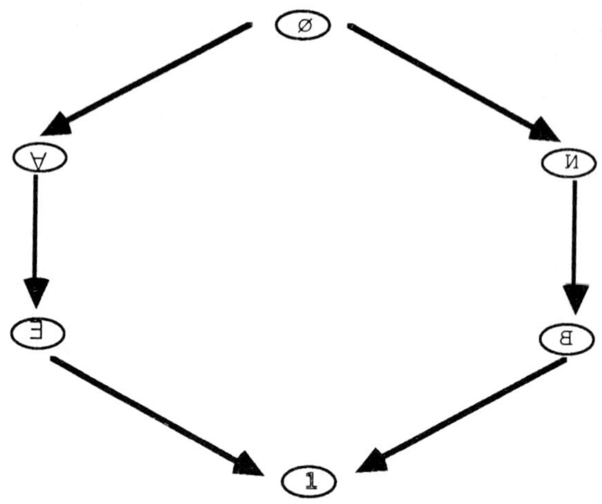

Fig. 3

The aim of the further discussion is to generalize this structure. Under the generalization we mean introducing new quantifiers in such a way that the structure of lattice is preserved and also hold all relations named as opposition.

3. APPROACHES TO GENERALIZATION

The key notion for the generalization considered here is the operation of "selecting" those elements from the given set X wich satisfy the property P.
Suppose that

$$X_P = \{x \in X \mid P(x)\}$$

then the set X_P will be used for truth valuation of the proposition $Qx \in X : P(x)$.

Roughly speaking, the truth value of this proposition will be obtained by measuring the set X_P and by subsequent comparing the obtained result with some patterns prepared in advance. These patterns are considered as some measuring sticks labelled by the quantification words like "all", "most", "exists" etc.

There are two ways to perform such a program. One of them consists in introducing some measure μ on the set X. Suppose the measure is normalized to unit interval [0,1]. In this case as above mentioned measurement sticks, we can choose some subset of [0,1]. Suppose for example that the following subsets are given:

$$r_1 = (0,1], \quad r_2 = [1,1], \quad r_3 = (0.5,1],$$

then valuation of $Qx:P(x)$ consists in measuring of X_P i.e. determination the number $\mu(X_P)$ and comparing it with given subsets.

For example if $\mu(X_P)\in r_1$, then the sentence "$\exists x\in X:P(x)$" has the truth value equal to "true", so that subset $r_1 \subseteq [0,1]$ can be encoded as "exists". The proposition $\forall x\in X:P(x)$ is true if $\mu(X_P)\in r_2$, while the sentence "for most x is P(x)" is considered as a true if $\mu(X_P)\in r_3$.

We will follow however the other way of measuring, which is more general and more suitable for generalization (see [1,7,8]).

For introducing the general paradigm, we consider now the reformulation of traditional quantifiers in the language of calculus of classes (see [6]).

It is clear that the universally quantified statement $\forall x\in X:P(x)$ can be defined as follows:

$$\forall x\in X:P(x) \Leftrightarrow X_P=X,$$

or in equivalent form:

$$\forall x\in X:P(x) \Leftrightarrow X_P\in K_\forall$$

where $K_\forall = \{X\}$.

For existentional statement we have:

$$x\in X:P(x) \Leftrightarrow X_P\neq\varnothing$$

but this is the same as

$$x\in X:P(x) \Leftrightarrow X_P\in K_\exists$$

where $K_\exists=2^X\backslash\varnothing$

Analogously we have:

$$\text{ᗺ}\ x\in X:P(x) \Leftrightarrow X_P\in K_{\text{ᗺ}},\quad K_{\text{ᗺ}}=2^X\backslash X$$

$$\text{И}\ x\in X:P(x) \Leftrightarrow X_P\in K_{\text{И}},\quad K_{\text{И}}=\{\varnothing\}$$

$$\mathbf{1}x\in X:P(x) \Leftrightarrow X_P\in K_{\mathbf{1}},\quad K_{\mathbf{1}}=2^X$$

$$\theta x\in X:P(x) \Leftrightarrow X_P\in K_\theta,\quad K_\theta=\varnothing$$

So that we have the following general scheme of definition:

$$Qx\in X:P(x) \Leftrightarrow X_P\in K_Q \qquad (1)$$

where K_Q is some family of subsets of 2^X.

This means that quantified statement $QxP(x)$ is true if and only if the set X_P belongs to the family K_Q.

First of all it must be observed that this definition is based on the two-valued logic: statement $QxP(x)$ is either true or false, and the same is valid for statement $X_P\in K_Q$.

In some situation it is convenient to treat predicate P as a fuzzy predicate.

Quantifier words of natural language sometimes are hardly modelled by usual crisp subset so that for modelling them we can also try to use fuzzy sets. For using fuzzy sets, the general

scheme (1) should be modified as follows.
Suppose that the truth value of the statement q is denoted by T(q) or by [q], then (1) can be generalized as:

$$[Qx\in X:P(x)]=[X_P\in K_Q] \tag{2}$$

For both cases we assume the following definition of the main relationships for quantifier phrases:

1) Q_1 contraries Q_2 iff $K_{Q1}\cup K_{Q2} \subseteq 2^X$ & $K_{Q1}\cap K_{Q2} \subseteq \emptyset$

2) Q_1 contradicts Q_2 iff $K_{Q1}\cup K_{Q2} = 2^X$ & $K_{Q1}\cap K_{Q2} = \emptyset$

3) Q_1 sub-contraries Q_2 iff $K_{Q1}\cup K_{Q2} \supseteq 2^X$ & $K_{Q1}\cap K_{Q2} \supseteq \emptyset$

4) Q_1 entails Q_2 iff $K_{Q1} \subseteq K_{Q2}$

5) Q_1 opposites Q_2 iff $K_{Q1}=K_{\overline{Q2}}$

where $K_{\overline{Q}}=\{A\subseteq X\,|\,\overline{A}\in K_Q\}$

4. TWO-VALUED LOGIC APPROACH

In this section it is assumed that predicate $P:X\to\{0,1\}$ is a usual predicate of two-valued logic, K_Q is some family of subsets of the power set 2^X i.e. $K_Q:2^X\to\{0,1\}$.

Before following on to classification of quantifiers let us first consider the problem of negations.
Based on (1) it is possible to distinguish the following basic negations:

$$\overline{X}_P\in K_Q \quad X_{\overline{P}}\in K_Q,\ X_P\in\overline{K}_Q,\ X_P\notin K_Q,\ X_P\in K_{\overline{Q}}.$$

Within the framework of classical set theory we have $\overline{X}_P=X\backslash X_P$ and $X_{\overline{P}}=\{x\in X\,|\,\overline{P(x)}\}=X\backslash\{x\in X\,|\,P(x)\}=\overline{X}_P$, i.e. $X_{\overline{P}}=\overline{X}_P$ hence $\overline{X}_P\in K_Q \Leftrightarrow X_{\overline{P}}\in K_Q$.
Since for any set A holds $a\notin A \Leftrightarrow a\in\overline{A}$, then $X_P\in K_Q \Leftrightarrow X_P\in\overline{K}_Q$.
Finally it means that there are the following three kinds of negation:

1. $X_P\in\overline{K}_Q$ which corresponds to full sentence negation $\overline{QxP(x)}$,

2. $\overline{X}_P\in K_Q$ which corresponds to predicate negation $Qx\overline{P(x)}$,

3. $X_P\in K_{\overline{Q}}$ which corresponds to quantifier negation $\overline{Q}xP(x)$.

Now we are ready to classify all quantifiers with respect to two criteria: quantitative and qualitative.
First we define a class of affirmative quantifiers which will be denoted by AQ. This class is defined as follows:

1. $\{X\}\in AQ$

2. $A\in AQ,\ A\subseteq B \Rightarrow B\in AQ$

3. $A\in AQ \Leftrightarrow \overline{A}\notin AQ$

This class forms the segment $[\{X\},2^X\backslash\emptyset]$ in the usual lattice of all possible subsets of the power set 2^X.
It can be seen that the above definition generalizes two classical

affirmative quantifiers, since the two element family $AQ=\{\{X\},2^X\backslash\varnothing\}$ satisfies all three conditions. It means that end points of segment $[\{X\},2^X\backslash\varnothing]$ corresond to classical quantifiers and all others represent different quantifiers.

Traditionally negative quantified statements are defined from affirmatives by predicate negation. This leads the way to generalization: for any affirmative quantifier Q we must assign a negative Q' such that

$$Q'xP(x) \Leftrightarrow Qx\overline{P(x)}\ .$$

To determine Q' corresponding to Q means to resolve the equation

$$[X_P\in K_{Q'}]=[X_P\in K_Q] \qquad (3)$$

with respect to unknown $K_{Q'}$.
Since this equation is equivalent to

$$X_P\in K_{Q'}\Leftrightarrow(X\backslash X_P)\in K_Q,$$

then we have the solution:

$$K_{Q'}=\{A \subseteq X \mid A=X\backslash B, B\in K_Q\} \qquad (4)$$

For the finite family we can write explicitly: if $K_Q=\{B_1,B_2,\ldots,B_n\}$ then

$$K_{Q'}=\{X\backslash B_1,X\backslash B_2,\ldots,X\backslash B_n\}.$$

If we denote by NQ the class of all negative quantifiers corresponding to the affermative one, then it is not difficult to verify that:

1. $\varnothing \in NQ$,
2. $B \in NQ,\ \ A \subseteq B \Rightarrow A\ e\ NQ$,
3. $A \in NQ \Rightarrow \overline{A} \notin NQ$,
4. $A \in AQ \Rightarrow A \notin NQ$,
5. $A \in NQ \Rightarrow A \notin AQ$,
6. There is biunivocal mapping $f:AQ \rightarrow NQ$ such that

   $A \subseteq B \Rightarrow f(B) \subseteq f(A)$

   this mapping is given by

   $f(B)=\{a \subseteq X \mid a \in 2^X\backslash A\}$
7. No pair of quantifiers belonging to the same class are in contradiction
8. For any affermative quantifier Q_a there is contradictory negative Q_n
9. Class NQ forms the segment $[\{\varnothing\},2^X\backslash X]$ in the lattice 2^X.

Before giving the further properties of classes AQ and NQ let us consider classification according to a second coordinate.
In classical quantification theory the following implications are

fulfilled:

$$\overline{\forall x P(x)} \Leftarrow \forall x \overline{P(x)}$$

$$\overline{\exists x P(x)} \Rightarrow \exists x \overline{P(X)}.$$

We use them as a definition of general and particular quantifiers respectively in our generalized case.
Namely, quantifier Q is called a *general* one if

$$\overline{QxP(x)} \Leftarrow Qx\overline{P(x)},$$

and it will be called *particular* if

$$\overline{QxP(x)} \Rightarrow Qx\overline{P(x)}.$$

Taking into account (1) the above conditions are equivalent to the following:

$$X_P \in \overline{K}_Q \Leftarrow \overline{X}_P \in K_Q,$$

$$X_P \in \overline{K}_Q \Rightarrow \overline{X}_P \in KQ.$$

It can happen that there are such quantifiers for which the following condition is fulfilled:

$$X_P \in \overline{K}_Q \Leftrightarrow \overline{X}_P \in K_Q.$$

Such quantifiers are called here *neutral*. From the above definitions it follows that for any general quantifier Qg is fulfilled:

$Q_n xP(x) \Rightarrow Q_g xP(x)$ for all neutral quantifiers Q_n,

and for any particular quantifier Q_P the following implication

$$Q_P xP(x) \Rightarrow Q_n xP(x)$$

is true for all neutral quantifiers.
Roughly speaking, general quantifiers are less than neutral and particular are bigger. From the definition of the classes AQ and NQ it follows that for any affermative quantifier there is a contradictory negative quantifier. The opposite is also true: for any negative there is a contradictory affermative. This means that for any Q1 there is Q2 so that the following equality is satisfied:

$$[\overline{Q_1 xP(x)}] = [Q_2 xP(x)]$$

where $Q_1 \in AQ$ and $Q_2 \in NQ$ or vice-verse $Q_1 \in Q$ and $Q_2 \in AQ$. In terms of family of subsets this equality is equivalent to

$$X_P \in \overline{K}_{Q_1} \Leftrightarrow X_P \in K_{Q_2}$$

Since we can solve the equation

$$[Q_2 xP(x)] \doteq [Q_3 x\overline{P(x)}]$$

where either $Q_2, Q_3 \in NQ$ or $Q_2, Q_3 \in AQ$, hence we can conclude the following.
For any given general quantifier Qg there is a particular Qp so that

$$\overline{Q_g xP(x)} \Leftrightarrow Q_p x\overline{P(x)} \quad (5)$$

and

$$\overline{Q_p xp(x)} \Leftrightarrow Q_g x\overline{P(x)} \quad (6)$$

It is obvious that for any given quantifier Q we can easily "calculate" quantifier Q' so that equivalence:

$$\overline{QxP(x)} \Leftrightarrow Q'x\overline{P(x)}$$

is satisfied, namely Q' is given by

$$K_{Q'}=\{A \subseteq X \mid A \in 2^X \backslash K_Q\}.$$

The above relations (5) and (6) are considered here as a generalized De Morgan laws.
It is clear that the pair of quantifiers which satisfies the De Morgan law is a contradictory pair.
Neutral quantifiers satisfy also the De Morgan law, it means that for any neutral quantifier Qn holds

$$\overline{Q_n xP(x)} \Leftrightarrow Q_n x\overline{P(x)} \quad (7)$$

The generalized square of opposition we can now represent as is shown an Fig. 4.

As an illustration let us consider a simple example. Suppose X={a,b}. There are thus four possibilities to define the class of affermative quantifiers, they are listed below.
1) AQ1={{a,b},{{a},{b},{a,b}}},
2) AQ2={{a,b},{{a,b},{b}},{{a},{b},{a,b}}},
3) AQ3={{a,b},{{a,b},{a}},{{a},{b},{a,b}}},
4) AQ4={{a,b},{{a,b}},{a}},{{a,b},{b}},{{a},{b},{a,b}}}.
All these possible families and corresponding families of negatives quantifiers NQ1,...,NQ4 are shown in Fig. 5 as four squares Fi={AQi,NQi}, i=1,...,4. Arrows in these representations indicate the relation of entailment, and dotted lines indicate the contradictions. For semplicity, subsets of X are denoted as strings of elements, for example family {{a},{b},{a,b}} is represented by the string a,b,ab.
Families AQ1,...,AQ4 and corresponding NQ1,...,NQ4 form the segments in the lattice of all subsets of the set X which is presented in Fig. 6.

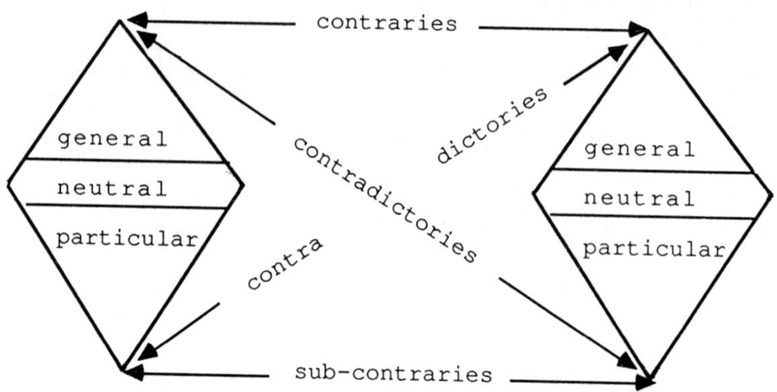

Fig. 4 - Generalized square of opposition.

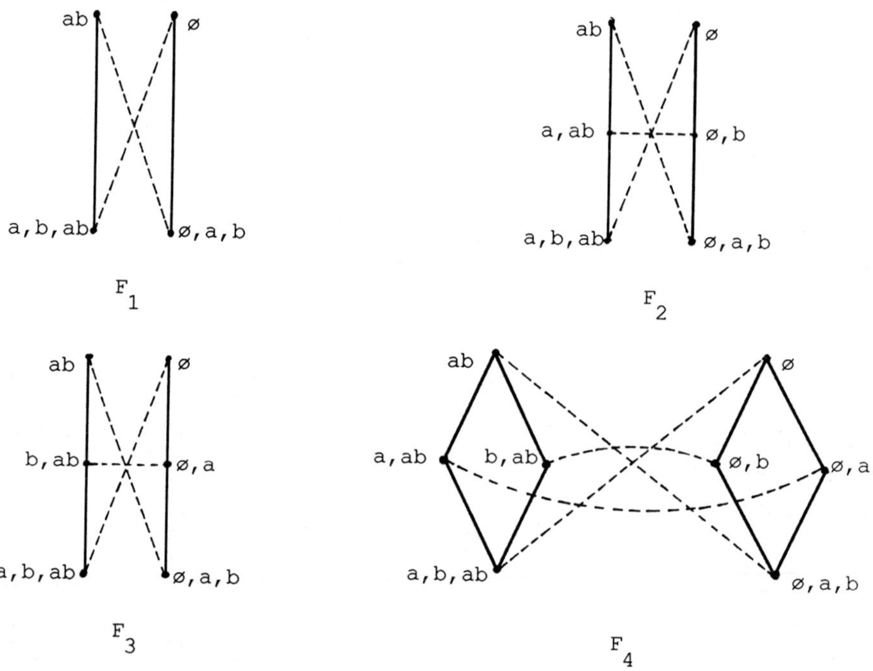

Fig. 5 - Possible sguares of opposition

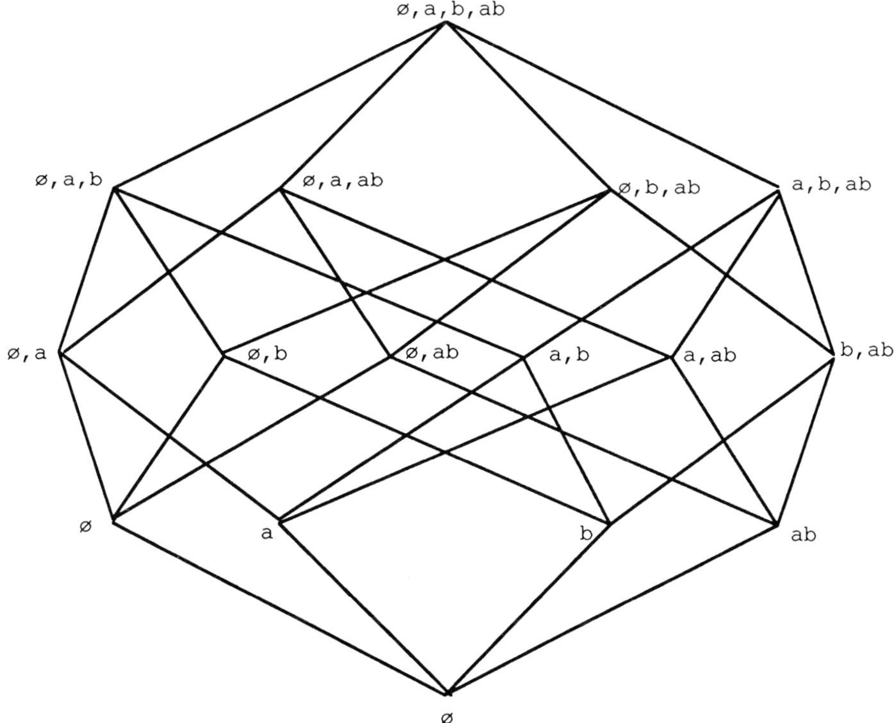

Fig. 6 - Lattice of all subsets

5. USING FUZZY SETS APPROACH

According to the main scheme (1) the proposition "$Qx\in X:P(x)$" is considered to be equivalent to the proposition "$X_P\in K_Q$", where K_Q is some family of subsets of the set X.

Both X_P and K_Q are some sets, and this observation leads the way to generalization: to use fuzzy sets. Following this way gives the hope of formalizing more adequately natural language quantifiers.

Let us recall the necessary notation of fuzzy sets.

Fuzzy subset A of the set X is meant here as a mapping

$$A:X \rightarrow L$$

where L is the same complete lattice ordered semigroups (closg) (see[3]).

Same important examples of closgs are the following:

B={0,1} which corresponds to two Boolean values: false and true,

$Bn=\{0,1\}^n$ n-fold Cartesian product of B,

I=[0,1] unit interval,

$In=[0,1]^n$ n-dimensional unit cube,

$FB=\{f|f:\{0,1\} \rightarrow [0,1]\}$ fuzzy subsets on $\{0,1\}$,

$FI=\{f|f:[0,1] \rightarrow [0,1]\}$ fuzzy subsets on I.

Fuzzy unary predicate P is meant also as a mapping

$$P:X \rightarrow L,$$

Two-argument predicate $\in$ is usually treated as a two valued predicate

$$[x\in X] = \begin{cases} \text{true, if x is an element of X} \\ \text{false, if x is not an element of X} \end{cases}$$

It is also possible to consider $x \in X$ as a fuzzy relation i.e. $[x\in X]\in L$, where L is some closg.

For the family K_Q there are many possibilities of treatments.

Same reasonable variants are listed below.

1. $K_Q:2^X \rightarrow B$ - crisp family of crisp subsets
2. $K_Q:2^X \rightarrow I$ - fuzzy family of crisp subsets
3. $K_Q:I^X \rightarrow B$ - crisp family of fuzzy subsets
4. $K_Q:I^X \rightarrow I$ - fuzzy family of fuzzy subsets

There are of course many other variants; the more "fuzzy" the less reasonable. Yet variant $P:X \rightarrow FI$, $K_Q:F(X) \rightarrow FI$, where $F(X)=\{A|A:X \rightarrow FI\}$ and $[X_p\in K_Q] \in FI$ seems to be too "fuzzy".

6. SAME REASONABLE MODELS

Let us consider two following variants:

1. $P:X \rightarrow I$, $K_Q:F(X) \rightarrow B$,
2. $P:X \rightarrow I$, $K_Q:2^X \rightarrow I$,

where $F(X)=\{A|A:X \rightarrow I\}$.

For each of these variants there are two possibilities of the valuation of the statement $Qx:P(x)$, one of them is binary valuation: $T(Qx:P(x))\in B$, and the second one is fuzzy valuation: $T(Qx:P(x))\in I$. We shall discuss them separately.

The two above mentioned variants seem to have simple interpretations, and they are also attractive from the formal point of view. Namely it is possible to treat them in such a way that all properties discussed for the crisp variant will be preserved.

In order to achieve this end operations for fuzzy sets are needed which satisfy complementary laws: excluded middle and non-contradiction.

We assume that union, intersection and negation satisfy the following conditions:

$$\overline{\emptyset} = X, \quad \overline{X} = \emptyset, \quad \overline{\overline{A}} = A, \quad A \cup \overline{A} = X, \quad A \cap \overline{A} = \emptyset,$$

and the De Morgan laws are also satisfied.
For the simplicity of notation fuzzy set $A:X \to [0,1]$ such that $A(x)=1$ for all $x \in X$ is denoted by X and similarly Ø indicates fuzzy set $A(x)=0$ for all $x \in X$.
One of the best known examples of operations satisfying these conditions are so called bold conncetives. The others are studied in detail by Weber [9].
Let us first discuss variant $P:X \to [0,1]$, $K_Q:[0,1]^X \to \{0,1\}$. A quite natural way of defining the truth value of the proposition $Qx:P(x)$ is the following:

$$T(X_P \in K_Q) = \begin{cases} 1, & \text{if there is } A \in K_Q \text{ s.t. } A \subseteq X_P \\ 0, & \text{otherwise} \end{cases}$$

The word "natural" is used here because by such a definition discussed here variant $P:X \to [0,1]$, $K_Q:[0,1]^X \to \{0,1\}$, $T(QxP(x)) \in \{0,1\}$ can be considered as a generalization of the crisp variant $P:X \to \{0,1\}$, $K_Q:2^X \to \{0,1\}$, $T(QxP(x)) \in \{0,1\}$.

Therefore it is superfluous to interprete here in fuzzy set language all conclusions derived in the previous paragraph.
The case when $T(QxP(x)) \to [0,1]$ is less natural because of arbitrary of the definition of the truth of the sentence $QxP(x)$. This truth can be defined (for examples) as follows:

$$T(QxP(x) = M(h(x)) \tag{8}$$

where M is some mean operator applied to fuzzy set $h:X \to [0,1]$, which, again quite arbitrary can be defined for examples as follows:

$$h = \cup\{A e K_Q \mid A \subseteq X_P\},$$

or

$$h = \cap\{A e K_Q \mid A \subseteq X_P\},$$

where $\cup$ can be conceived as a operator of union, supremum, lowest upper bound etc. (The same is valid for $\cap$).
The mean operator can be defined for example as follows:

$$M(f(x)) = \max_{x \in X} f(x)$$

$$M(f(x)) = \int_X f(x)\, d\mu(x)$$

Axiomatics of mean operators are fully discussed in [2].

Let us now consider the variant $P:X \to I$, $K_Q:2^X \to [0,1]$. For which again there are two possibilities: either $T(QxP(x)) \in \{0,1\}$ or

$T(QxP(x)) \in I$. The first case is hard to define formally with the reasonable motivation. This follows from the fact that there is no natural way to compare fuzzy set X_P with crisp subset of X "weighted" by some grade of numbership to the family K_Q. It will be considered therefore the case $T(QxP(x)) \in I$.
For the truth valuation on the sentence $Qx:P(x)$ it can be used, again rather arbitrary, formula (8) with

$$h(x) = \perp \{X_P(x) \ \mathrm{T} \ K_Q(A)\}$$

where $\perp$ is same t-conorm which is extended for all subsets A of the set X, and T is some t-norm.
R. Yager (see [11]) in his substitution method uses

$$h(x) = \max_{A \subseteq X} \{\min(X_P(x), K_Q(A))\}. \qquad (9)$$

7. CONCLUDING REMARKS

Quantified statements are frequentely used by operation researches for formulation decisional processes and algorithms. For example, so called optimal decision is usually defined as a maximal element of same prespecified admissible set of alternatives with respect to the same order relation $\geq$. The set X_{opt} of all optimal elements in X is defined as:

$$X_{opt} = \{x_o \in X \mid \forall x: P(x)\},$$

where

$$P(x) = \begin{cases} \text{true, if } x_o \geq x \\ \text{false, if } x_o < x. \end{cases}$$

Very often it is more justificable to determine a satisfactory instead of optimal element.
Element x_s might be defined satisfactory if:

for *many* $x \in X$ is fulfilled $x_s > x$

or if

for *most* $x \in X$ holds $x_s > x$, etc.

A lot of so called heuristic algorithms are constructed for determining such a element x_o that

for *most* x it is not true that x is better than x_o.

The quantifier word *most* is conceived in this case very differently. It might mean for example *all* examined elements, the *majority* of all available elements, or it could also mean that x_o is *far enough* from the worst element, etc.

Algorithm is meant here as a constructively defined mapping from the data D to the results R.
In the case of fuzzy mapping $A: D \rightarrow R$, for any given input $d \in D$ corresponds some fuzzy subset of R, which is called "right result".

This fuzzy subset is caused mainly by the execution of branchings i.e. conditional instructions.
Conditional instruction of the type:

if $Qx \in X:P(x)$ then I1 else I2

is treated to be equivalent to the execution of both I1 and I2, but with a different degree of membership to the "wright result". These degrees are equal to $T(Qx \in X:P(x))$ and $T(Qx \in X:P(x))$ respectively for the instructions I1 and I2.
In the case of $T(QxP(x)) \in B$ we arrive at usual branching i.e. only *one* instruction is executed: either I1 or I2.

For the case of so called collective decision making K_Q is referred to as a majority space, and the elements of this space are called coalitions.
Suppose that $d_1, \ldots, d_m$ are individual decisions, for example such as $d_i \in B^n$ for the same fixed $n \geq 1$.
Having fixed family $K_Q: 2^m \to \{0,1\}$ the aggregated social decision $d^s = (d_1^s, \ldots, d_n^s)$ can be obtained by the majority rule (see [1,8]):

$$d^s = \bigcup_{c \in K_Q} \left(\bigcap_{i \in c} d_i \right)$$

It means that $d_j^s = b$, where $b \in \{0,1\}$ holds only in the case when there exists same coalition $c = \{i1, i2, \ldots, ik\}$ in the family K_Q such that all members of it agree with decision $d_j^s = b$.

ACKNOWLEDGEMENTS

The paper was written while I was visiting the Institute of Statistics and Operations Research at the University of Trento in Italy in 1985. I am very grateful for the financial support and for providing excellent conditions and stimulating atmosphere.
I am much indebted to Professors M. Fedrizzi and A. Pollastri for their encouragement and support. My particular thanks are due to Prof. Z. Hellwig from Academy of Economics in Wroclaw and to Prof. M. Zenga from University of Milan for making my visit to Italy possible and for their continuous support and encouragement.

BIBLIOGRAPHY

[1] V.I. Danilov, Group choice models, Tech. Kibernetika, 1.198b, 143-164 (in Russian).

[2] D. Dubois, H. Prude, A review of fuzzy set aggregation connectives, Inf. Sciences, 36, 1985, 85-121.

[3] J.A. Goguen, The logic of inexact concepts, Synthese, 19,

1968, 325-373.

[4] R.J. Hintikka, Quantifiers in logic and quantifiers in natural languages, in: S. Korner (ed.), Philosophy of Logic, Basil Blackwell, Oxford, 1978.

[5] P.L. Peterson, On the logic of "few", "many" and "most", Notre Dame Journal of formal Logic, 1, 1979.

[6] H. Reichenbach, Elements of symbolic Logic, The Free Press, 1966.

[7] Y.A. Shreider, Voting systems and their representation as a majority structures, Kibernetika, 4, 1979, 57-61 (in Russian).

[8] N.Y. Vilenkin, Y.A. Shreider, Majority spaces and quantifier "majority", Semiotika i Informatika, Vol. 8, 1977, 45-82 (in Russian).

[9] S. Weber, $\perp$ - decomposable masures and integrals for Archimedean t-conorms $\perp$, Journal of Math. Anal. and Appl. N. 1, 1984, 114-138.

[10] R.R. Yager, Quantifiers in formulation of multiple objective decision functions, Informations Sciences 31, 1983, 107-139.

[11] R.R. Yager, Aggregating Evidence using quantified statements, Information Sciences, 36, 1985, 179-206.

[12] L.A. Zadeh, A computational approach to fuzzy quantifiers in natural languages, Comp. and Maths. with Appls. Vol. 9, N. 1, 1983, 149-184.

PART II

APPLICATIONS IN KNOWLEDGE-BASED SYSTEMS, DECISION AND CONTROL

Fuzzy Logic in Knowledge-Based Systems,
Decision and Control
M.M. Gupta and T. Yamakawa (Editors)

HEURISTIC LEARNING EXPERT SYSTEMS
General Principles

A.F. Rocha*, M. Theoto** and P. Torasso***

* Departamento de Fisiologia, UNICAMP, 13100 Campinas
** Escola de Enfermagem USP, 05403 São Paulo
*** Dipartimento di Informatica, Univ. Torino

KEY WORDS: Fuzzy Set Theory; Diagnostic Expert Systems; Heuristic Learning; Frame Knowledge Representation; Neural Net Theory.

ABSTRACT

The present paper discuss the general organization and the main components of an intelligent system having some capability of heuristic learning and reasoning. The structure of the system is quite innovative since it integrates knowledge representation formalism (and in particular frames) with results on human reasoning and learning obtained in studying and modelling neural nets. The final goal of this kind of approach is to provide a diagnostic expert system with the ability of modifying and optimizing its knowledge on the basis of the experience the system can gain by solving particular cases. This particular form of learning (heuristic learning) is a keystone for human knowledge acquisition and it is based on the refinement of knowledge by means of repeated observations of the external world. It corresponds of some basic neural mechanisms known as sensory adaptation, attenuation, conditioning, etc. The same rules determining these neural mechanisms are used here to guide the heuristic learning.

INTRODUCTION

The current generation of expert systems lacks one of the most significant properties of the human brain: The capability to adapt control strategies and knowledge sources by observing the external environment. This particular form of learning (called here, heuristic or inductive learning) may result in developing and refining heuristic by solving particular cases. The brain mechanisms involved on heuristic learning are the same ones responsible for classic and operant conditioning, sensory adaptation, etc.

Knowledge is represented in the brain through the association of neurons, and it is dependent on both the structure of this association as well as on individual properties of each neuron (Rocha, 1983). Stimulus pairing results on modifications of the cellular coupling (some linkages are strengthened, some others reduced) and on alteration of the stimulus enconding (frequent stimuli are encoded by the most efficient length and energy codewords) (Rocha and Bassani, 1982). These processes are the most important ones for heuristic acquisition.

Heuristic learning preceeds the deductive learning both when phylogeny and ontogeny is considered. The repeated observation of the same pattern shapes the cellular connectivity of the human brain and engraves the neural nets responsible for the processing of our behavior. This initial modelling may be improved by further logical analysis.

All the attempts to provide computers with some reasoning and/or learning capability privileged logics (Barr and Feigenbaun, 1982; Michelski, Carbonell & Mitchell, 1983; Michaelsen, Michie and Boulanger, 1985). Because of this, all models had to take for granted a initial knowledge base. named heuristic. Biology, in turn, neglected logics, and has been studying in deep details the rules to shape the animal behavior by controlling the external stimulation (Adey, 1966; Allen and Tsukahara, 1974; Buño and Velluti, 1977; Buzsaki et al, 1979; Hubel and Wiesel, 1970; Kleinamn and John, 1975; sdiman and Rakic, 1973). The time is ripe to a step in the direction of a total understanding of the human reasoning, by combining the knowledge acquired by these two different areas of the human endeavour.

The purpose of the present paper is to discuss the general structure of an Expert System based on frame, neural nets and fuzzy logic theories, and having some capability of heuristic learning taking into account the knowledge acquired by Biology about the brain functioning.

THE INITIAL REMARKS

The behavior of each neuron may be described using fuzzy automata theory (Rocha et al, 1980, Rocha, 1981). Briefly, any neuron N is assumed to be a fuzzy automaton:

$$N = (S, SC, Q, \phi, \beta)$$

to which

1) S is the set of inputs or stimuli;
2) SC is the set of outputs or classes or neural responses;
3) Q is the set of neural states;
4) β is the output function;
5) ϕ is the family of state membership ($\mu(q,q')$) matrices

$$\mu(q,q') = \mu(\Omega,t).f(r).\mu(s,s').\mu(sc,sc')$$

dependent on:

5a) The time Ω the neuron is able to hold information - The function $\mu(\Omega,t)$ controls the state transition depending on the relation between the neural memory and the time of the stimulus occurence. States of high entropy are favoured by repeated application of excitatory (inhibitory) stimuli during (outside) the retention period Ω.

5b) The number of stimuli received in the retention time Ω. States of high entropy are favoured by frequent (unfrequent) excitatory (inhibitory) stimuli.

5c) The stimulus membership ($\mu(s,s')$) - This function controls the state transition depending on the relations holded by the stimuli. States of high entropy are favoured by frequent paired stimuli.

5d) the control ($\mu(sc,sc')$) exercised over the neuron by other neural structures.

The coupling ($\mu(Na,Nj)$) between two neurons Na and Nj depends on the capacity Na have to trigger different responses on Nj. Thus, $\mu(Na,Nj)$ is dependent on both the entropy $h(SCa,j)$ transmitted from Na to Nj, as well as on the entropy ($h(SCj)$) of Nj:

$$\mu(Na,Nj): SCa \times SCj \longrightarrow [\,0,1\,]$$

$$\mu(Na,Nj) = h(SCa,j) \,/\, h(SCj)$$

$$SCa,j = \beta j(q,SCa)$$

The heuristic knowledge is composed by those sets of actions providing the best environment adaptation and its acquisition is guided by the information patterns existing in the surrouding environment. These external patterns shape specific neural nets in the brain, by changing the neuronal coupling, according with their frequency and association of their components. This process is the responsible for most of the knowledge imprinting during the training of the expert.

Any neural model Mj represented in the neural net NNj, becomes fully characterized if its structural description is provided by the neural graph NGj underlying NNj and if its interpretation is furnished by the language L(NGj) defined over this NGj (Rocha et al, 1980; Rocha, 1983). Well learned models are represent on neural nets that (Rocha et al, 1987):

1) have a small number of nodes;
2) are more likely to be a tree;
3) handle context free languages,and
4) require less processing times.

The neural nets NG are organized in the Nervous Systems (NS) depending on the neuronal coupling:

$$NG = (\{SCj\}_{NS} \;,\; \{\mu(Na,Nj)\}_{NS} \;)$$

The coupling power of theses nets are calculated as

$$cp(NG) = |M|_{p}\; |m|_{s}\; u(Na,Nj)$$

where |M| and |m| stand for the operations realized over parallel (p) and serial (s) pathways, respectivelly. |M| could be a max-operation, and |m| could be a min-operation as preconized by classical fuzzy set theory in the case of logic operators |or| and |and|, respectivelly. However, other functions are more likely to replace these max and min, in order to turn formalization closer to reality (Greco and Rocha, 1986).

The fuzzy grammars realized over such nets are fuzzy relation grammar, with productions of the type

$$SCa \xrightarrow[\alpha(p)]{} SCj$$

$$\alpha(p) = \mu(SC_{a,a+1}, SC_{a+1,j})$$

where $\alpha(p)$ expresses the dependence on the context.

The fuzzy language L(NG) defined by these grammars is composed by productions of the type

$$L(NG)\ (n) = |M|\ (\ (p1)\ |m|\ (p2)\ \ldots\ |m|\ (pk)$$

|M| being taken over all derivation chains from the point basis to the node |n| (Rocha et al, 1980).

If new pieces of informations are in disagreement with a model Mj, they may reduce the strength of its neural net NNj, and may increase the connectivity on its complementar neural nets rNNj (Rocha, 1982a). In such situation disagreement and error may trigger a procedure to find whether a new and more complete model may be developed from NNj or from rNNj to accomodate the new knowledge (Rocha, 1983). This procedure searchs for α-cut level sets on NNj and rNNj being able to maintain the highest possible connectivity and the smallest disagreement with the new data. If this α-cut level sets exists, it will be a germ or frame for the development of an Expert Model.

Any information disagreeing with the model Mj currently in use may be accepted as such or may be questioned as bad observation. This will depend on the autonomy of Mj, or in other words, on the capacity of Mj to be maintained or to be build again in the brain (Rocha, 1982b). The autonomy of Mj will depend, among others, on the personality of each brain or on different kinds of learning strategies (Rocha, 1987). Some ones are conservative and prefer to maintain Mj while questioning observation; some others are non-conservative and prefer to challenge Mj to fit or not the new data. Con- flicting information in the case of conservative systems, and data about old models in the case of non-conservative models may be stored on memory systems. These memory systems may differ on their retention times (e.g, Addey,1966). Besides this, the memory system may store data used to instantiate the neural nets or frames.

The inductive learning imprints the external environment in the brain, such that the most important external patterns become represented many times in the central nervous systems. Each time a pattern is represented in the brain, it shares not only a different neighborhood but also distinct memberships with these neighbours nets (e.g., Dreyer et al, 1975; Loe et al, 1977). Any neural frame shares, therefore, many and distinct relations with other patterns (Rocha, 1985).

The brain processing may be viewed as a navigation throughout a set of neural nets representing external patterns (sensory areas) and a set of neural nets representing actions (motor, behavior, autonomic areas, logic reasoning) to be exercized upon both the external and/or internal environments. This navigation is dependent on excitations and inhibitions triggered over the synapsis (the site of connection between neurons) by both external information or by internal activated neural nets (frames). Specific and momentary pathways are activated for the actual processing dependent not only on the actual incoming messages but also by the previous received and processed data. These momentary nets (called operational nets (OG), Rocha et al, 1980) are being built and rebuilt constantly, and correspond to the actual working hypothesis. Each working hypothesis may trigger specific observations of both the external (consultation) and internal (recalling) environment necessary for its validation.

The traffic of messages ($\pi(Na,Nj)$) between any pai of neurons Na,Nj is determined by the balance between inhibitory (I) and excitatory (E) actions exercised over them by the incoming messages:

$$\pi(Na,Nj) = \mu(E,Nj) - \mu(I,Nj)$$

The operation nets NG are dependent on $\pi(Na,Nj)$:

$$OG = (\{SCj\}_{NS} , \{\pi(Na,Nj)\}_{NS})$$

Navigation is controlled by specific neural systems and rules (Coquery, 1972; Kleiman and John, 1975; Massion, 1973; Rocha, 1978)

The brain may observe itself to study the behavior of the models imprinted on it. Self observation is the keystone to the development of any system having the capacity to model its own models, the first step on creating a deductive learning. Logics may be assumed as a model developed by man during his historical development, to analise and create models (Rocha and Rocha, 1985).

Message traffic control, housecleaning, memory control, choice of learning strategy, etc. will demand the supervision of specialized neural systems or nets. These high order systems are developed throughout phylogeny and ontogeny (Luria, 1974).

THE GENERAL STRUCTURE

Any diagnostic expert system having some similarit with the human way of thinking, has to be composed of three main modules:

1) CONSULTATION SYSTEM: whose task is to carry on the actual consultation on particular cases working the knowledge base relative to a particular human (e.g.medical) specialization. It its the responsible for the navigation in the set of frames used to represent findings (observed data) and hipotheses (structured knowledge about diagnosis). It uses two different kinds of frames to accomodate the difference between the characterization of findings and prototypical knowledge associated with the diagnostic hypothesis. Frames share also distinct relations:

a) Hierarchical Relations of the type ancestral and descendent linkages, and

b) Associative Relations describing the membeship among frames representing chunks of knowledge having some features in common. It is very important to take this kind of relations in consideration in order to increase the similarity between the Expert System and the human brain, once the major characteristic of the Nervous System is association of its cells.

2) MEMORY SYSTEM: is composed of three different memories short, medium and long term memories. Its task is to store the experience of the system and data used to instantiate prototypical knowledge. Because of this, it is organised into:

2a) Knowledge Memory System: whose task is to store information about new states of knowledge which may be triggered if conflicting data is proved to be true (conservative systems), or about old states of knowledge which have to be restored if some working hypothesis were prooved to be false. The system itself is not conservative in order to avoid a "reductio ad infinitum".

2b) Record File System: to store the most significant dat gathered during specific consultations. It allows the system to build again old models, to fill in missing data at the first consultation, and/or to correct data.

3) SUPERVISOR: takes care of the most complex tasks in the system: moves information to and from the memory systems; keeps track of the time in the system by adding a unit of time to the actual time at the end of each consultation; organises the record files; controls learning; decides for the learning strategy to be used, etc. The supervisor is also in charge of the relations between the entire system and the external world.

All of this means that by analyzing cases and by receiving after some time a confirmation or disconfirmation of the correctnes of chosen diagnosis, the supervisor is able to improve the behavior of the system. This is obtained by strengthening some links between frames and reducing the importance of some other associations. The result of this is a constant updating of the knowledge. To perform this job the supervisor may decide for conservative or non-conserva- tive approaches by determining, for instance, the number of repeated observations necessary to activate the mechanism that updates the knowledge base according with the experience.

THE CONSULTATION SYSTEM

The knowledge about a particular domain is represented by frames (Aikins, 1983; Cravetto et all, 1985; Fikes, 1985) at different levels of detail: the most general knowledge is represented at the level of contexts, viewed here a set of families of diagnosis. Theses families, called here gross diagnosis classes, are a step deeper on the refinement of the working hypothesis toward the most detailed level, the level of diagnosis. This organization allows, as in the brain, (Allen and

Tsukahara, 1974; Rocha, 1982a), two different kinds of strategy to be used during the consultation:

1) quick approach: if there is a strong correlation betwee data and diagnosis, the consultation is simplified by inquiring only those diagnosis whose frames were activated by the provided external data,

2) sequential approach: on the contrary, if the correlation between data and hypothesis is low, then consultation starts activating frames at the level of the context which may guide the process toward the level of gross diagnosis classes, by taking into consideration both the external data as well as previous knowledge about frame association. The hypothesis gathered at this level will be used, in turn, to orient the inquiry to obtain the necessary data to refine the process toward the diagnosis level.

Frames are activated by means of two different mechanisms:

1) because some important data, which could act as triggers are provided by the external world or by the File Record System, and or

2) because some its parent and/or or relative frames were activated.

The Consultation System carries out its job in two main phases:

PASSIVE PHASE: The system is given informantion from the external world to initiate the consultation. These pieces of information are used to activate some initial working hypothesis, and

ACTIVE PHASE: Once the confidence on some activated working hypothesis reaches a predefined threshold established by the supervisor, or the external world stops to spontaneously provide data, the system starts to actively inquire the external world or the File Record System for filling the slots of the activated frames.

The Consultation system may organize data according with some procedure and temporal guidelines, called Procedure Steps, separating clear different activities of the inquiring, as for example, anamnesis, physical examination, complementary tests, etc. in the case of a medical expert system.

The Knowledge Base is organised as a family of frames describing findings and hypothesis.

FINDINGS: are atomic pieces of information about hypothesis and obtained from the external world or from the Record File System. Findings include, for example in the case of a medical expert system, information about signals, symptons and complementary laboratory tests.

All findings share the same frame structure, named here Frame type 1:

Identification

Slot name: Name and specification

(Fever Periodic)
Slot confidence: the confidence [0,1] assigned to the received information

Structural description

Slot Procedure Step: describes the steps on which the findings has to be furnished or asked for.

Slot Relatives: contains the degree of membership [0,1] with other findings in the Knowledge Base

Prototypical Description

A set of slots providing the best characterization of th data to be handled by the system. In the case of an Expert Medical System this description involves 6 slots:

Slot Type
Slot Intensity
Slot Duration
Slot Frequency
Slot Location
Influencing Factors

Frames describing findings are not required to contain all these slots if the corresponding features are not applicable to the findings under consideration.

DIAGNOSTIC HYPOTHESIS: Structured knowledge about a particular diagnosis. All the diagnostic hypothesis are described by the Frame Type 2:

Identification

Slot Name: the name of the Diagnosis.
Slot Evidence: measured on the closed interval [0,1] and dependent on both the confidence on the its findings and on the correlation between these findings and the actual diagnosis.

Structural Description

Slot Ancestors: describes the linkage with the Gross Diagnostic Classes. It cannot be empty.

Slot Relatives: contains the list of the diagnosis sharing the highest membership with the current diagnosis.

Slot Descendents: empty

Control Knowledge Description

Slot Triggers: the list of the findings that activate the frame to be taken into consideration by the Consultation System. The list has the following general form:

$$\{ Fi=Ln, \int_{Dj} (Fi=Ln) \}$$

Fi - the name of the finding;
Ln - the Generic Instantiation of Fi concerning the

diagnosis Dj, or in other words, the list of the allowed values assumed by the slots of Fi to associate it to Dj;
f_{Dj} – the membership function for Fi concerning the diagnoses Dj;

Prototypical Description

Slot Essential Components: Contains the findings that have to occur in order the diagnoses Dj may be taken into consideration as a candidate for the final diagnosis Triggers are essential components, too.

Slot Complementary Components: Contains the findings that provide additional support for the diagnoses Dj

The structure of the elements of these two later slots is that of a fuzzy logic graph (Greco and Rocha, 1987):

Terminals: List of essential findings ordered by the importance they have to elicit the specific diagnoses Dj.

Arcs: membership fucntions between findings and diagnoses;

Non-terminal nodes: logic connectives associating findings;

The strenght between findings and second order nodes are the values for f_{Dj} (Fi=Ln), whereas the strength of the arcs between the second order nodes and the upper nodes are dependent on f and on the logic connective assigned to the upper node (Greco and Rocha, 1987; Rocha et al, 1987).

Essential and complementary components are discriminated by means of the power of the arcs of their fuzzy logic graphs. The discriminating threshold is the same for all frames and it is part of the knowledge of the supervisor.

GROSS DIAGNOSTIC CLASSES: A family of diagnosis sharing a common subset of triggers and used to organise knowledge at a more generic level. All diagnostic classes share the same type of frame:

Identification

Slot Name
Slot Evidence

Structural description

Slot Ancestors: just one context
Slot Relatives: contains the list of the gross classes sharing the highest membership with the actual diagnostic class.
Slot Descendents: the list of the diagnosis sharing the highest correlations (measured by means of a membership funtion, Greco and Rocha, 1987) with the actual diagnostic class.

Control Knowledge and Prototypical descriptions

As described above for Hypothesis or Diagnosis.

CONTEXT OF EXPERTISE: A family of Gross Diagnostic Classes sharing a common set of triggers and used to provide some loose organization of knowledge to guide the initial steps of very fuzzy consultations. They share the following frame structure:

Identification

Slot Name
Slot Evidence

Structural description

Slot Ancestors: empty
Slot Relatives: list of contexts sharing the highest membership with the actual context.
Slot Descendents: the list of the diagnostic classes sharing the highest correlations with the actual context.

Control Knowledge Description

Slot Triggers

Prototypical Description

Slot Components: A fuzzy logic graph of findings.

Each consultation begins with the external world providing some initial data about findings used to instantiate some thype 1 frames. These findings start the process of activation of the frames of type 2 (high order frames) used to organise the knowledge at the level of context, gross diagnostic classes and diagnosis.

The starting level of the consultation depends on the correlation between the initial data and the Diagnosis at the Knowledge Base. At one extreme, if the provided data are strongly correlated to some Diagnosis, then they may be sufficient to activated the frames at the level of Diagnosis and to end the passive phase of the consultation. On the contrary, if the initial data are poorly associate with the known Diagnosis, then they will first activate one context of expertise, wich may guide the choice of some gross diagnostic classes to be investigated. Intermediate degrees of correlation may trigger the consultation at the level of Gross Classes.

The next step of the consultation will be to rank the activated frames according with their degree of evidence before to begin the inquiry of the active phase.

The degree of evidence of a hypothesis is evaluated by taking into account the confidence on the findings and on the correlation between findings and frames and/or the membership between frames. The logic connectives used to combine data is also relevant for the calculus of the evidence of a frame (Torasso &

Console, 1987). The functions to be used for the calculus of the evidence may be chosen among the classical propositions in the literature, or they may be established by means of experimentation (Greco and Rocha, 1987).

Since only partial information may be provided by the esternal world during the passive phase, the initial ranking of gross diagnostic classes may be obtained using just part of the above procedure.

After selecting the most promissing gross classes the system may starts the inquiry of their descendents or close relatives. The analysis of descendents begins exploring triggers; procedes analysing essential components, and finishes studying complementary components. Relative frames may be activated at the level of Gross Diagnostic Classes and Diagnosis.

Whenever an essential component is not verified, the frame it belong to is disregarded.

Any finding provided by the external world to the system that are not a component of any of the activated frames, will be associated to a list of unaccounted findings.

At the end of the consultation, the remaining activated frames are reordered according with their degrees of evidence, and all diagnosis with an evidence degree below an acceptance threshold will be disregarded. (Rocha, 1983). All findings that are accounted for only by disregarded frames, as well as those remaining on the unaccounted list will be associated to the empty context, representing the ignorance of the system.

CHANGING THE KNOWLEDGE

The repeated observation of some external patterns will modify the Knowledge of the system by changing the membership function between Findings and Context, Gross Diagnostic Classes and Diagnosis, as well as the membership relations between these high order frames. The way these values will change is determined by the rules established by Biology to classic and operant conditioning, sensory adaptation, etc, and formalized with the aid of fuzzy automata and fuzzy logics theories (Rocha 1982 a,b, 1983, 1985).

The idea is that repeated pairing between findings and Diagnosis, Gross Diagnostic Classes, etc. as well as repeated association between these high order frames will strength their membership functions and relations in the same way the cellular connectivity is enhanced in the Brain by stimulus pairing. The values of these fuzzy functions or relations concerning competitive frames may be reduced obeying the same rules.

In this condition, the membership function $f\,(Fj=L)_{Hi}$ between findings Fj and hypothesis Hi is dependent on (Rocha et al, 1980) the frequency of association between L and Hi. This frequency is dependent on the number |x| of associations of this type occurring at the time |t|, within the retention time interval (|[rt,RT]|)

defined for the system memory, and on an ideal number |r| of repetitions to change the system:

$$f_{Hi}(Fj=L) \rightarrow \begin{cases} 1 \rightarrow \begin{cases} x \rightarrow |r| \\ rt < t < RT \end{cases} \\ 0 \text{ otherwise} \end{cases}$$

The evaluation of the actual value of f (Fj,L) requires, therefore, the following data from the supervisor:

|r| the ideal number of observations

the retention time interval [rt,RT]

the initial value $f^{o}_{Hi}(Fj=L)$

and it uses the following types of functions:

1) $f^{t}_{Hi}(Fj=L) = f^{t-1}_{Hi}(Fj=L) + (1 - f^{t-1}_{Hi}(Fj=L))*\Omega$

2) $f^{t}_{Hi}(Fj=L) = f^{t-1}_{Hi}(Fj=L) - f^{t-1\ 1)}_{Hi}(Fj=L)*\Omega$

3) $f^{t}_{Hi}(Fj=L) = f^{o}_{Hi}(Fj=L) + [(f^{t-1}_{Hi}(Fj=L)) - f^{t-1}_{Hi}(Fj=L)*\Omega)$
$*(1-f^{o}_{Hi}(Fj=1))]$

and the following rules:

a) Ω is a function of |r|, being small if |r| is large;

b) if L occurs within rt<t<RT then use function 1;

c) if L occurs within t<rt or t>RT then if the system has

c1) to recover the previous state then uses function 3;

c2) to be inhibited by a competing model then uses function 2;

d) if L does not occur within rt<t<RT then if the system has

d1) to recover then use function 3;

d2) to be inhibited then uses function 2.

In the same line of reasoning, the membership re- lation $\mu(Hi,Hj)$ between two hypothesis Hi and Hj will be considered dependent on:

1) use: $f_{Hj}(Hi)$ calculated in the same way as $f_{Hi}(Fj{=}L)$;

2) the structure of the knowledge $\mu s(Hi,Hj)$ calculated with the aid of the following procedures:

   2a) Procedure to analyse frames: the similarity between the frames increases as the similarities between triggers, essential and complementar components are enhanced;

   2b) Procedure to analyse Triggers: the similarity between the triggers increases as the intersection of their generic instantiation lists augments;

   The generic instantiation of a finding is a set of constraints about the admissable values on their slots;

   2c) Procedure to analyse Components: the similarity between essential or complementary components increases as the similarity of their fuzzy graphs augments;

   2d) Procedure to analyse Fuzzy Graphs: the similarity between two fuzzy graphs increases as the intersection of their production languages enhances;

3) the hierachy (ancestors) organizing the knowledge in the system $\mu(Ai,Aj)$, calculated with the aid of the following rules:

   3a) Take the strongest/weakest relation between ancestors and their descendants, depending the system to be non-conservative/conservative, to chose the fathers that will influence the membership betweem hypothesis;

   3b) Take the membership between the chosen fathers as the actual value of $\mu(Ai,Aj)$.

The actual value $\mu(Hi,Hj)$ will be calculated as:

$$z = f_{Hj}(Hi) + (1 - f_{Hj}(Hi)) * f_{Hj}(Hi) * \mu s(Hi,Hj)$$

$$\mu(Hi,Hj) = z + (1 - z) * z * \mu(Ai,Aj)$$

The calculations will be executed by the supervisor at the end of the consultation:

1) first, for those hypothesis chosen as possible diagnosis, and for those data used to reached such conclusions;

2) for unaccounted findings and the chosen hypothesis, and

3) for unaccounted findings and the empty context.

The association between unaccounted findings and chosen hypothesis will permit to modify the previous knowledge by adding new informations to it, whereas the association between findings and the empty context may generate questions (Rocha, 1983) to challenge the actual knowledge.

When a finding is disconnected from a disregarded hypothesis, a trace of this linkage will be maintained in the system by increasing the membership between the disregarded frame and their relative frames considered as possible. By this way, frequent disregarded hypothesis may be fused into frequent accepted ones in order to accomodate new patterns in the external world resulting from modification of old ones.

THE SUPERVISOR AND THE MEMORY SYSTEM

Each time a particular case is submitted to the Expert System, the consultation system will provide a list of possible diagnosis by analysing the initial data and by asking for additional support for some selected hypothesis. Once the system has obtained some final possible diagnosis, the supervisor has to take track of this decision and of the findings supporting it, in order to take some decision about the consequences these conclusions may have upon the knowledge structuring in the system.

In solving particular cases, the consultation system points to findings supporting final diagnosis. Whenever this association becomes frequent and if the diagnostic hypothesis is confirmed, the supervisor has to strengthen the membership between the diagnosis and the findings supporting it. On the contrary, if hypothesis is disconfirmed then the reduction of such relations should be the case to be considered. In a similar way, the membership between frequently co-occurring hypothesis should be increased if the hypothesis are confirmed, or reduced if they are reject.

This kind of heuristic reasoning is part of the task of the supervisor, which has to decide about the knowledge restructuring taking into consideration not only the confirmation /disconfirmation of the decision at the end of the consultation, but also the conservative/non-conservative strategies declared in the system.

The impact of the confirmation/disconfirmation of a diagnotisc hypothesis decreases in the long run, since many details about the knowledge restructuring are lost if the time between consultation and the acknowledge of the result of this consultation increases. Forgetting is fundamental to the human reasoning, not only as a mechanism to disregard unconfirmed and low frequent hypothesis, but also to summarize data to be transferred among systems. If the strenght of data and frame association is time dependent, then summarizing may be easily processed by the same rules proposed in the previous section to determine the knowledge restructuring in consequence of the use of the system.

The human memory is assumed to be divided into short, medium and long term memories (e.g, Addey, 1966), each one playing different and complementary roles in the system. Information is passed from short toward long term memory if it is proved to be useful to system.

The supervisor will evaluated the possible modification of the knowledge base at the end of the consultation using the mechanisms discussed in the previous section. The supervisor will chose one of the following actions, depending on the kind of learning strategy declared in the system:

1) non-conservative strategy: it will change the Knowledge Base according with the results of the calculation, and will transfer the old counterpart of the modified structure to the short term memory.

2) conservative strategy: it will maintain the actual Knowledge Structure, and will put the results of the calculation in the short term memory.

Each time the supervisor stores information in the memory system, it attachs a time label to the transferred data to keep track of the forgetting. At the same time, it will verify the time label of all stored information. Those data having its memory time reaching the superior limit RT of the retention time will have their membership functions and relations recalculated using the formula and rules proposed in the previous section and taking in consideration the confirmation/disconfirmation process. After this, the data:

1) will be used to modify the conservative system or they will be removed from the memory of non-conservative system in the case of hypothesis confirmation. Otherwise, they

2) will be moved from short to medium term memory;

3) will be moved from medium to long term memory, and

4) will be removed up from long term memory of conservative systems or they will determine the recover of the previous learning state in the case of non-conservative systems.

Only those data having their calculated membership functions and relations greater a defined threshold (part of the knowledge of the supervisor) will be moved according with the strategy above. This procedure corresponds to the heuristic summarizing process discussed before.

The recovering of past learning states in the case of non-conservative systems, and the calculation of future learnings in the case of conservative systems may require the consultation to the File Record System, because the Memory System itself is not conservative. This means that, no trace of the old memory knowledge is kept in the memory systems. This could prevent the system from fully recovering any previous state or from having all data necessary to perfectly define future transitions. In this situation, the supervisor will ask the File Record System about some specific pieces of informations.

The File Record System is an auxiliary memory system similar to the files maintained by the (medical) expert about previous consultations. The supervisor is the responsible for its maitenance, too. The idea is that the supervisor will put in the File Record the necessary information to recover old knowledge or to guide the future knowledge modifications. This means that a summary of the consultation and of the resultant conclusions will be stored in the File Record. This summary will consist of the chosen diagnostic hypothesis and the findings supporting them.

CONCLUSION

The two most striking features of the system presented so far are:

1) the hierachical relations between diagnosis, gross diagnostic classes and contexts, and the associative relations among relative frames at each of these levels of knowledge organization, because they mimic the way knowledge is stored in the brain by means of association of neurons and neural nets (Rocha, 1985), and

2) the possibility to modify the membership functions correlating findings and hypothesis and the membership relations associating hypothesis, according with the experience gained by the system in solving particular cases.

The repetion of stimulus patterns modify the coupling of neurons and shapes preferential neural nets for knowledge representation in the brain, because the neural processing is dependent on message navigation in the brain (e.g., Coquery, 1972; Massion, 1973; Rocha et al, 1980).

Consultation in the Expert System proposed here, is dependent on navigation in its Knowledge Base. The frame activation and the maintenance of this activation are dependent on the strength of the membership shared by fidings, diagnosis, etc. Repetition of similar observations will change this membership in a similar way the neural coupling is modified by repeated stimulus pattern, and, thus, it will reorganize the Knowledge Base in the same way the expert's knowledge is modified by heuristic learning.

An improvement of this process is to represent the same knowledge many times in the Knowledge Base, each of its frames sharing distinct memberships with the frames representing other models. This process will mimic the cerebral representation of the expert knowledge (Rocha, 1985) and it will enhance the system's ability to cope with the external pattern variability. However, the multiple knowledge representation and association may eventually provoke a combinatorial explosion, if not controlled. It is necessary, therefore, to reduce and even remove associations and representations that turn to be useless due to changes in the external environment. This is the role played by forgetting in the proposed Expert System. The implementation of a Multiple Knowledge Representation may be also a complex task on the actual serial processing machines and it will be a necessity on the future paralell processor computers.

REFERENCES

Adey, W. R. - Neurophysiological correlates of information transaction and storage of brain tissue. In: Progess in Physiological Psychology, E. Stellar and J. M. Sprage, Academic Press, 1966/1-43

Aikins, J. - Prototypical knowledge for expert systems. Artificial Intelligence, 1983/23:163-210

Allen, G. I. and Tsukahara, N. - Cerebrocerebelar communication systems. Physiological Review, 1974/54:957-106

Barr, A. and e. A. Feigenbaun - The Handbook of Artificial Intelligence. Heuristech Press/Willian Kaufman, 1982

Buño, W. Jr and J. C. Velluti - Relations of hippocampal theta cycles with bar pressing during self-stimulation. Physiology & Behavior, 1977/615-621

Buzsaki, G., E. Grastyan, I. N. Tveritskaya and J. Czopf - Hippocampal evoked potentials and eeg changes during classical conditioning in the rat. Electroencephalography and Clinical Neurophysiology, 1979/47:64-71

Clancey, W. J. - The epistemology of a rule-based expert system - a framework for explanation". Artificial Intelligence, 1983/20:215-251

Clancey, W. J. - Heuristic classification. Artificial Intelligence, 1985/27:280-350

Coquery, J. M. - Fonctions motrices et controle des messages sensoriels d'origine somatique. J. Physiologie (Paris), 1972/6:321-333

Cravetto, C., Lesmo L., Molino G., Torasso P. - LITO2: a frame based expert systgem for medical diagnosis in Hepatology. In De Lotto, I., Stafanelli M. (eds). Artificial Intelligence in Medicine, North-Holland, 1985:107-119

Dreyer, D. A., R. J. Schneider, C. B. Metz and B. L. Whitsel - Representation of head and face in postcentral gyrus of the macaque. J. Neurophysiology, 1975/38:714-733

Fikes, R. - The role of frame-based representation in reasoning. Comm of the ACM, 1985/28:904-920

Greco, G. and A.F. Rocha - The fuzzy logic of a text understanding. Fuzzy Sets and System, 1987 (in press)

Hubel, D. H. and T. N. Wiesel - Period of susceptibility to the physiological effects of unilateral eye closures in kittens. J. Physiology 1970/206:419-436

Kleinman, D. and E. R. John - Contradiction of auditory and visual information by brain stimulation. Science, 1975/271-273

Kulikowsky, C. Artificial intelligenge methods and system for medical consultation. IEEE Trans. Pattern Anal.Machine Intell.; PAMI 1980/2:464-474

Loe, P. R., B. L. Whitsel, D. A. Dreyer and C. B. Metz - Body representation in ventral basal complex of thalamus. J. Neurophysiology, 1977/40:1339-1355

Luria, A. R. - Cerebro y lenguage, Paidos, Buenos Aires, 1974

Massion, J. - Intervention des coies cérébello-corticales et cortico-cérébelleuses dans l'organization et la régulation du movement. J. Physiologie (Paris), 1973/61:117A-170A

Michaelsen, R. H., D. Michie, A. Boulanger - The technology of exptert system - Tansplanting expert knowledge to machines. Byte 1985/10:303-312

Micholski, R. S., J. G. Carbonell, T. M. Mitchell - Machi Learning. Tioge Publ. Comp., 1983

Rocha, A. F. - Temporal influences of the reticular formation on sensory processing. In The Reticular Formation Revisited, Hobson J. A. andBrazier, M. A. B. (eds), Raven Press, 1980/105-115

Rocha, A. F.; E. Françozo, M.I. Hadler and M.A. Bailduino - Neural languages. Fuzzy Sets and System, 1980/3:11-35

Rocha, A.F. - Neural fuzzy point processes. Fuzzy Sets and Systems, 1981/5:127-140

Rocha, A.F. - Basic properties of neural circuits. Fuzzy Sets and Systems, 1982a/7:109-121

Rocha, A.F. - Toward a theoretical and experimental approach of fuzzy learning. In Approximate Reasoning in Decision Analysis, M.M. Gupta and E. Sanchez (eds.), North-Holland, 1982b:191-200

Rocha, A. F. and J. W. M. Bassani - Information theory applied to the study of neural codes. In a General Survey of Systems Methodology, Troncale, L. (ed), Society for General Systems Research, 1982/2:528-533

Rocha, A.F. - Expert sensory systems: Initial considerations In Approximate Reasoning in Expert Systems, M.M. Gupta, A. Kandel, W. Blander and J.B. Kiska (eds), Elsevier, 1985:549-570

Rocha, A.F., F. Giorno and B. Leão - The physiology of the expert system. In Expected System in the Fuzzy Age, A. Kandel (ed), Addison Wesley, 1987 (in press)

Sidman, R. L. and P. Kakic - Neuronal migration, with special reference to developing human brain: a review. Brain Research, 1973/62:1-36

Torasso, P., L. Console - Approximate reasoning and Prototypical knowledge. Proc. Second IFSA Congress, Tokyo, 17:686-689

Fuzzy Logic in Knowledge-Based Systems,
Decision and Control
M.M. Gupta and T. Yamakawa (Editors)

A NEW APPROACH TO FUZZY CONTROL

Tu Xiangchu   Tu Chengyuan

Beijing Polytechnic University*

The authors made a new approach to the artificial intelligent (AI) control, combining fuzzy control with the conventional PID regulation and utilizing a self-optimizing technique. The regulation strategy of this new regulator developed by the authors, the method of fuzzication of the variables involved—the expressions of their membership functions, and the way of on-line self-optimization of operating parameters were described in detail. Also, some results of the emulation, such as the data and response curves were given as examples, showing the excellence of our regulator.

Keywords: Fuzzy control, Process control, Artificial intelligence, PID control, on-line self-optimization

It is well known that fuzzy language is an effective means for expressing human's thought, thus came out many kinds of fuzzy controller, provided with its related artificial intelligence (AI).

The fuzzy controllers so far belonged to either one of the following types: non-linear proportional or non-linear PD controller. Some of them, which been more or less effectively used in practice, are of the latter type [1,2]. S. Z. Long and P.Z. Wang succeeded in realizing some kind of relation between them —a simple and effective relation[3]. It's easily seen that, this kind of relation implies just a PD control function, of course, again with the characterizing non-linearity of fuzzy control—a valuable properity which has been used to conveniently simulate the human's thought as well as his operation skill.

Nevertheless, none of the up to date fuzzy controller, even that one developed by S.Z. Long and P.Z. Wang, has been provided with the I-function of the conventional PID controller so as to minimize it's static error.

We, now, develop a strategy of fuzzy control as follows:

$$\underset{\sim}{Y} = K\underset{\sim}{E} + K_i \int \underset{\sim}{E}\, dt + K_d \underset{\sim}{D} \tag{1}$$

where

$$E = UO - U, \quad D = \frac{dE}{dt} \tag{2}$$

U......the real value of the quantity to be controlled;
UO.....it's set-point value;
E......the deviation of U from UO, or the input to the fuzzy controller;
Y......output from the controller, or the magnitude of the controlling function.

* Prof. Tu Xiangchu, Department of Radioelectronics, Beijing Polytechnic university, the Eastern Suburb of Beijing, China

A new kind of controller, the fuzzy PID controller thus been developed. It holds simultaneously the conventinal PID as well as the fuzzy control function as it's excellent features. Furthermore, it has been provided with self-optimization function—self-optimization of K, K_i and K_d. During the first period of our research, we merely, however, optimize K, for the purpose of simplization of our design. Also for this purpose, the cooresponding quantities in eq. are fuzzified merely to sever grades: NB (negative big), NM (negative medium), NS (negative small), ZE (zero), PS (positive small), PM (positive medium), PB (positive big), as so did in paper[3].

The membership functions used in our regulator are as follows:

$$\mu_S(|X|) = \mu_M(|X|) = \mu_B(|X|) = 0 \qquad |X| = 0$$

$$\mu_S(|X|) = \begin{cases} 1 & 0 < |X| \leqslant X_1 \\ \exp\left[-\left(\frac{|X| - X_1}{\sigma_X}\right)^4\right] & |X| > X_1 \end{cases}$$

$$\mu_M(|X|) = \exp\left[-\left(\frac{|X| - X_2}{\sigma_X}\right)^4\right] \qquad |X| > 0 \tag{3}$$

$$\mu_B(|X|) = \begin{cases} \exp\left[-\left(\frac{|X| - X_3}{\sigma_X}\right)^4\right] & 0 < |X| \leqslant X_3 \\ 1 & |X| > X_3 \end{cases}$$

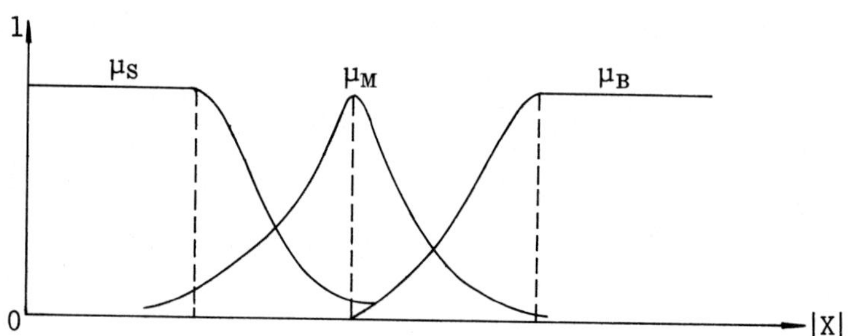

Here, X may denote any of the variables E, D and Y (we let $D = \frac{dE}{dt}$ for the purpose of convenience to deal with BASIC II programming). Moreover, all these expressions do deal with the absolute value of X, thus subscripts used are smiplified into S (Small), M (Medium) and B (Big), and the letter P (Positive) or N (Negative) before them as generally used may be omitted.

Substituting the real parameter E, D or Y for X into (3), we have:

$$\mu_S(|E|) = \mu_M(|E|) = \mu_B(|E|) = 0 \qquad |E| = 0$$

$$\mu_S(|E|) = \begin{cases} 1 & 0 < |E| \leqslant E_1 \\ \exp\left[-\left(\frac{|E| - E_1}{\sigma_E}\right)^4\right] & |E| > E_1 \end{cases} \tag{4}$$

$$\mu_M(|E|) = \exp\left[-\left(\frac{|E| - E_2}{\sigma_E}\right)^4\right] \qquad |E| > 0$$

$$\mu_B(\ |E|\ ) = \begin{cases} \exp\left[-\left(\dfrac{|E| - E_3}{\sigma_E} \right)^4 \right] & 0 < |E| \leqslant E_2 \\ 1 & |E| > E_3 \end{cases}$$

$$\mu_S(\ |D|\ ) = \mu_M(\ |D|\ ) = \mu_B(\ |D|\ ) = 0 \qquad |D| = 0$$

$$\mu_S(\ |D|\ ) = \begin{cases} 1 & 0 < |D| \leqslant D_1 \\ \exp\left[-\left(\dfrac{|D| - D_1}{\sigma_D} \right)^4 \right] & |D| > D_1 \end{cases}$$

$$\mu_M(\ |D|\ ) = \exp\left[-\left(\frac{|D| - D_2}{\sigma_D} \right)^4 \right] \qquad |D| > 0 \tag{5}$$

$$\mu_B(\ |D|\ ) = \begin{cases} \exp\left[-\left(\dfrac{|D| - D_3}{\sigma_D} \right)^4 \right] & 0 < |D| \leqslant D_2 \\ 1 & |D| > D_3 \end{cases}$$

$$\mu_S(\ |Y|\ ) = \mu_M(\ |Y|\ ) = \mu_B(\ |Y|\ ) = 0 \qquad |Y| = 0$$

$$\mu_S(\ |Y|\ ) = \begin{cases} 1 & 0 < |Y| \leqslant Y_1 \\ \exp\left[-\left(\dfrac{|Y| - Y_1}{\sigma_Y} \right)^4 \right] & |Y| > Y_1 \end{cases}$$

$$\mu_M(\ |Y|\ ) = \exp\left[-\left(\frac{|Y| - Y_2}{\sigma_Y} \right)^4 \right] \qquad |Y| > 0 \tag{6}$$

$$\mu_B(\ |Y|\ ) = \begin{cases} \exp\left[-\left(\dfrac{|Y| - Y_3}{\sigma_Y} \right)^4 \right] & 0 < |Y| \leqslant Y_2 \\ 1 & |Y| > Y_3 \end{cases}$$

In our emulation experiments, the following values have been used: $E_1 = 1$, $E_2 = 3$, $E_3 = 5$, $\sigma_E = 1.6$; $D_1 = 0.05$, $D_2 = .1$, $D_3 = .2$, $\sigma_D = 2.8$; $Y_1 = 101$, $Y_2 = 110$, $Y_3 = 120$, $\sigma_Y = 18$.

The on-line self-optimization of K was implemented with a target function somewhat like $Q = \int |E| t\, dt$, an useful criterion for the optimization of the control systems. And the method of on-line optimization used is similar to the limited Depth-first Search in AI. We practise this kind of optimization in an onedimension space of the parameter K. As for our design idea, it is evident from the following fig.

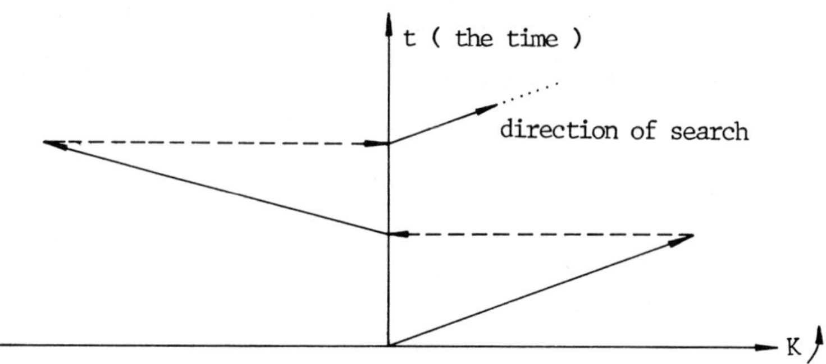

FIGURE 1
The method of limited Depth-first Search for K

The generalized plant to be controlled is either a 1'st order inertial element 1/ (TS+1) only, or together with a time-lag element $e^{-\tau s}$ serially connected between them, which been usually recommended as the typical simplized model for most technical thermodynamic processes.

Our simulation experiments were carried out on microcomputer TP803, an equivalent to TRS-80, being programmed in BASIC II. The sampling rate choosed was 1 time per $\Delta T = 0.01$, as compared with $T = 10$, or again together with $\tau = 0.05$ in our simulation system.

Transient responses of this system were obtained for 50 % setpoint disturbance (i.e. starting from U = 50 % when UO = 100 %), as well as for a δ-disturbance in the interval t = 9 ~ 9.5. Figure 3 and 5 show the results, with the curve-poltting rate choosed as 40 ΔT, i.e., 40 times the sampling rate. For recognition's sake the ordinates in these two fig. have been replaced by another quantity-the deviation, and been twice doubled to 4E, as shown in Figure 4 and 6, where the listing rate was 50 ΔT, and the plotting rate 20 ΔT. The influence of δ-disturbance is easily seen.

Evidently, the results obtained, even if from our preliminary and rough experiments, are quite satisfactory: the overshoot < 0.1 % for the system 1/(10s+1), and < 0.8 % for a system 1/(10s+1) in series with $e^{-0.05s}$.

However we are sure that, the foregoing results can be much more improved by means of increasing the number of grades of $\underset{\sim}{E}$, $\underset{\sim}{\frac{dE}{dt}}$ and $\underset{\sim}{Y}$ to some tens, as well as by means of carefully adjustment of the parameters of the fuzzy controller. This will be just the furthermore work of the author's research.

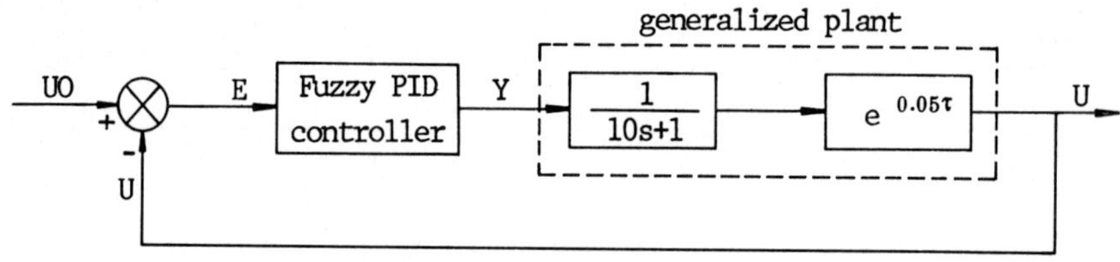

FIGURE 2
Block diagram of the simulation system

Fig. 3. Transient Responce for 1/(10s+1), -30% Disturbance

U=f(N)

U0=100

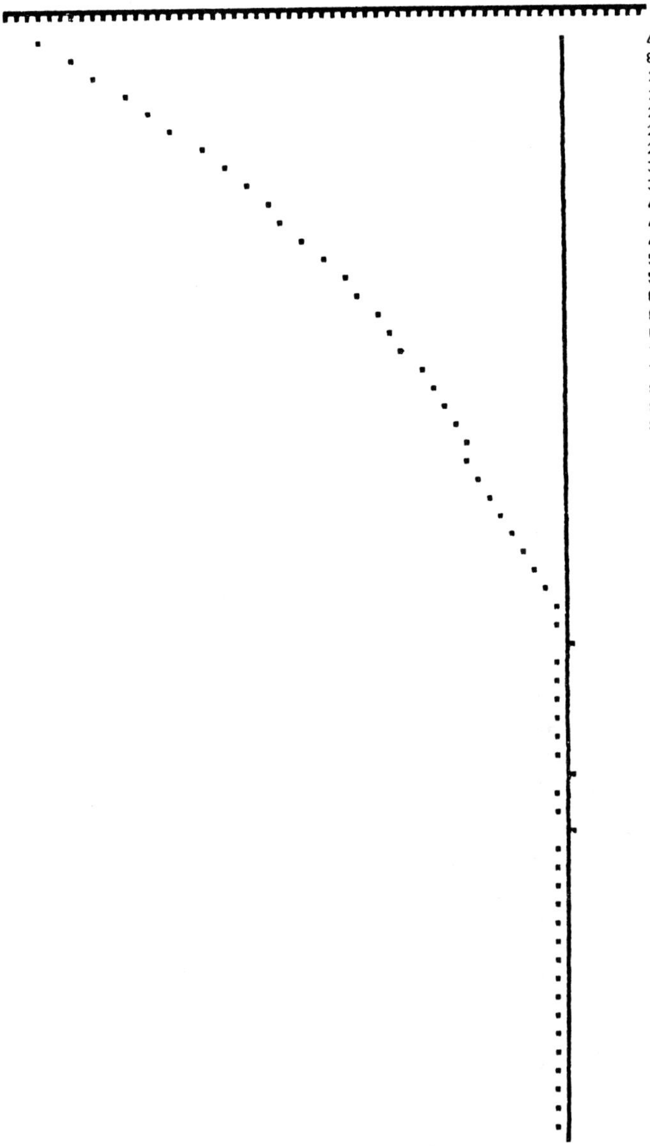

U	N
52.7448	40
55.3819	80
57.9157	120
60.3501	160
62.689	200
64.9363	240
67.0955	280
69.1699	320
71.163	360
73.078	400
74.9179	440
76.6856	480
78.384	520
80.0158	560
81.5837	600
83.09	640
84.5373	680
85.9279	720
87.2639	760
88.5476	800
89.7809	840
90.9658	880
91.3122	920
91.2668	960
92.3934	1000
93.4759	1040
94.516	1080
95.5152	1120
96.4753	1160
97.3978	1200
98.2845	1240
99.6359	1280
99.8081	1320
100.099	1360
99.8752	1400
99.8173	1440
99.9058	1480
99.9897	1520
99.8337	1560
99.9176	1600
100.001	1640
99.8456	1680
99.9297	1720
100.013	1760
99.8574	1800
99.904	1840
99.7548	1880
99.7853	1920
99.9714	1960
99.872	2000
99.7948	2040
99.9689	2080
99.9652	2120
99.903	2160
99.9055	2200
99.6801	2240
99.9216	2280
99.8051	2320
99.7731	2360
99.9933	2400

Fig. 4. Transient Responce for 1/(10s+1), -30% Disturbance (with List)

E	U	D	N
46.586	53.414	-.0666199	50
43.3385	56.6615	-.0633698	100
40.2494	59.7506	-.0602799	150
37.311	62.689	-.0573425	200
34.5158	65.4842	-.0545425	250
31.857	68.143	-.0518875	300
29.3278	70.6722	-.0493546	350
26.922	73.078	-.0469437	400
24.6336	75.3664	-.0446548	450
22.4567	77.5433	-.0424805	500
20.386	79.614	-.0404053	550
18.4164	81.5837	-.0384369	600
16.5427	83.4573	-.0365601	650
14.7604	85.2396	-.0347824	700
13.0651	86.9349	-.0330811	750
11.4525	88.5476	-.0314713	800
10.8296	89.1704	-.0308456	820
10.2191			840
9.62076			860
9.03421			880
8.45928			900
8.68778			920
8.91173			940
8.73325			960
8.16428			980
7.60658			1000
7.05991			1020
6.52408			1040
5.99885			1060
5.48402			1080
4.97939			1100
4.48476			1120
3.99991			1140
3.52467			1160
3.05884			1180
2.60221			1200
2.15446			1220
1.7155			1240
1.28249			1260
.364182			1280
.225288			1300
.191872			1320
.200417			1340
-.0993118			1360
.164269			1380
.124847			1400
.251381			1420
.182716			1440
.0165253			1460
.0942459			1480
.172295			1500
.0103378			1520
.0881424			1540
.166306			1560
4.47845E-03			1580
.0823975			1600
.160667			1620
-1.05286E-03			1640
.0764999			1660
.154419			1680
-7.41577E-03			1700
.0702744			1720
.148315			1740
-.0133972			1760
.064415			1780
.142571			1800
-.0190277			1820
.0960083			1840
3.96729E-04			1860
.245178			1880
.178383			1900
.214706			1920
.126541			1940
.0285873			1960
.147827			1980
.128006			2000

Fig. 5 Transient Response for 1/(10s+1) & EXP(-.05s), -30% Disturbance

U=f(N)

U0=100

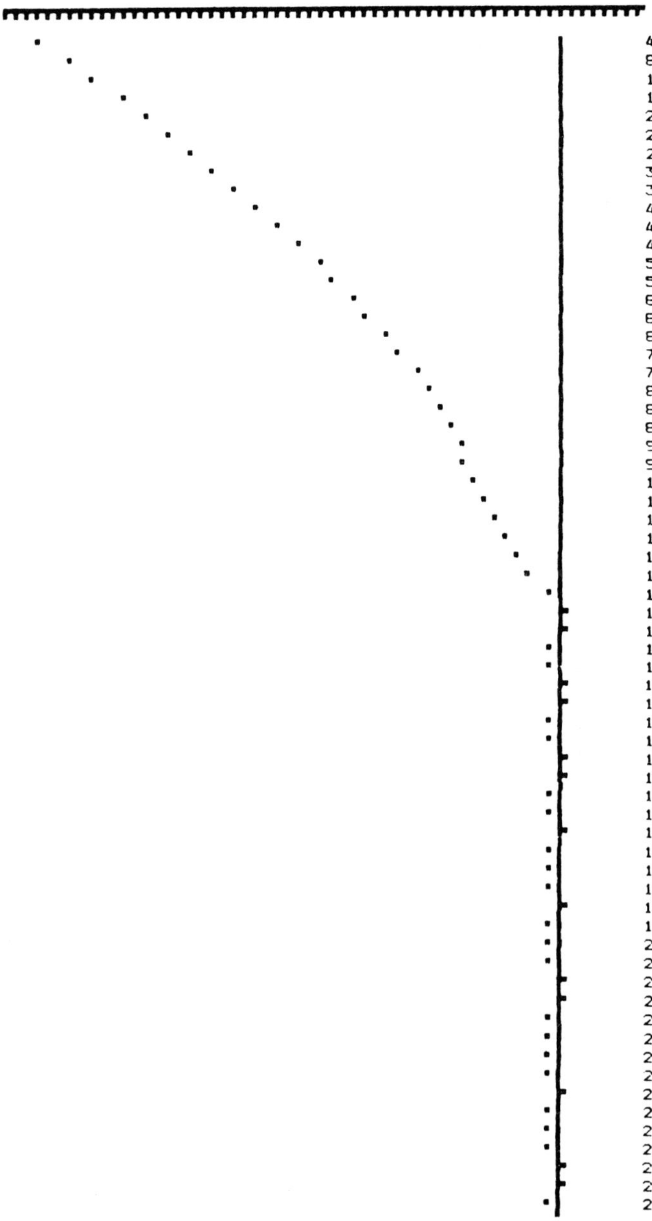

U	N
52.4077	40
55.058	80
57.6045	120
60.0511	160
62.4018	200
64.6603	240
66.8303	280
68.9151	320
70.9182	360
72.8428	400
74.6919	440
76.4685	480
78.1754	520
79.8154	560
81.3911	600
82.905	640
84.3596	680
85.7571	720
87.0998	760
88.3899	800
89.6294	840
90.8203	880
91.3689	920
91.1227	960
92.255	1000
93.343	1040
94.3882	1080
95.3925	1120
96.3574	1160
97.3547	1200
99.7135	1240
100.8	1280
100.033	1320
99.3962	1360
99.3039	1400
100.502	1440
100.243	1480
99.5271	1520
99.2495	1560
100.619	1600
100.324	1640
99.4707	1680
99.0341	1720
100.584	1760
99.9642	1800
99.2872	1840
99.1944	1880
100.592	1920
99.9729	1960
99.2958	2000
99.0748	2040
100.62	2080
100.358	2120
99.6888	2160
99.3463	2200
99.2845	2240
99.9329	2280
100.452	2320
99.5999	2360
99.2914	2400
99.2297	2440
100.234	2480
100.594	2520
99.5994	2560

Fig. 6 Transient Response for 1/(10S+1) & EXP(-.05*S), -30% Disturbance

Curve : 4E=f(N)

E	U	D	N
46.9198	53.0802	-.0669556	50
43.656	56.344	-.0636902	100
40.5514	59.4486	-.0605812	150
37.5982	62.4018	-.0576286	200
34.7891	65.2109	-.0548172	250
32.1169	67.8831	-.0521469	300
29.5751	70.4249	-.0495987	350
27.1572	72.8428	-.0471802	400
24.8573	75.1427	-.0448837	450
22.6695	77.3305	-.0426941	500
20.5885	79.4115	-.0406113	550
18.6089	81.3911	-.0386278	600
16.7259	83.2741	-.0367432	650
14.9347	85.0653	-.0349503	700
13.2309	86.7692	-.0332489	750
11.6101	88.3899	-.0316238	800
10.9842			820
10.3706			840
9.76923			860
9.17975			880
8.60194			900
8.63109			920
8.85616			940
8.87727			960
8.30545			980
7.74496			1000
7.19555			1020
6.65703			1040
6.12917			1060
5.61176			1080
5.10461			1100
4.60749			1120
4.12022			1140
3.64258			1160
3.17442			1180
2.64533			1200
1.26843			1220
.286522			1240
-.222702			1260
-.800072			1280
-.471504			1300
-.0334168			1320
.40741			1340
.601783			1360
.739349			1380
.696121			1400
.720902			1420
-.502159			1440
-.609558			1460
-.24337			1480
.196198			1500
.47287			1520
.877236			1540
.750519			1560
.362473			1580
-.618744			1600
-.756638			1620
-.323525			1640
.0883484			1660
.52932			1680
.614098			1700
.905914			1720
.654121			1740
-.583702			1760
-.394646			1780
.0357895			1800
.448402			1820
.712837			1840
.669609			1860
.805595			1880
.120621			1900
-.591949			1920
-.403282			1940
.0270844			1960
.439735			1980
.704239			2000

REFFERENCE

[1] M. Gupta et al., Fuzzy Automata and Decission processes, Elssevier North-Holland Inc., 1977. §14, 19 and 20.

[2] D. Dubois et al., Fuzzy Sets and Systems, Academic Press, Inc. (London), 1980, pp. 297-305.

[3] S.Z. Long and P.Z. Wang, Fuzzy Math. (published in China), 1982, No. 3, pp. 105-112.

Fuzzy Logic in Knowledge-Based Systems,
Decision and Control
M.M. Gupta and T. Yamakawa (Editors)

FUZZY CONCEPTS IN HUMAN RELIABILITY

Takehisa ONISAWA

Department of Basic Engineering, Kumamoto University, 2-39-1, Kurokami, Kumamoto, 860 Japan

ABSTRACT This paper proposes the concept "error possibility" in place of the error rate in order to represent human reliability. The error possibility can be expressed by a fuzzy subset of the unit interval [0, 1]. The fuzzy expression represents our feeling that human error may exist even if the error rate is estimated to be very small.

Moreover this paper deals with a model of the relationships between human reliability and the performance shaping factors. The present model enables one to select necessary and/or important performance shaping factors in order to estimate human reliability of a given task. This model uses the fuzzy measures and the fuzzy integrals. The validity of the present model is verified by the data of the basic experiment.

Key Words: Human reliability, error possibility, performance shaping factors, fuzzy measures, fuzzy integrals

1. INTRODUCTION

Even in a highly automated system there inevitably will be instnaces that the computer will not be programed to handle. A human being is still indispensable to the system in this situation. However he always does not make an error. Even though not all human errors cause troubles in the system, some of them lead to a big accident. A human being is not a little concerned in the accident. These facts have been well known by the incidents at Three Mile Island in 1979 and at Chernobyl in 1986. There have been many studies concerning human errors, Gelder, [1], Lind [2], Melchers [3]. It has been said that the reliability study which does not include human factors tends to be optimistic. It is necessary to include human reliability in the reliability analysis. However human behaviour is affected by many factors, e.g., his adaptability and flexibility, task environment, etc. These factors are called performance shaping factors (abbreviated as PSF's), Swain and Guttmann [4]. Furthermore the assessment of probability distributions for human error rates in probabilistic risk assessment is largely based on experts' judgement. Since risk assessments usually deal with rare events for which meaningful statistical records do not exist, it is very difficult to calibrate experts using their performance, Apostolakis [5]. So it is not adequate to express human reliability in terms of the error rate.

The fuzzy theory has been suggested to be an effective tool for human reliability, Terano et al. [6]. This paper proposes the concept "error possibility", Onisawa [7] to represent human reliability instead of the error rate which is used in the conventional analysis. The error possibility can be expressed by a fuzzy subset of the unit interval [0, 1]. The error possibility represents our feeling that a human error may exist even if the error rate is estimated to be very small. The error possibility is evaluated from many points of view as human reliability.

Moreover this paper deals with a model of the relationships between human reliability and PSF's, Onisawa [8]. Human behaviour is affected by many PSF's as abovementioned. It is difficult to assign weighting factors to the PSF's or to develop an equation for the relative influence of all the PSF's on the performance of a given task, Swain and Guttmann [4]. The present model enables one to select necessary and/or important PSF's in order to estimate human reliability of a given task. This model uses the fuzzy measures and the fuzzy integrals, Sugeno [9]. The validity of the present model is verified by the data of the basic experiment on a microcomputer.

2. ERROR POSSIBILITY (Onisawa [7])

2.1 Error Possibility

Let us consider a fuzzy set E of [0, 1] associated with a possibility distribution E(e) of the type

$$E(e) = \frac{1}{1 + 20 \times |e - e_0|^m}, \qquad (1)$$

where $m = m_L$ for $e \leqq e_0$ and $m = m_U$ for $e \geqq e_0$.

The parameter m is related to the fuzziness and e_0 gives the maximal grade of E(e), i.e., $E(e_0) = 1$. E is called an "error possibility" and e is called a "likelihood of error". Eq. (1) implies that even if e_0 is very small, there exists a possibility that a human operator may make an error since $E(e) > 0$ for all $e \in [0, 1]$. There may be many ways to derive the error possibility, Onisawa [10]. However in this paper e_0 and m are assumed to be derived from the triplet of the error rate. The error rate is defined as

$$\text{human error rate} = \frac{\text{frequencies of error}}{\text{total chances which a human operator makes an error.}} \qquad (2)$$

Usually the error rate is given by the triplet $[Er_L, Er_M, Er_U]$ where Er_M is the recommended value of the error rate, Er_L is its lower bound and Er_U is its upper bound.

i) e_0 from Er_M

The error rate is very small, e.g., 10^{-2}, 10^{-3}, etc. These values seem to be very small from the viewpoint of our daily life. However these values are not necessarily small from the viewpoint of human reliability. Eq. (3) is defined in order to fit the very small value to our feeling.

$$e_0 = f(Er_M) = \begin{cases} \dfrac{1}{1 + (K \times \log(1/Er_M))^3}, & Er_M \neq 0, \\ 0, & Er_M = 0, \end{cases} \qquad (3)$$

where K is a constant.

So Eq. (1) implies that even if the error rate is estimated to be very small, there is a possibility that a human operator may make an error. This consideration fits our feeling. According to Swain and Guttmann [4], $10^{-2} - 10^{-3}$ are considered as the error rates of a routine and 5×10^{-5} as the lowest lower bound of the error rate. For example, K is defined as

$$K = \frac{1}{\log (1/(5 \times 10^{-3}))} \tag{4}$$

so that the error rates of a routine correspond to about 0.5 and the lowest lower bound of the error rate corresponds to the small value. When K is given by Eq. (4), the recommended value of the error rate is compared with 5×10^{-3}. **Table 1** shows the classification of e_0 when K is given by Eq. (4).

Table 1 Classification of e_0

Class	Bounds of e_0	Bounds of the error rate
C_1	1.0	1.0
C_2	0.9 - 1.0	7.830×10^{-2} - 1.0
C_3	0.7 - 0.9	1.841×10^{-2} - 7.830×10^{-2}
C_4	0.5 - 0.7	5.0×10^{-3} - 1.841×10^{-2}
C_5	0.3 - 0.5	8.872×10^{-4} - 5.0×10^{-3}
C_6	0.2 - 0.3	2.225×10^{-4} - 8.872×10^{-4}
C_7	0.1 - 0.2	1.636×10^{-5} - 2.225×10^{-4}
C_8	0.05 - 0.1	7.243×10^{-7} - 1.636×10^{-5}
C_9	0.0 - 0.05	0.0 - 7.243×10^{-7}
C_{10}	0.0	0.0

ii) m_L and m_U from $[Er_L, Er_M, Er_U]$

(1) Define $k_L = Er_M / Er_L$ and $k_U = Er_U / Er_M$. The parameters k_L and k_U are rewritten as k together. The parameter k implies uncertainty bounds for the estimated error rate.
(2) Four uncertainty bounds are defined such as $k \leqq 3$, $3 < k \leqq 5$, $5 < k \leqq 10$ and $10 < K$. These uncertainty bounds are referred to **Table 2**, Swain and Guttmann [4]. Table 2 shows the general guidelines for estimating uncertainty bounds for estimated error rates.
(3) In the class C_5 define $m = 2.0$ for $k \leqq 3$, $m = 2.5$ for $3 < k \leqq 5$, $m = 3.0$ for $5 < k \leqq 10$ and $m = 3.5$ for $10 < k$, respectively.
(4) In each class the representative value of e_0 is defined as shown in **Table 3**. The representative value is the middle point of the bounds of e_0 in each class. The error possibilities in the class C_1 and C_{10} are defined later.

Let $E_i(e)$ be the error possibility in the class C_i and e_{0i} be the representative value of e_0 in the class C_i, i.e., $E_i(e_{0i}) = 1$ (i = 2, 3, ---, 9).

(5) Within the same uncertainty bounds m is obtained so as to satisfy Eq. (5)

$$E_i(e_i) = E_5(e_5), \quad i = 2, 3, ---, 9 \tag{5}$$

where $e_i = f(10 \times Er_{Mi})$ when m_U is obtained and $e_i = f(Er_{Mi} / 10)$ when m_L is obtained.

Table 2 General Guidelines for Estimating Uncertainty Bounds for Estimated Error Rates

Error Rate	Lower Bound	Upper Bound
0.01 < Er	Er / 5	2 × Er − 5 × Er
0.001 < Er < 0.01	Er / 3	3 × Er
Er < 0.001	Er / 10	10 × Er

Table 3 Representative Value of e_0

Class	Representative Value of e_0
C_2	0.95
C_3	0.8
C_4	0.6
C_5	0.4
C_6	0.25
C_7	0.15
C_8	0.075
C_9	0.025

The value of m in each class is compared with that of m in the class C_5 in the sense of Eq. (5). **Fig. 1** shows the example of obtaining m within the uncertainty bounds $3 \leqq k < 5$ in the class C_6 when K is given by Eq. (4).

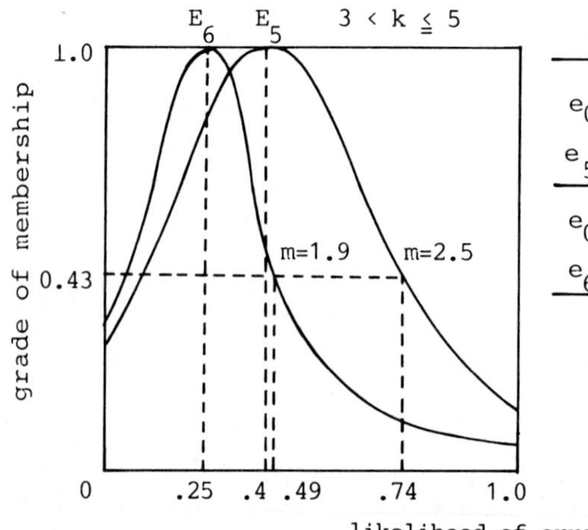

e_{05}	0.4 → $Er_{M5} = 2.32 \times 10^{-3}$ ↓
e_5	0.74 ← 2.32×10^{-2}
e_{06}	0.25 → $Er_{M6} = 4.80 \times 10^{-4}$ ↓
e_6	0.49 ← 4.80×10^{-3}

Fig. 1 Example of Obtaining m

Table 4 shows the results obtined by the above procedure when K is given by Eq. (4). However the values with the boldface are used by referring to Table 2. The parameter m for k > 10 is used when the error rate of a given task is assumed to be fuzzier.

Table 4 Parameter m

Class	m_L, m_U			
	k ≦ 3	3 < k ≦ 5	5 < k ≦ 10	10 < k
C_2 (0.01 < Er)	2.7, 2.7	3.5, 3.5	4.0, 4.0	4.7, 4.7
C_3 (0.01 < Er)	1,3, 3.1	1.7, 3.8	2.0, 4.6	2.3, 5.4
C_4 (0.001 < E < 0 .001)	1.9, 2.6	2.4, 3.3	2.9, 3.9	3.4, 4.6
C_5 (0.001 < Er < 0.001)	2.0, 2.0	2.5, 2.5	3.0, 3.0	3.5, 3.5
C_6 (Er < 0.001)	1.6, 1.5	1.9, 1.9	2.3, 2.3	2.7, 2.7
C_7 (Er < 0.001)	1.1, 1.2	1.4, 1.5	1.7, 1.8	1.9, 2.1
C_8 (Er < 0.001)	0.8, 0.9	1.0, 1.1	1.1, 1.4	1.3, 1.6
C_9 (Er < 0.001)	0.5, 0.7	0.6, 0.8	0.7, 1.0	0.9, 1.2

The error possibility in the class C_1 is defined as

$$E(e) = \begin{cases} 1, & e = 1, \\ 0, & e \neq 1, \end{cases} \tag{6}$$

and that in the class C_{10} is defined as

$$E(e) = \begin{cases} 0, & e \neq 0, \\ 1, & e = 0. \end{cases} \tag{7}$$

These error possibilities are non-fuzzy.

The error possibility is a relative concept in the sense of Eqs. (3), (4) and (5).

2.2 Evaluation of Error Possibility

The error possibility E given in **Fig. 2** is evaluated with respect to the following three points. The reliability is considered in these aspects. Let $(E)_\alpha = (e_1(\alpha), e_2(\alpha))$ be α-cut of E.

(1) likelihood of error e_0 which gives the maximal grade of E

$$J_1 = e_0. \tag{8}$$

J_1 is the center of the error possibility. This evaluation J_1 shows which class in Table 1 the error possibility E belongs to.

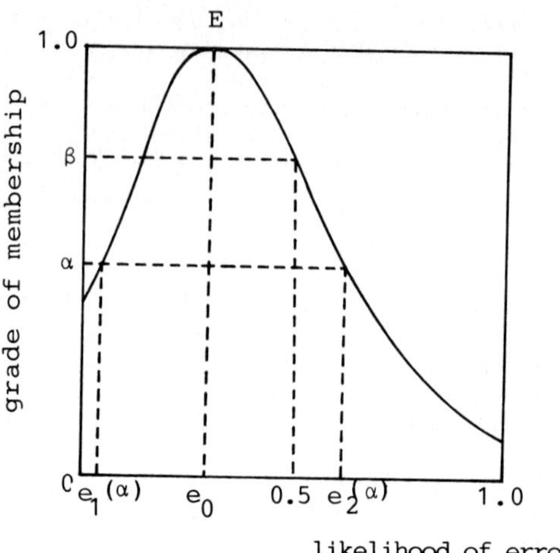

Fig. 2 Error Possibility

(2) potentiality for error

Define

$$J_2 = \frac{\int_0^{\beta} (e_2(\alpha) - 0.5)\alpha d\alpha}{\int_0^{1} (1 - 0.5)\alpha d\alpha}, \tag{9}$$

where $e_2(\beta) = 0.5$.

J_2 is evaluated when $J_1 \leqq 0.5$ since $J_1 > 0.5$ shows that the likelihood of error e_0 is large. This evaluation J_2 is interpreted as the potentiality for error since J_2 evaluates the part of the likelihood of error such as $e_2(\alpha) \geqq 0.5$ for $\alpha \in [0, \beta]$. It can be considered that the larger J_2 becomes, the higher the potentiality for error becomes even if e_0 is small. The high potentiality for error implies that it is necessary to improve the task environment or to practice a human operator in the task.

(3) fuzziness of error possibility

Define

$$J_3 = \frac{\int_0^{1} (e_2(\alpha) - e_0)\alpha d\alpha}{\int_0^{1} (1 - e_0)\alpha d\alpha}, \tag{10}$$

where $e_2(\alpha) \geqq e_0$ for $\alpha \in [0, 1]$.

J_3 evaluates the fuzziness of the error possibility. It can be considered that if J_3 is large, J_1 is not evaluated with confidence.

In the reliability analysis using the error possibility, human reliability can be estimated from many aspects, not only the likelihood of error which gives the maximal grade of the membership but also the potentiality for error and the fuzziness of the error possibility.

3. PERFORMANCE SHAPING FACTORS MODELLING (Onisawa [8])

Human reliability is affected by many PSF's. It is difficult to assign weighting factors to PSF's or to develop an equation for the relative influence of all PSF's on the performance of a given task since 1) there are many PSF's, 2) PSF's are dependent on one another, 3) PSF's are not always necessary and sufficient factors for human error and 4) the relationships between them are ambiguous.

The fuzzy measures and the fuzzy integrals are suitable for a model of human evaluation process and there have been some applications, Onisawa [8], Sugeno [9], Sugeno [11], Ishii and Sugeno [12], Onisawa et al.[13]. It is not necessary to assume the linearity and the independence of items in modelling an evaluation process by the use of the fuzzy measures and the fuzzy integrals. The description of the fuzzy measures and the fuzzy integrals is omitted here.

3.1 Fuzzy Integral Model (Onisawa et al. [13])

3.1.1 Identification of Fuzzy Measures

Let A be an object, $Z = \{z_1, z_2, ---, z_n\}$ be a set of its attributes and $h_A : Z \rightarrow [0, 1]$ is the characteristic function of the object A. Let g be a fuzzy measure to express the grade of the subjective importance of the attributes in overall evaluation of the object. Then the overall evaluation of A is given by

$$E_A = \oint h_A(z) \circ g. \tag{11}$$

Eq. (11) is called a fuzzy integral model of the evaluation with respect to an object with various attributes. In this paper the obeject is human reliability and a model of human reliability, on an average, of n persons is structured. Fuzzy measures are identified so as to minimize the performance index.

$$J = [(1/n) \sum_{k=1}^{n} (E_k^* - E_k)^2]^{1/2}, \tag{12}$$

where E_k^* is a subjective evaluation of an object by the k-th person and E_k is the output of the fuzzy integral model of the k-th person such as

$$E_k = \oint h_k(z) \circ g, \qquad 1 \leqq k \leqq n, \tag{13}$$

where h_k is the subjective characteristic function of the object given by the k-th person and g's are the mean fuzzy measure among n persons in the sense that g's minimize J.

The identification algorithm is referred to Ishii and Sugeno [12].

3.1.2 Structure Identification

Let us discuss how to identify the structure of a model. Suppose that all the possible attributes of an object can be listed up. Some notions are defined to identify the structure of a model.

Definition : **degree of overlapping**

$$\mu_{ij} = \begin{cases} \dfrac{g(\{z_i, z_j\}) - (g(\{z_i\}) + g(\{z_j\}))}{g(\{z_i\}) \wedge g(\{z_j\})}, & i \neq j, \\ 0, & i = j, \end{cases} \qquad (14)$$

where $\wedge$ stands for min.

$\mu_{ij} \in [-1, +\infty)$ is called a degree of overlapping of g between the i-th and the j-th attributes. If $\mu_{ij} \geq 0$, then g is super-additive for $\{z_i\}$ and $\{z_j\}$, i.e., $g(\{z_i, z_j\}) \geq g(\{z_i\}) + g(\{z_j\})$. This implies that $\{z_i\}$ and $\{z_j\}$ are necessary to be evaluated together. It can be considered that $\{z_i\}$ and $\{z_j\}$ support each other. On the other hand if $\mu_{ij} < 0$, then g is sub-additive. It can be considered that $\{z_i\}$ and $\{z_j\}$ cast a doubt on each other in the evaluation of the object when the both are evaluated. As an extreme case, if $\mu_{ij} = -1$, then $g(\{z_i, z_j\}) = \max(g(\{z_i\}), g(\{z_j\}))$, i.e., one of the subjective importance of $\{z_i\}$ and that of $\{z_j\}$ has no influence on the subjective importance of $\{z_i, z_j\}$. One of $\{z_i\}$ and $\{z_j\}$ may be eliminated in the evaluation of the object.

Definition : **normalized μ_{ij}**

$$W_{ij} = \begin{cases} \mu_{ij}, & \mu_{ij} < 0, \\ 1 - \dfrac{1}{1 + \mu_{ij}}, & \mu_{ij} \geq 0. \end{cases} \qquad (14)$$

$W_{ij} \in [-1, +1)$ is the normalization of $\mu_{ij} \in [-1, +\infty)$.

Definition : **overlapping coefficient**

$$\eta_i = \sum_{j=1}^{J} \frac{W_{ij}^3}{J - 1}, \qquad (16)$$

where J is the number of attributes.

η_i is the value expressing the mean degree of overlapping between the i-th attribute and others. Eq. (16) is assumed in order to indicate greater contribution of W_{ij} to η_i when W_{ij} is large, and in order to reflect the signs of W_{ij} on η_i. $\eta_i \in [-1, +1)$ is called an overlapping coefficient. If $0 \leq \eta_i \leq 1$, then the j-th attribute does not overlap with others on an average, i.e., this attribute is relevant to evaluate an object.

Definition : **necessity coefficient**

$$\xi_i = 1 + \eta_i \times (1 - g(\{z_i\})), \qquad -1 \leq \eta_i < 0. \qquad (17)$$

ξ_i is called a necessity coefficient expressing the degree of necessity of the i-th attribute in the structure of a model. The larger η_i becomes and the larger $g(\{z_i\})$ becomes, the larger ξ_i becomes. The attributes with $\eta_i < 0$ may be considered whether they are relevant or irrelevant in a model by referring to the value of ξ_i. The attribute $\{z_i\}$ may be eliminated if the following inequality is satisfied:

$$\xi_i \,/ \max_{i,\, \eta_i < 0} \xi_i < 0.7, \qquad (18)$$

where $\{z_i\}$ is not eliminated if $\eta_i \geqq 0$.

In other words if $\eta_i \geqq 0$, then $\{z_i\}$ may be regarded as no overlapping on an average and therefore this attribute may be considered relevant. If $\eta_i < 0$, then Eq. (17), which takes $g(\{z_i\})$ into consideration, will be used. However, a threshold 0.7 is introduced in this case so that the necessity coefficient of the relevant attribute may not have too widely different values. It is called a fuzzy measure analysis, Onisawa et al. [13] to select the relevant attribute to evaluate an object as well as to build a fuzzy integral model.

3.2 Distinction between Error Possibilities

It is necessary to express the distinction between error possibilities in order to consider the influence of the task conditions on human error. The distinction between error possibilities is defined by the following steps.

(1) According to Swain and Guttmann [4], the logarithmic distribution of human error rates may be often considered as a normal distribution. The lognormal

Table 5 Means and Standard Deviations of Lognormal Numbers

Class	Mean Error Rate $\overline{Er}$	Standard Deviation σ	Likelihood of Error
C_2	1.364×10^{-1}	0.335	0.950
C_3	3.555×10^{-2}	0.332	0.800
C_4	9.552×10^{-3}	0.253	0.597
C_5	2.392×10^{-3}	0.246	0.404
C_6	4.987×10^{-4}	0.404	0.253
C_7	4.958×10^{-5}	0.401	0.150
C_8	4.931×10^{-6}	0.396	0.075
C_9	1.585×10^{-7}	0.401	0.037

random numbers are so generated that the mean error rate in each class is the value as shown in **Table 5.** The random numbers are so picked up that they belong to $[Er / 5, Er \times 5]$ in the classes C_2 and C_3, $[Er / 3, Er \times 3]$ in the classes C_4 and C_5 and $[Er / 10, Er \times 10]$ in the classes C_6, C_7, C_8 and C_9, respectively.

(2) Define

$$L = \log \overline{Er}, \tag{19}$$

$$Ln = L + n\sigma, \tag{20}$$

where $n = 0, \pm 1, \pm 2, \pm 3$ and σ is the standard deviation of logarithm,

$$Er_{Mn} = 10^{Ln}. \tag{21}$$

(3) Eq. (21) is substitued into Eq. (3),

$$e_n = \frac{1}{1 + (K \times \log (1/Er_{Mn}))^3}, \tag{22}$$

where $K = 1/(\log (1/5 \times 10^{-3}))$.

(4) The representative error possibility in each class

$$E(e) = \frac{1}{1 + 20 \times |e - e_0|^m}, \tag{23}$$

where m is referred to Table 4,

is drawn. The parameter e_0 is the likelihood of error for n = 0 in each class.

(5) The grade $E(e_n)$ is obtained in each class.

The distinction between likelihoods of error e_0 and e_n is expressed by referring to the grade as shown in **Table 6.**

Table 6 Distinction between Likelihoods of Error based on their Grades

Grade (+)	Grade (-)	Qualitative Expression
1 - 0.9	1 - 0.95	no distinction
0.9 - 0.7	0.95 - 0.85	a little distinction
0.7 - 0.5	0.85 - 0.7	ordinary distinction
0.5 -	0.7 -	wide distinction

grade(+) means $E(e_n)$ for n = +1, +2, +3 and
grade(-) means $E(e_n)$ for n = -1, -2, -3.

Fig. 3 shows the example in the class C_5.

The distinction between error possibilities is defined as follows. Let $E_1(e)$ and $E_2(e)$ be two error possibilities and let e_{01} and e_{02} be the likelihoods of error such as

$$\begin{cases} E_1(e_{01}) = 1, \\ E_2(e_{02}) = 1. \end{cases} \tag{24}$$

The distinction between error possibilities is expressed as shown in Table 6 by referring to the grade such as max $(E_1(e_{02}), E_2(e_{01}))$. **Fig. 4** shows the example of the distinction between error possibilities. In Fig. 4(a)

$$\begin{cases} E_1(0.4) = 0.80 \text{ (grade (+))}, \\ E_2(0.25) = 0.69 \text{ (grade (-))}. \end{cases} \tag{25}$$

The distinction between E_1 and E_2 is expressed by referring to the grade 0.80 (= max(0.80, 0.69)), that is to say, there exists a little distinction. In Fig. 4(b)

$$\begin{cases} E_1(0.4) = 0.80 \text{ (grade (+))}, \\ E_2'(0.25) = 0.98 \text{ (grade(-))}. \end{cases} \tag{26}$$

The distinction between E_1 and E_2' is expressed by referring to the grade 0.98 (= max (0.80, 0.98)), that is to say, there is no distinction.

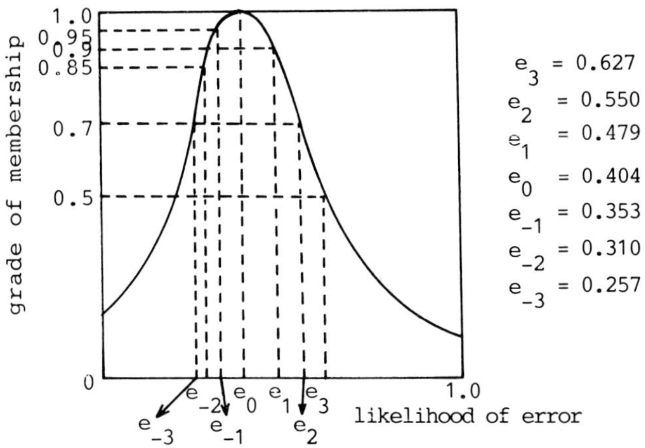

$e < e_{-3}$, $e_3 < e$; wide distinction

$e_{-3} < e < e_{-2}$, $e_2 < e < e_3$; ordinary distinction

$e_{-2} < e < e_{-1}$, $e_1 < e < e_2$; a little distinction

$e_{-1} < e < e_1$; no distinction

Fig. 3 Distinction between Likelihoods of Error Based on their Grades (Class C_5)

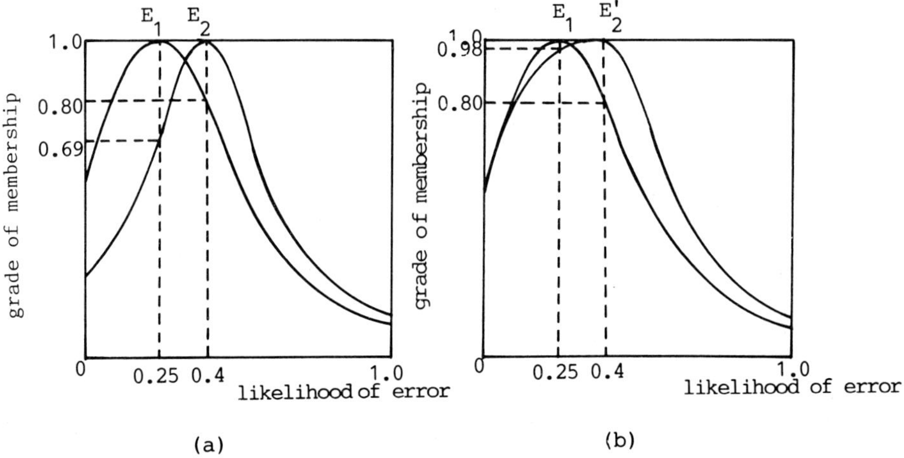

Fig. 4 Distinction between Error Possibilities

3.3 Experiments

Four kinds of tests are prepared as the following. Each test is consisted of 100 consecutive trials. These tests are performed by using a microcomputer and a color CRT. The subject is required to

(1) arrange five numbers less than 1000 in order of size,
(2) add up or subtract mentally two numbers less than 100,
(3) answer the remainder when the number less than 1000 is divided by 3,
(4) input the nonsense syllable precisely which is displayed on the CRT. The nonsense syllable is consisted of 10 alphabets and numerics.

Each trial of the test is displayed on the CRT. **Table 7** shows the various task conditions.

Table 7 Task Conditions

Item	Detail of Conditions
Equipment	* Normal * The display is detached and placed slantwise. * The color of the display is changed. * The subject must use the alphabetical keys instead of the numerical keys. * The size of alphabets and numerics on the CRT is small. * The location of alphabets and numerics on the CRT varies.
Environment	* Normal * Room temperature is high. * Some persons go in and out of the room where the experiment is performed. * The radio is on during the experiment.
Time Limit per One Trial	Test 1:* The time limit is fixed 25 secs. * The time limit is varied from 11 to 16 secs at random. Test 2:* The time limit is fixed 10 secs. * The time limit is varied from 3 to 5 secs at random. Test 3:* The time limit is fixed 10 secs. Test 4:* The time limit is fixed 15 secs.
Subjects	* Students (male 14, female 9) * Non-Students (male2, female 8)

* Normal implies that the task conditions are not made artificially.

Table 8 Statements in Questionnaire

A. Equipment Factor
1. Was it easy to operate the keyboard which you used ?
2. Was it easy to see the CRT which you used ?
3. Did you carefully answer each trial ?

Please evaluate as a whole the adequacy of the equipment which you used ?

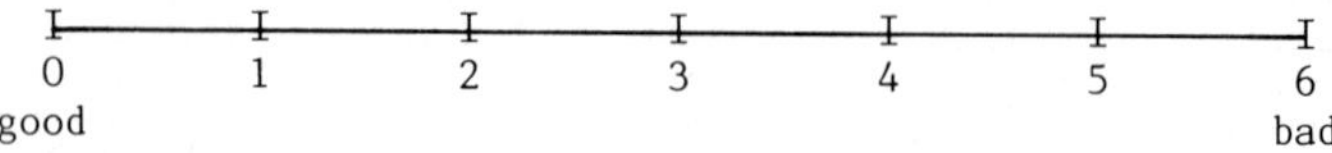

B. Environment Factor
4. Was the room temperature kept at a comfortable temperature ?
5. Was the room noisy ?
6. Were you disturbed by the comings and goings of people ?
7. Did you concentrate on the test in given environmental conditions ?

Please evaluate as a whole the environmental conditions.

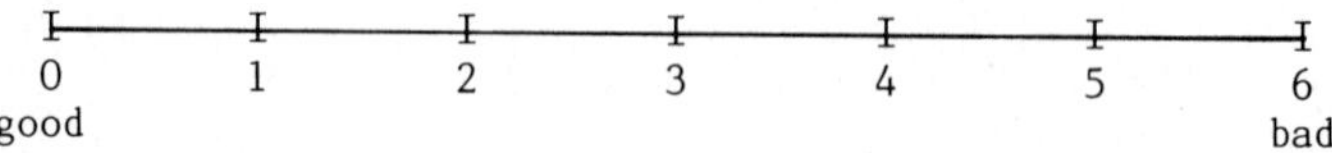

Table 8 Statements in Questionnaire (contd.)

C. Personality Factor
  8. How extent are you familiar with this test ?
  9. Are you good or not at this test ?
 10. How extent did you feel the difficulty of this test ?
 Please evaluate your aptitude for this test ?

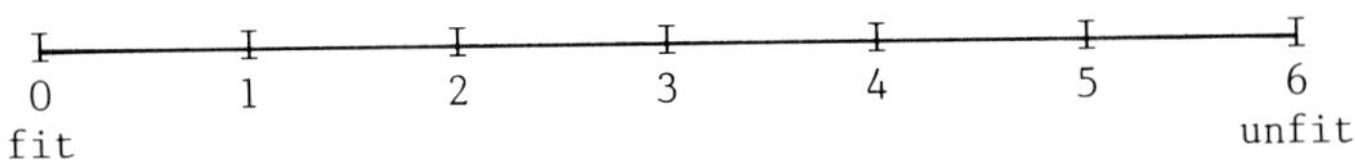

D. Fatigue FActor
 11. Did you feel that you took a long time to complete this test ?
 12. Was your posture comfortable for you during the test ?
 13. How extent was the degree of your fatigue before starting this test ?
 Please evaluate as a whole the degree of your fatigue during this test.

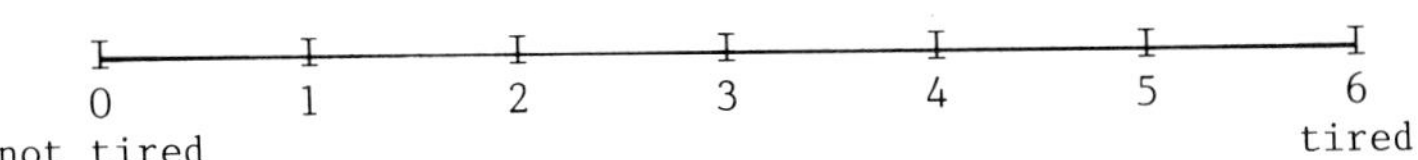

E. Motivation Factor
 Were you willing to perform this test ?

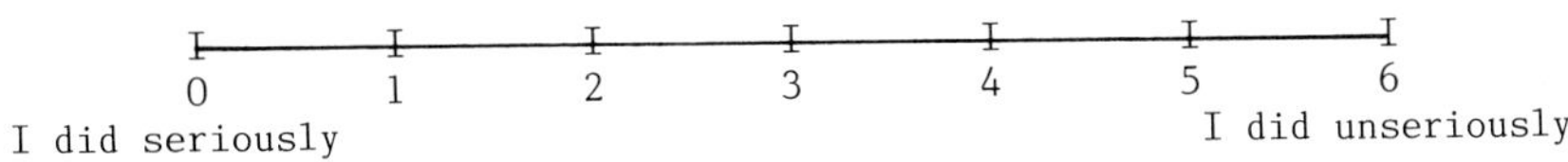

Finally please arrange from A to E in order of importance of PSF's for human reliability.

The subject is required to give a correct answer to each test under the various task conditions. After the test he is also required to answer the questionnaire orally. **Table 8** shows the statements in the questionnaire which are selected by referring to Swain and Guttmann [4].

Let us explain how to use the scales taking an example of the overall evaluation of the adequacy of equipment:
If the subject feels that the adequacy of equipment is very good or very bad, then he answers 0 or 6. If the subject feels that the adequacy of equipment is good or not, then he answers 1 or 5. If the subject feels that the adequacy of equipment is slightly good or slightly bad, then he answers 2 or 4. If the subject feels that the adequacy of equipment is neutral, that is to say, bad or good to the same extent, then he answers 3.

3.4 Results

The number of the data is 145. Of which: 75 for the test 1, 34 for the test 2, 23 for the test 3 and 13 for the test 4.

Define

$$L = \sum_{i=1}^{N} \log Er_i \,/\, N, \tag{27}$$

where Er_i is the error rate of the i-th subject and N is the number of the data.

$$Er = 10^L, \tag{28}$$

$$\sigma = \left(\sum_{i=1}^{N} (L - \log Er_i)^2 / N \right)^{1/2}. \tag{29}$$

* experiments under normal conditions ----- 80 cases

$$Er = 4.3 \times 10^{-2}, \quad \sigma = 0.328.$$

* experiments under abnormal conditions ----- 65 cases

$$Er = 8.4 \times 10^{-2}, \quad \sigma = 0.429.$$

The normal conditions mean that the task conditions are not made artificially and that the fixed time in Table 7 is used as the time limit in each test. On the other hand the abnormal conditions mean that at least one of the task conditions is made bad artificially.

From the above results, K in Eq. (3) is defined as

$$K = 1 / \log(1/5 \times 10^{-2})). \tag{30}$$

The results by the questionnaire are normalized in [0, 1] as the following: $0 \rightarrow 0$, $1 \rightarrow 0.167$, $2 \rightarrow 0.333$, $3 \rightarrow 0.5$, $4 \rightarrow 0.667$, $5 \rightarrow 0.833$ and $6 \rightarrow 1$.

Fuzzy measures are identified by using 90 data which are picked up at random from 145 data. Numbers in the following tables correspond to the questionnaire numbers.

(1) from Basic items to PSF

Let $h_{Zi}(z_j)$ be the result of the questionnaire for the j-th basic item by the i-th subject and e_{Zi} be the result of the questionnaire for the PSF Z by the i-th subject. The fuzzy measures g_Z's in Eq. (31) are identified.

$$e_{Zi} = \int h_{Zi}(z) \circ g_Z. \tag{31}$$

A. Equipment Factor

Table 9 shows the result of the fuzzy measure analysis for the equipment factor. It is found that the importance of the item 2 is the highest and that imporntance of the item 1 is the second highest. It is also found that the importance of the item 3 is not so high as that of the items 1 and 2. The basic item 3 is eliminated by Eq. (18). **Table 10** shows the final result. The performance index becomes smaller than that in Table 9. Hereafter only the final result of the fuzzy measure analysis is shown in tables.

Table 9 Result of Fuzzy Measure Analysis Equipment Factor

Basic Item	1	2	3
g	0.445	0.853	0.171
η	- 0.128	- 0.663	- 0.462
ξ	0.929	0.902	0.617
J		0.212	

Table 10 Final Result of Fuzzy Measure Analysis
Equipment Factor

Basic Item	1	2
g	0.753	0.822
η	- 0.446	- 0.446
ξ	0.890	0.921
J	0.201	

B. Environment Factor

Table 11 Result of Fuzzy Measure Analysis
Environment Factor

Basic Item	4	5	7
g	0.432	0.432	0.137
η	- 0.383	- 0.360	+ 0.257
ξ	0.782	0.796	-
J		0.179	

Table 11 shows the result of the fuzzy measure analysis for the environment factor. It is found that the basic item 7 is necessary for the environment factor since its overlapping coefficient is positive in spite of the low importance. This table also shows that the basic item 6 is eliminated. This result reflects that there were not many the comings and goings of people which disturbed the subject.

C. Personality Factor

Table 12 Result of Fuzzy Measure Analysis
Personality Factor

Basic Item	8	9	10
g	0.265	0.743	0.265
η	- 0.128	- 0.663	- 0.462
ξ	0.527	0.926	0.527
J		0.196	

Table 12 shows the result of the fuzzy measure analysis for the personality factor. It is found that the importance and the necessity coefficient of the item 9 is much higher than those of the items 8 and 10. This table shows that the basic item 9 represents this PSF, the personality factor.

D. Fatigue Factor

Table 13 Result of Fuzzy Measure Analysis

Basic Item	11	12	13
g	0.405	0.074	0.784
η	+ 0.262	+ 0.423	+ 0.121
ξ	-	-	-
J		0.273	

Table 13 shows the result of the fuzzy measure analysis for the fatigue factor. It is found that the importance of the item 12 is low. However it is also found that all basic items in this table are necessary for the fatigue factor since the overlapping coefficients of all items are positive.

(2) from the PSF's to the likelihood of error

Let $H_i(Z)$ be the evaluation of the PSF Z by the i-th subject, which is the output of the fuzzy integral model, that is,

$$H_i(Z) = \int h_{Zi}(z) \circ g_Z, \tag{32}$$

where g_Z's are the identified fuzzy measures in step (1).

However, as for the factor E, the result by the questionnaire is used. Let e_i be the likelihood of error of the i-th subject which is obtained by his error rate and Eq. (3). The fuzzy measures g's in Eq. (33) are identified.

$$e_i = \int H_i(Z) \circ g. \tag{33}$$

Table 14 shows the result.

Table 14 Result of Fuzzy Measure Analysis PSF's

PSF	A	B	C	D	E
g	0.373	0.316	0.106	0.328	0.328
η	- 0.031	- 0.151	+ 0.031	- 0.288	- 0.350
ξ	0.981	0.897	-	0.807	0.765
J			0.218		

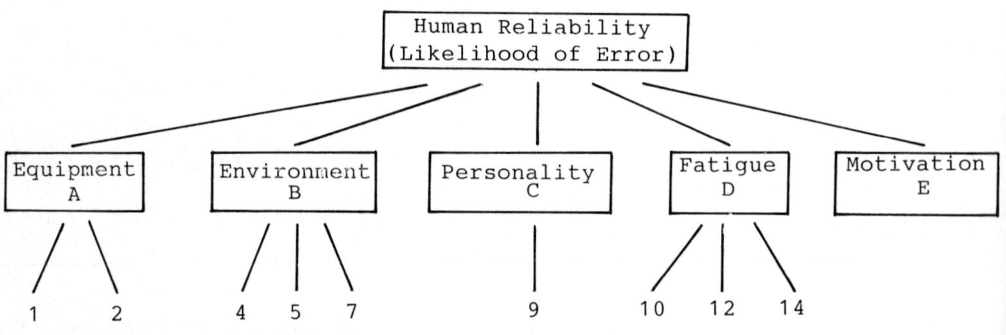

Fig. 5 Model of Structure of Human Reliability

It is found that the order of the importance of each PSF is the following: 1. equipment factor(A), 2. fatigue factor(D), 2. motivation factor(E), 4. environment factor(B) and 5. personality factor(C). This table also shows that the importance of the personality factor is not so high as others but that this factor is necessary for human reliability since the overlapping coefficient of this factor is positive. **Fig.** 5 shows the structure of human reliability.

The score is given to the result of the questionnaire on the order of the importance of PSF's as the following: the first = 10 points, the second = 5 points, the third = 0, the fourth = -5 points, the fifth = -10 points. The result is on an average A = 1.14, B = 0.45, C = -1.08, D = -0.45, E = -0.06. This order of the importance are not completely consistent with the result in Table 14. However, roughly speaking, the order is A > (B, D, E) > C and the importance of the equipment factor is the highest and that of the personality factor is the lowest in the both results.

Likelihood of error for the remaining 55 data are compared with those which are obtained by g (identified in (2)) and the fuzzy integral model. The performance index J is 0.222 which is approximately equal to J in Table 14. This result suggests that the present PSF model can estimate human reliability in the given task conditions.

3.5 Distinction between Two Error Possiblities

The mean error rate of the tests under the normal conditions is 4.3×10^{-2}. The likelihood of error is 0.465 and belongs to the class C_5 in Table 1. E_5 in **Fig.** 6 is the error possibility under the normal conditions. On the other hand the mean error rate under the abnormal conditions is 8.4×10^{-2}. The likelihood

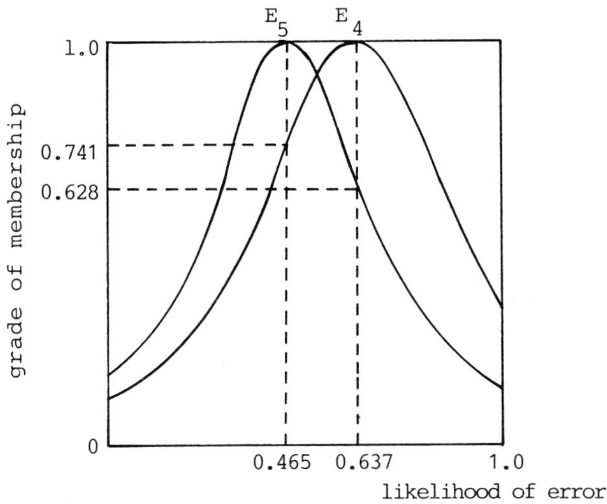

Fig. 6 Distinction between Error Possibilities E_4 and E_5

of error is 0.637 and belongs to the class C_4. E_4 in Fig. 6 is the error possibility under the abnormal conditions. It is found that the grade of 0.637 in E_5 is 0.628 and that the grade of 0.465 in E_4 is 0.741. This means that there exists the distinction between E_4 and E_5 and that E_4 is greater than E_5. The task conditions for E_4 should be improved in order to decrease the error possibility to the same extent as E_5.

Table 14 shows that the personality factor is necessary for human reliability since the overlapping coefficient of this factor is positive. The subjects are picked up who answer any of 0, 1 and 2 in the questionnaire on the personality factor after the test under the abnormal conditions. There are 29 cases. Their mean error rate of the tests under the abnormal conditions is 5.1×10^{-2} and the likelihood of error is 0.506. **Fig.** 7 shows that the grade of 0.506 in E_5 is 0.967 and that there does not exist the distinction between them. It is found that the subjects, who answer in the questionnaire that they are good at the test, can perform the test with the same reliability as E_5 even under the

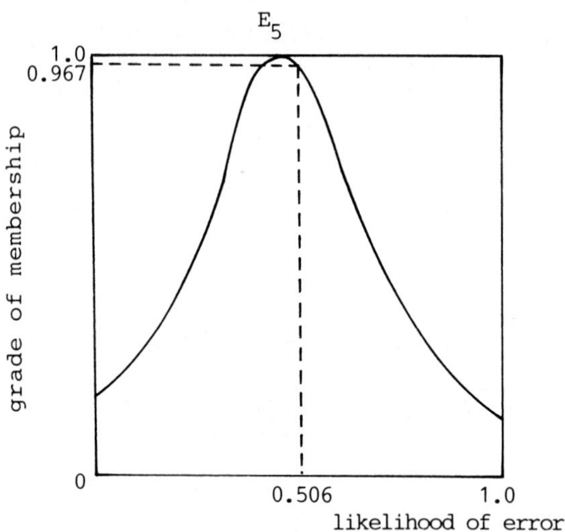

Fig. 7 Necessity of Personality Factor for Human Reliability

abnormal conditions. This result shows that the reliability is kept high to a certain extent by the apt disposition of a human operator.

4. CONCLUSIONS

The concept "error possibility" is proposed instead of the error rate in human reliability. By using the error possibility in human reliability the reliability can be considered from many satndpoints, not only the likelihood of error which gives the maximal grade of the membership but also the potentiality for error and the fuzziness of the error possibility. It is difficult to see the above point directly from the result of the probabilistic analysis since the uncertainty of the result is subjectively considered by experts on their engineering judgements in the probabilistic method.

A fuzzy set is generally a relative concept. The error possibility is also a relative concept. It is necessary to justify the parameter of the membership function. The derivation of the error possibility is founded on the following consideration:

(1) Although the error rate seems to be small from the viewpoint of our daily life, it is never small from the viewpoint of human reliability.

(2) Even if the error rate is estimated to be very small from the viewpoint of human reliability, there is a possibility that a human operator may make an error.

(3) The larger Er_M/Er_L or Er_U/Er_M is, the fuzzier the error rate is.

The algorithm of deriving the parameter of the membership function in this paper is applicable to the case that there is a possibility of happening even if its rate is estimated to be small.

The model of the relationships between human reliability and PSF's is also proposed. It is important to analyze the relations between human reliability and PSF's from the practical point of view such as

(1) the selection of PSF's to be considered in human reliability
(2) the improvement of PSF's to increase human reliability,
(3) the estimation of human reliability under the given task conditions, etc.

Fuzzy concepts are necessary for the analysis of the relations between human reliability and PSF's since these relationships as well as human reliability are ambiguous. In this paper these relations are analyzed by using the fuzzy measures and the fuzzy integrals.

This paper shows that the fuzzy concepts are useful for the analysis of human reliability.

REFERENCES

[1] K. van Gelder: Human Error with a Blending-Process Simulator, IEEE Transactions on Reliability, Vol R-29, No. 3, (1980) pp. 258 - 264.

[2] N. C. Lind: Models of Human Error in Structure Reliability, Structural Safety, 1, (1983) pp. 167 - 175.

[3] R. E. Melchers: Human Error in Structure Reliability Assessments, Reliability Engineering, 7, (1984) pp. 61 - 75.

[4] A. D. Swain and H. E. Guttmann: Handbook of Human Reliability with Emphasis on Nuclear Power Plant Applications, NUREG/CR-1278, (1980).

[5] G. Apostolakis: On the Assessment of Human Error Rates Using Operational Experiences, Reliability Engineering, 12, (1985) pp. 93 - 105.

[6] T. Terano, Y. Murayama and N. Akiyama: Human Reliability and Safety Evaluation of Man-Machine Systems, Automatica, Vol. 19, No. 6, (1983) pp. 719 - 722.

[7] T. Onisawa: An Approach to Human Reliability in Man-Machine System Using Error Possibility, Fuzzy Sets and Systems (to appear).

[8] T. Onisawa: Performance Shaping Factors Modelling Using Human Error Possibility and Fuzzy Integrals, The Japanese Journal of Ergonomics, Vol. 22, No. 2, (1986) pp. 81 - 89 (in Japanese).

[9] M. Sugeno: Theory of Fuzzy Integrals and its Applications, Doctoral Dissertation, Tokyo Institute of Technology, (1974).

[10] T. Onisawa: A Representation of Human Reliability Using Fuzzy Concepts, Information Sciences (to appear).

[11] M. Sugeno: Fuzzy Measures and Fuzzy Integrals: A Survey, in: M. M. Gupta, G. N. Saridis and B. R. Gaines, (eds.), Fuzzy Automata and Decision Process, (North-Holland, Amsterdam, 1977) pp. 89 - 102.

[12] K. Ishii and M. Sugeno: A Model of Human Evaluation Process Using Fuzzy Measures, International Journal of Man-Machine Studies, Vol. 22, No. 1, (1985) pp. 19 - 38.

[13] T. Onisawa, M. Sugeno, Y. Nishiwaki, H. Kawai and Y. Harima: Fuzzy Measure Analysis of Public Attitude towards the Use of Nuclear Energy, Fuzzy Sets and Systems, Vol. 20, No. 3, (1986) pp. 259 - 289.

Fuzzy Logic in Knowledge-Based Systems,
Decision and Control
M.M. Gupta and T. Yamakawa (Editors)

FUZZY ANALOGICAL REASONING AND ITS APPLICATION

XU Chunxi

Editorial Office, Journal of Fuzzy Mathematics,
Huazhong University of Science & Technology,
Wuhan, P. R. China

CHEN Shiquan*

Guizhou Institute for the Prevention & Control
of Industrial Hygiene & Occupational Diseases,
Guiyang, P. R. China

In this paper the fuzzy analogical reasoning be discussed quantitatively. First three inferential methods be suggested, namely Near degree operation, Addition comprehensive method and Fuzzy maximum path calculation. Then these methods be used to infer the standards of maximum allowable concentration of industrial toxicant vinylbenzene (styrene). The experimental reuslts have showed that these methods be usually effective in practive.

Keywords: Fuzzy analogical reasoning, Near degree operation, Addition power comprehensive method, Fuzzy maximum path calculation.

1. INTRODUCTION

Given two objects X and Y, we now investigate what conclusion can be inferred when all facts known are:

(1) There are n different qualities $q_1, q_2, \ldots, q_n$ such that for each $1 \leq i \leq n$, X and Y possess quality q_i in degrees a_i and b_i, respectively.

(2) There is another quality q_{n+1}, such that X possesses quality q_{n+1} in degree a_{n+1}. More precisely, we want to ask in what degree does Y possess quality q_{n+1}? Clearly, such a reasoning is a fuzzy conditional reasoning and can be denoted as:

---

* Present Address: Department of Mathematics, Guizhou Normal University, Guiyang, . P. R. China

$$
\begin{array}{ll}
\text{X possesses } a_1,\ a_2,\ \ldots,\ a_n \text{ and } a_{n+1} & \\
\text{Y possesses } b_1,\ b_2,\ \ldots,\ b_n & \\
\hline
\text{Y possesses } \qquad\qquad b_{n+1} &
\end{array}
\qquad (1)
$$

Now let us analyze some features of reasoning of the form (1). First, because the concepts $a_i(1 \leq i \leq n+1)$ and $b_j(1 \leq i \leq n)$ contained in the premises are usually imprecise or fuzzy, and hence so is the concept b_{n+1} contained in the conclusion, the reasoning is fuzzy in essence. Thus it seems appropriate to find inferential methods from fuzzy mathematics, which is just one of the aims of this paper. Hence, from now on, it is assumed that a_i, $b_i \in [0, 1]$, i.e., we use a number in [0, 1] to describe the degree in which an object possesses a quality.

Secondly, if $a_i \approx b_i$ for some i, it may be said that X and Y resemble each other in quality q_i. Hence if $a_i \approx b_i$ for all i $1 \leq i \leq n$, the form (1) can be regarded as a model for analogical reasoning. But we would like to call the reasoning of the form (1) as fuzzy analogical reasoning since it is a fuzzy reasoning.

This paper suggests three inferential methods for the reasoning of the form (1) without assumption of $a_i \approx b_i$, i=1,2,...,n., and presents their application to inferring reliable standards of maximum allowable concentration of industrial toxicant. It is necessary to point out that our experiments have showed that our methods are usually effective in practice.

2. NEAR DEGREE OPERATION

A vector $(a_1, a_2, \ldots, a_n)$ is called a fuzzy vector if it satisfies $a_i \in [0, 1]$, i=1, 2,..., n. The set of all fuzzy vectors $(a_1, a_2, \ldots, a_n)$ is called a fuzzy vector space (Wang Peizhang [3]) and denoted by V.

Let N: $V \times V \to [0, 1]$ be a mapping. If N satisfies the following conditions, then N is called a near degree or near measure in V.

$$
\begin{array}{lll}
\text{(i)} & N(A,A)=1 & \\
\text{(ii)} & N(A,B)=N(B,A) & \qquad (A,B,C \in V) \\
\text{(iii)} & C \subseteq B \subseteq A \rightrightarrows N(A,C) \leq N(A,B). &
\end{array}
$$

N(A,B) is called the *near degree* of A and B with respect to N. An important example of near degree is as follows:

$$N(A,B) = 1 \pm \frac{1}{n} \sum_{i=1}^{n} (a_i - b_i), \qquad i=1,2,\ldots,n. \qquad (2)$$

Where $A=(a_i,\ldots,a_n)$ $B=(b_i,\ldots,b_n)$.

We can get following two formulas to obtain inferential result according to form (2).

If $(a_i - b_i) \geq 0$, then

$$b_{n+1} = a_{n+1} \cdot (1 - \frac{1}{n} \sum_{n=1}^{n} (a_i - b_i)) \qquad i=1,2,\ldots,n, \tag{2a}$$

If $(a_i - b_i) < 0$, then

$$b_{n+1} = a_{n+1} / 1 + \frac{1}{n} \sum_{n=1}^{n} (a_i - b_i) \qquad i=1,2,\ldots,n. \tag{2b}$$

Clearly, the above formulas expresses the idea that greater the value of N(A,B) is, the more similar X and Y are. Of course, in estimating the similarity between X and Y by using the (2), it should be guaranteed that the choice of referential qualities $q_1,\ldots,q_n$ is appropriate. If not, the obtained result will be of no use.

3. ADDITION POWER COMPREHENSIVE METHOD

All the qualities are treated without distinction of importance by using *near degree operation*. In fact, some are important, others are secondary, and still others are insignificant among them. We can use the so-called *Addition power comprehensive method* to stick out main qualities for overcoming the defect. First according to the degrees of importance of the qualities $q_1,\ldots,q_n$ we give them power weights $d_1,\ldots d_n$, respectively, thus getting a power weight set or power vector D.

$$D = (d_1, d_2, \ldots, d_n)$$

where $d_i \in [0, 1]$ $\sum_{i=1}^{n} d_i = 1$, $i=1,2,\ldots,n.$

Then we can establish a comprehensive model in the form:

$$1 \pm ((D \circ A^T) - (D \circ B^T)) \tag{3}$$

where $\circ$ is the composition operation of fuzzy matrices. We can get following two formulas to obtain inferential result according to form (3)

If $(D \circ A^T) - (D \circ B^T) \geq 0$, then

$$b_{n+1} = (1 - ((D \circ A^T) - (D \circ B^T))) \cdot a_{n+1} \tag{3a}$$

If $(D \circ A^T) - (D \circ B^T) < 0$, then

$$b_{n+1} = a_{n+1} / 1 + ((D \circ A^T) - (D \circ B^T)) \tag{3b}$$

4. FUZZY MAXIMUM PATH CALCULATION

These qualities of X and Y are considered separably by the two methods mentioned above. If we consider not only the degrees of qualities themselves, but also the relations between these qualities, the reliable standard of reasoning will be greatly raised.

So we can use the fuzzy matrix to repersent the relation between these qualities.

If A is regarded as a 1×a matrix, the direct product A×A can be denoted by a fuzzy symmetrical matrix or a fuzzy graph. We may look for the interreationship of these qualities from the matrix or graph, and find out a path that manifests a intimatest relation among the qualities of A. We can find out the path among the qualities of B, too. Then we can calculate their a different value and draw a conclusion.

First we introduce the following definitions:

Definition 4.1 Soppose W is a finite vertex set. $W \in \mathscr{F}(W)$ is a fuzzy vertex set. $E \in \mathscr{F}(W) \times \mathscr{F}(W)$ is a fuzzy edge set. If $E(w_i, w_j) \leq W(w_i) \wedge W(w_j)$, $(\forall w_i,\ w_j \in W)$, the pair (W,E) is called a fuzzy graph.

Definition 4.2 Soppose $C \subset E$, if C is a set connecting vertices $w_1, w_2, \ldots, w_n$, the w_1, w_i, w_{i+1}, w_n do not coincide, and initial vertex is identical with the terminal vertex, then C is called a circuit. If the initial vertex and the terminal vertex are different, then C is called a path P.

Definition 4.3 The sum q of the membership degree of all elements in P is called the connected degree of P.

Definition 4.4 $h = \frac{n}{n-1}$ is called the average connected degree of P.

Definition 4.5 A path whose everage connected degree h is the highest in E is called a fuzzy maximum path.
Let E consist of l elements, then there are l! paths in E. How we find out a fuzzy maximum path? First we can show a fuzzy graph (W,E) with a symmetrical matrix (Minieka [1]). For instance when there are n vertices in W, we have

$$\begin{array}{c} \\ W_1 \\ W_2 \\ \vdots \\ W_n \end{array} \begin{array}{c} \begin{array}{ccc} W_1 & W_2 & \cdots W_n \end{array} \\ \begin{bmatrix} e_{11} & e_{12} & \cdots e_{1n} \\ e_{21} & e_{22} & \cdots e_{2n} \\ \vdots & \vdots & \vdots \\ e_{n1} & e_{n2} & \cdots e_{nn} \end{bmatrix} \end{array}$$

We can find out the fuzzy maximum path using the following:

(i) Let the master diagonal elements all zero.

(ii) Suppose $C_k=(x_1,x_2,\ldots,x_n)$ is a circuit from vertex w_k to w_k (k=1,2,..., n, $w_k \in W$) in fuzzy graph $(W,E)=(e_{ij})_{n\times n}$, where x_i (i=1,2,...n) is an edge of membership degree in E. $x_i \in E$, i=1,2,..., n.

(iii) To make $x_1=e_{i_1k}=\max e_{ik}$ i=1,2,...,n.

Let $e_{i_1j}=0$, ($j\neq k$, j=1,2,...,n) and $e_{ki_1}=0$

To make $x_2-e_{i_2i_1}=\max c_{ii_1}$ i=1,2,...,n

Let $e_{i_2j}=0$ ($j\neq i_1$, j=1,2,...,n) and $e_{i_1i_2}=0$.

.
.
.

To make $x_n=e_{i_ni_{n-1}}=\max e_{ii_{n-1}}$ i=1,2,..., n.

If the maximum elements are not unique in identical contingency, we can take out an element such that the other elements are small in an identical row.

(iv) According to (iii) we can find out $C_1,C_2,\ldots,C_n$ by turns.

(v) Cut out an edge x_{i_0} whose membership degree is minimum in C_k, (k=1,2,..., n). If the minimum elements are not unique in C_k, cut out one of them arbitrarily. The C_k has become a path P_k.
(vi) We can find a P_{i_0} whose connected degree q is the highest in P_1. $P_2,\ldots,P_n$ using the following:

$$\mathrm{Max}(q_1,q_2,\ldots,q_n)$$

where $q_i=x_1+x_2+\ldots+x_n-x_{i_0}$. (i=1,2,...,n)

(vii) Calculate the average connected degree h using the following:

$$h=\frac{q}{n-1}, \qquad q=\sum_{i=1}^{n} x_i. \qquad i\neq i_0,\ i=1,2,\ldots,n.$$

It is the fuzzy maximum path that the average connected degree h are the highest in $q_1, q_2,\ldots, q_n$.
we can draw a conclusion by as follows:

$$1\pm(h_1-h_2) \tag{4}$$

where h_1 and h_2 are the average connected degree of fuzzy maximum path among A and B, respectively.

If $h_1-h_2\geq 0$, then

$$b_{n+1}=(1-(h_1-h_2))\cdot a_{n+1} \tag{4a}$$

If $h_1-h_2<0$, then

$$b_{n+1}=a_{n+1}/1+(h_1-h_2). \tag{4b}$$

5. USING FUZZY ANALOGICAL REASONING TO INFER THE STANDARDS OF MAXIMUM ALLOWABLE CONCENTRATION OF INDUSTRIAL TOXICANT

There are about five-handred thousands of toxic substances in the world. How are obtained exactly the standards of maximum allowable concentration (MAC) of toxicant in modern production? How are the standards set up fast? The fuzzy analogical reasoning method is a good one.

Industrial toxicant vinylbenzene and vinyl chloride are very resemblance in many qualities. We can establish the analogical reasoning model between them. We investigate their qualities in three respects which are properties of physics and chemistry, toxicity experiment and clinical manifestation. The seven qualities which are colour, excretory quantity, dissolve, slow toxicity, acute toxicity, slow toxicosis and acute toxicosis are compared, respectively. They can be listed as Table 1 (cf.[2]).

According to Table 1, we can give the corresponding fuzzy vectors A, B, which are the qualities in degree a_i and b_i, $i=1,2,...,7.$, respectively, representing styrene and vinyl chloride.

$$A=(1,\ 0.4,\ 1,\ 0.6,\ 0.6,\ 0.6,\ 0.6\ ) \qquad \text{(St.)}$$

$$B=(1,\ 0.7,\ 1,\ 0.9,\ 0.8,\ 0.9,\ 0.8\ ) \qquad \text{(VC)}$$

First we use Near degree operation to calculate the result.
Because of $(a_i-b_i)<0$, the form (2b) be used. When A,B be substituted into (2b), we have

$$b_{n+1}=(30\text{mg/m}^3)/1-\frac{1}{7}\ (0.3+0.3+0.2+0.3+0.2)=36.9\text{mg/m}^3,$$

Where because MAC of VC is 30mg/m^3, the $a_{n+1}=30\text{mg/m}^3$.

From above operation we get that MAC of St. is 36.9mg/m^3.

Then we use Addition power comprehensive method so that the seven qualities be treated with distinction of importance. The slow toxicity, acute toxicity, slow toxicosis and acute toxicosis are important, the dissolve is secondary, the colour and excretory quantity are insignificant among these qualities.

Table 1

		styrene (St.)	vinyl chloride (VC)
physical and chemical properties	color	colorless	colorless
	excretory quantity	40%	69.4%
	dissolution	dissolve in organic solvent but dissolve little in water	idem
toxicity experiment	slow toxicity	breathe with difficulty blurred consciousness	muscle tic short of breath
	acute toxicity	weak stimulation in respiratory tract	lose weight react slowly
clinical manifestation	slow toxicosis	headache sickness	headache coma
	acute toxicosis	abdominal distension cough	abdominal distension dry skin

The seven qualities of St. and VC can be added power in turns as follows:

$$D=(0.05,\ 0.05,\ 0.1,\ 0.2,\ 0.2,\ 0.2,\ 0.2\ )$$

Instituting D, A, B into the expression (3b), we have

$$b_{n+1}=(30\text{mg/m}^3)/1+((0.05\ 0.05\ 0.1\ 0.2\ 0.2\ 0.2\ 0.2\ )\circ\begin{bmatrix}1\\0.4\\1\\0.6\\0.6\\0.6\\0.6\end{bmatrix}$$

$$-(0.05\ 0.05\ 0.1\ 0.2\ 0.2\ 0.2\ 0.2)\circ\begin{bmatrix}1\\0.7\\1\\0.9\\0.8\\0.9\\0.8\end{bmatrix})$$

$$=(30\text{mg/m}^3)/0.795=37.7\text{mg/m}^3$$

For composition operation of fuzzy matrices, there are many formulas which can be used (cf.[3]), Here we use the operation of arithmetic ×,+. and get that the MAC of St'. is 37.7mg/m^3.

Finally we use Fuzzy maximum calculation to infer MAC of St. The interrelationship between the seven qualities of St. and VC can be listed as follows, respectively.

Table 2. (St)

	color	excretory quantity	disso-lution	slow toxi-city	acute toxi-city	slow toxi-cosis	acute toxi-cosis
color	1	0.1	0.3	0.2	0.15	0.2	0.15
excretory quantity	0.1	1	0.4	0.35	0.38	0.3	0.34
disso-lution	0.3	0.4	1	0.55	0.58	0.5	0.6
slow toxicity	0.2	0.35	0.55	1	0.45	0.6	0.36
acute toxicity	0.15	0.38	0.58	0.45	1	0.42	0.6
slow toxicosis	0.2	0.3	0.5	0.6	0.42	1	0.52
acute toxicosis	0.15	0.34	0.6	0.36	0.6	0.52	1

Table 3. (VC)

	color	excretory quantity	disso-lution	slow toxi-city	acute toxi-city	slow toxi-cosis	acute toxi-cosis
color	1	0.1	0.3	0.2	0.15	0.2	0.15
excretory quantity	0.1	1	0.7	0.65	0.68	0.6	0.58
disso-lution	0.3	0.7	1	0.85	0.88	0.8	0.9
slow toxicity	0.2	0.65	0.85	1	0.75	0.8	0.7
acute toxicity	0.15	0.58	0.88	0.75	1	0.85	0.9
slow toxicosis	0.2	0.6	0.8	0.8	0.85	1	0.78
acute toxicosis	0.15	0.85	0.9	0.7	0.9	0.78	1

Step 1. According to the (i) of part 4 we have matrix 1 from Table 2:

$$\begin{bmatrix} 0 & 0.1 & 0.3 & 0.2 & 0.15 & 0.2 & 0.15 \\ 0.1 & 0 & 0.4 & 0.35 & 0.38 & 0.3 & 0.34 \\ 0.3 & 0.4 & 0 & 0.55 & 0.58 & 0.5 & 0.6 \\ 0.2 & 0.35 & 0.55 & 0 & 0.45 & 0.6 & 0.36 \\ 0.15 & 0.38 & 0.58 & 0.45 & 0 & 0.42 & 0.6 \\ 0.2 & 0.3 & 0.5 & 0.6 & 0.42 & 0 & 0.52 \\ 0.15 & 0.34 & 0.6 & 0.36 & 0.6 & 0.52 & 0 \end{bmatrix}$$

matrix 1

Step 2. According to the (ii) and (iii) of part 4, when k=1,

$C_1=(x_1, x_2, \ldots, x_7)$

$x_1=e_{i_1 1}=\max e_{i1} \qquad i=1,2,\ldots, 7,$

$=\max \{0, 0.1, 0.3, 0.2, 0.15, 0.2, 0.15\}$

$=0.3$

Let $e_{3j}=0$, $(j\neq 1,\ j=2,3,\ldots,7)$ and $e_{13}=0$, then we have matrix 2 as follows:

$$\begin{bmatrix} 0 & 0.1 & 0 & 0.2 & 0.15 & 0.2 & 0.15 \\ 0.1 & 0 & 0.4 & 0.35 & 0.38 & 0.3 & 0.34 \\ 0.3 & 0 & 0 & 0 & 0 & 0 & 0 \\ 0.2 & 0.35 & 0.55 & 0 & 0.45 & 0.6 & 0.36 \\ 0.15 & 0.38 & 0.58 & 0.45 & 0 & 0.42 & 0.6 \\ 0.2 & 0.3 & 0.5 & 0.6 & 0.42 & 0 & 0.52 \\ 0.15 & 0.34 & 0.6 & 0.36 & 0.6 & 0.52 & 0 \end{bmatrix}$$

matrix 2

$x_2=e_{i_2 3}=\max\{0, 0.4, 0, 0.55, 0.58, 0.5, 0.6\}=0.6$

Let $e_{7j}=0$, $(j\neq 3,\ j=1,2,4,\ldots,7)$ and $e_{37}=0$, then we have matrix 3 as follows:

$$\begin{bmatrix} 0 & 0.1 & 0 & 0.2 & 0.15 & 0.2 & 0.15 \\ 0.1 & 0 & 0.4 & 0.35 & 0.38 & 0.3 & 0.34 \\ 0.3 & 0 & 0 & 0 & 0 & 0 & 0 \\ 0.2 & 0.35 & 0.55 & 0 & 0.45 & 0.6 & 0.36 \\ 0.15 & 0.38 & 0.58 & 0.45 & 0 & 0.42 & 0.6 \\ 0.2 & 0.3 & 0.5 & 0.6 & 0.42 & 0 & 0.52 \\ 0 & 0 & 0.6 & 0 & 0 & 0 & 0 \end{bmatrix}$$

matrix 3

$x_3=e_{i_3 7}=\max\{0.15, 0.34, 0, 0.36, 0.6, 0.52, 0\}=0.6$

Let $e_{5j}=0$, $(j\neq 7,\ j=1,2,\ldots,6)$ and $e_{75}=0$, then we have matrix 4 as follows:

$$\begin{bmatrix} 0 & 0.1 & 0 & 0.2 & 0.15 & 0.2 & 0.15 \\ 0.1 & 0 & 0.4 & 0.35 & 0.38 & 0.3 & 0.34 \\ 0.3 & 0 & 0 & 0 & 0 & 0 & 0 \\ 0.2 & 0.35 & 0.55 & 0 & 0.45 & 0.6 & 0.36 \\ 0 & 0 & 0 & 0 & 0 & 0 & 0.6 \\ 0.2 & 0.3 & 0.5 & 0.6 & 0.42 & 0 & 0.52 \\ 0 & 0 & 0.6 & 0 & 0 & 0 & 0 \end{bmatrix}$$

matrix 4

$$x_4=e_{i_4 5}=\max(0.15,\ 0.38,\ 0,\ 0.45,\ 0,\ 0.42,\ 0)=0.45$$

Let $e_{4j}=0$, $(j\neq 5,\ j=1,2,3,4,6,7)$ and $e_{54}=0$, then we have matrix 5 as follows:

$$\begin{bmatrix} 0 & 0.1 & 0 & 0.2 & 0.15 & 0.2 & 0.15 \\ 0.1 & 0 & 0.4 & 0.35 & 0.38 & 0.3 & 0.34 \\ 0.3 & 0 & 0 & 0 & 0 & 0 & 0 \\ 0 & 0 & 0 & 0 & 0.45 & 0 & 0 \\ 0 & 0 & 0 & 0 & 0 & 0 & 0.6 \\ 0.2 & 0.3 & 0.5 & 0.6 & 0.42 & 0 & 0.52 \\ 0 & 0 & 0.6 & 0 & 0 & 0 & 0 \end{bmatrix}$$

matrix 5

$$x_5=e_{i_5 4}=\max(\ 0.2,\ 0.35,\ 0,\ 0,\ 0,\ 0.6,\ 0\ )=0.6$$

Let $e_{6j}=0$, $(j\neq 4,\ j=1,2,3,5,6,7)$ and $e_{46}=0$, then we have matrix 6 as follows:

$$\begin{bmatrix} 0 & 0.1 & 0 & 0.2 & 0.15 & 0.2 & 0.15 \\ 0.1 & 0 & 0.4 & 0.35 & 0.38 & 0.3 & 0.34 \\ 0.3 & 0 & 0 & 0 & 0 & 0 & 0 \\ 0 & 0 & 0 & 0 & 0.45 & 0 & 0 \\ 0 & 0 & 0 & 0 & 0 & 0 & 0.6 \\ 0 & 0 & 0 & 0.6 & 0 & 0 & 0 \\ 0 & 0 & 0.6 & 0 & 0 & 0 & 0 \end{bmatrix}$$

matrix 6

$$x_6=e_{i_6 6}=\max(\ 0.2,\ 0.3,\ 0,\ 0,\ 0,\ 0,\ 0)=0.3$$

Let $e_{2j}=0$, $(j\neq 6,\ j=1,\ldots,5,7)$ and $e_{62}=0$, then we have matrix 7 as follows:

$$\begin{bmatrix} 0 & 0.1 & 0 & 0.2 & 0.15 & 0.2 & 0.15 \\ 0 & 0 & 0 & 0 & 0 & 0.3 & 0 \\ 0.3 & 0 & 0 & 0 & 0 & 0 & 0 \\ 0 & 0 & 0 & 0 & 0.45 & 0 & 0 \\ 0 & 0 & 0 & 0 & 0 & 0 & 0.6 \\ 0 & 0 & 0 & 0.6 & 0 & 0 & 0 \\ 0 & 0 & 0.6 & 0 & 0 & 0 & 0 \end{bmatrix}$$

matrix 7

$$x_7=e_{i_7 2}=\max(0.1,\ 0,\ 0,\ 0,\ 0,0,\ 0,)=0.1$$

So $C_1=(0.3\ 0.6\ 0.6\ 0.45\ 0.6\ 0.3\ 0.1)$.

Using same method we can find out $C_2, C_3, \ldots, C_7$, respectively.

$C_2=(0.38\ 0.6\ 0.6\ 0.55\ 0.6\ 0.2\ 0.1)$

$C_3=(0.6\ 0.6\ 0.45\ 0.6\ 0.3\ 0.1\ 0.3)$

$C_4=(0.6\ 0.52\ 0.6\ 0.58\ 0.4\ 0.1\ 0.2)$

$C_5=(0.6\ 0.6\ 0.55\ 0.6\ 0.3\ 0.1\ 0.15)$

$C_6=(0.6\ 0.55\ 0.6\ 0.6\ 0.38\ 0.1\ 0.2)$

$C_7=(0.6\ 0.58\ 0.55\ 0.6\ 0.3\ 0.1\ 0.15)$

Step 3. According to the (v), (vi) and (vii) of part 4, we have

$q_1 = \Sigma\ (0.3,\ 0.6,\ 0.6,\ 0.45,\ 0.6,\ 0.3) = 2.85$

$q_2 = \Sigma\ (0.38,\ 0.6,\ 0.6,\ 0.55,\ 0.6,\ 0.2) = 2.93$

$q_3 = \Sigma\ (0.6,\ 0.6,\ 0.45,\ 0.6,\ 0.3,\ 0.3) = 2.85$

$q_4 = \Sigma\ (0.6,\ 0.52,\ 0.6,\ 0.58,\ 0.4,\ 0.2) = 2.9$

$q_5 = \Sigma\ (0.6,\ 0.6,\ 0.55,\ 0.6,\ 0.3,\ 0.15) = 2.8$

$q_6 = \Sigma\ (0.6,\ 0.55,\ 0.6,\ 0.6,\ 0.38,\ 0.2) = 2.93$

$q_7 = \Sigma\ (0.6,\ 0.58,\ 0.55,\ 0.6,\ 0.3,\ 0.15) = 2.78$

$\text{Max}\ (q_1, \ldots, q_7) = \max\ (2.85,\ 2.93,\ 2.85,\ 2.9,\ 2.8,\ 2.93,\ 2.78) = 2.93$

$$h_1 = \frac{2.93}{6} = 0.488$$

Similarly, we can calculate h_2 from Table 3 and get

$$h_2 = \frac{4.4}{6} = 0.733$$

Because of $h_1 - h_2 < 0$, the form (4b) be used. When h_1 and h_2 be substituted into (4b), we have

$$b_{n+1} = (30\text{mg/m}^3)/1 + (0.488 - 0.733) = 39.7\text{mg/m}^3$$

We get that MAC of St. is 39.7mg/m^3 by the fuzzy maximum path calculation.

6. CONCLUSION

In this paper we have discussed three methods of *fuzzy analogical reasoning* and used these mehtods inferring the MAC of St.. In these methods *Near degree operation* is concise. *Addition power comprehensive method* and *Fuzzy maximum path calculation* may yield better inferentical results. In fact, that MAC of St. is 39.7mg/m^3, which be obtained by Fuzzy maximum path calculation, is close to the 40mg/m^3 which is provided for production now. If Addition power comprehensive method and Fuzzy maximum path calculation are combined to use, the result will be very effective.

ACKNOWLEDGEMENTS

The anthors would like to thank Prof. Yu Yandong as well as the referees for their suggestions and kind help.

REFERENCES

[1] E. Minieka, Optimization Algorithms for Network and Graphs, Marcel Dekker, Inc., New York and Basel, 1978.

[2] Labour hygiene and occupational disease, People hygiene publishing house, China, 1981.

[3] Wang Peizhuang, Theory and Applications of Fuzzy Sets, Shanghai Publishing House of Science and Technology, China, 1983.

Fuzzy Logic in Knowledge-Based Systems,
Decision and Control
M.M. Gupta and T. Yamakawa (Editors)

DYNAMIC TEAM DECISION PROBLEMS WITH ULTRAFUZZY INFORMATION STRUCTURES

Hideyuki NOJIRI

Department of Commerce,
Kumamoto University of Commerce,
2-5-1 Oe, Kumamoto 862, Japan

ABSTRACT

In this paper, the concept of fuzzy sets of type 2 is introduced to the dynamic team theory and the dynamic team decision processes which contain ultrafuzzy states, ultrafuzzy dynamic information structures, ultrafuzzy information signals, ultrafuzzy decision rules, and ultrafuzzy actions are formulated. After some definitions of fuzzy relations of type 2, fuzzy sets of type 2 induced by mappings, composition, Cartesian product, ultrafuzzy mappings, and ultrafuzzy functions, the ultrafuzzy dynamic team decision models are proposed.

Keywords: fuzzy team theory, ultrafuzzy sets, fuzzy information structure, fuzzy information, fuzzy decision, team decision.

1. INTRODUCTION

Since Zadeh's original papers [5, 7] which have proposed the concept of fuzzy sets of type 2 as an extension of fuzzy sets, fuzzy linguistic logic [7] and type 2 fuzzy automata and grammars [2] have been formulated. The algebraic structures of fuzzy sets of type 2 have been investigated by Mizumoto and Tanaka [3]. The ultrafuzzy team decision problems have been formulated by Nojiri [4].

In this work, we introduce the concept of fuzzy sets of type 2 [3, 5, 7] to the dynamic team theory proposed by Y.C. Ho and K.C. Chu [1] and study the ultrafuzzy dynamic information structures of organization.

In Sec.2, we give some definitions of fuzzy relations of type 2, fuzzy sets of type 2 induced by mappings, composition, Cartesian product, ultrafuzzy mapping, and ultrafuzzy function. In Sec.3, dynamic team decision problems are described. In Sec.4, dynamic team decision problems with ultrafuzzy information structures are formulated. In Sec.5, the concepts of an ultrafuzzy information structure and an ultrafuzzy decision rule are defined. The dynamic team decision problem under ultrafuzzy constraints is dealt with and the concepts of fuzzy expected utility of the dynamic team are given.

2. FUZZY SETS AND FUZZY RELATIONS WITH FUZZY MEMBERSHIP FUNCTIONS

In this section we list some preliminary definitions and properties of fuzzy sets of type

2 [3, 4] that will be used in the following. Then we shall define fuzzy relations of type 2, fuzzy sets of type 2 induced by mappings, composition, Cartesian product, ultrafuzzy mapping, and ultrafuzzy function by the use of the extension principle [7].

DEFINITION 2.1. Let X be a nonempty set. A *fuzzy set A of type 2* in X is characterized by a *fuzzy membership function* μ_A as

$$\mu_A : X \to [0,1]^J, \tag{2.1}$$

where the value $\mu_A(x)$ is called a *fuzzy grade* and is a fuzzy set in $J \subseteq [0,1]$.

NOTE. In this paper it is assumed that J is a finite subset of $[0, 1]$.

DEFINITION 2.2. Let $\mu_A(x)$ and $\mu_B(x)$ be two fuzzy grades (fuzzy sets in J) of fuzzy sets of type 2, A and B, respectively, and be represented as follows:

$$\mu_A(x) = \frac{f(u_1)}{u_1} + \frac{f(u_2)}{u_2} + \cdots + \frac{f(u_n)}{u_n}$$

$$= \sum_i \frac{f(u_i)}{u_i}, \quad u_i \in J, \tag{2.2}$$

$$\mu_B(x) = \frac{g(w_1)}{w_1} + \frac{g(w_2)}{w_2} + \cdots + \frac{g(w_m)}{w_m}$$

$$= \sum_j \frac{g(w_j)}{w_j}, \quad w_j \in J, \tag{2.3}$$

where the functions $f : J \to [0,1]$ and $g : J \to [0,1]$ are membership functions of fuzzy grades $\mu_A(x)$ and $\mu_B(x)$, respectively, and the values $f(u_i)$ and $g(w_j)$ in $[0,1]$ represent the grades for u_i and w_j in J, respectively. The operation $+$ stands for logical sum (or). Then the operations of fuzzy sets of type 2 are defined by using the extension principle as follows [7]. For all $x \in X$,

Union:

$$A \cup B \iff \mu_{A \cup B}(x) = \mu_A(x) \sqcup \mu_B(x)$$

$$= \left(\sum_j \frac{f(u_i)}{u_i} \right) \sqcup \left(\sum_j \frac{g(w_j)}{w_j} \right)$$

$$= \sum_{i,j} \frac{f(u_i) \wedge g(w_j)}{u_i \vee w_j}. \tag{2.4}$$

Intersection:

$$A \cap B \iff \mu_{A \cap B}(x) = \mu_A(x) \sqcap \mu_B(x)$$

$$= \left(\sum_i \frac{f(u_i)}{u_i} \right) \sqcap \left(\sum_j \frac{g(w_j)}{w_j} \right)$$

$$= \sum_{i,j} \frac{f(u_i) \wedge g(w_j)}{u_i \wedge w_j}. \tag{2.5}$$

Complement:

$$A^{C} \Longleftrightarrow \mu_{A^C}(x) = \neg\, \mu_A(x)$$

$$= \sum_i \frac{f(u_i)}{1-u_i} \quad . \tag{2.6}$$

where $\vee$ and $\wedge$ represent max and min, respectively.

DEFINITION 2.3. Let $J=\{u_1, u_2, \ldots, u_n\}$ be a subset of $[0,1]$ which satisfies $u_1<u_2<\cdots<u_n$; then a fuzzy grade $\mu_A(x)=\sum_i f(u_i)/u_i$ in J is said to be *convex* if for any integers i, k with $i \leqslant k$, the following is satisfied, i.e.,

$$f(u_j) \geqslant \min\{f(u_i),\ f(u_k)\}, \tag{2.7}$$

where j is any integer which satisfies $i \leqslant j \leqslant k$.

DEFINITION 2.4. A fuzzy grade $\mu_A(x)=\sum_i f(u_i)/u_i$ in J is said to be *normal* if

$$\max_i f(u_i)=1. \tag{2.8}$$

Otherwise it is subnormal.

NOTE. In this paper it is assumed that the grades of fuzzy sets of type 2 are normal and convex. Normal convex fuzzy grades in J form a distributive lattice under $\sqcup$ and $\sqcap$, where the greatest element is 1/1(=1) and the least element is 1/0(=0). Fuzzy sets of type 2 in a set X form a distributive lattice under the operations $\cup$ and $\cap$ defined in (2. 4) and (2. 5), respectively, where the grades characterizing these fuzzy sets of type 2 are normal convex fuzzy grades in $J \subseteq [0, 1]$[3].

DEFINITION 2.5. For all $x \in X$,

Containment:

$$\begin{aligned} A \subseteq B &\Longleftrightarrow \mu_A(x) \sqsubseteq \mu_B(x) \\ &\Longleftrightarrow \mu_A(x) \sqcap \mu_B(x) = \mu_A(x) \\ &\Longleftrightarrow \mu_A(x) \sqcup \mu_B(x) = \mu_B(x). \end{aligned} \tag{2.9}$$

Equality:

$$A=B \Longleftrightarrow \mu_A(x)=\mu_B(x). \tag{2.10}$$

Algebraic product:

$$\begin{aligned} AB &\Longleftrightarrow \mu_{AB}(x)=\mu_A(x)\cdot\mu_B(x) \\ &= \left(\sum_i \frac{f(u_i)}{u_i}\right)\cdot\left(\sum_j \frac{g(w_j)}{w_j}\right) \\ &= \sum_{i,j} \frac{f(u_i)\wedge g(w_j)}{u_i\cdot w_j}, \quad u_i,\ w_j \in J. \end{aligned} \tag{2.11}$$

Algebraic sum:

$$A \oplus B \iff \mu_{A \oplus B}(x) = \mu_A(x) \dot{+} \mu_B(x)$$

$$= \sum_{i,j} \frac{f(u_i) \wedge g(w_j)}{u_i \dot{+} w_j}$$

$$= \sum_{i,j} \frac{f(u_i) \wedge g(w_j)}{u_i + w_j - u_i \cdot w_j}, \qquad u_i, w_j \in J. \tag{2.12}$$

DEFINITION 2.6. For a family of fuzzy sets ot type 2, $Q = \{ A_i \mid i \in I \}$, the *union* $C = \cup_I A_i$ and the *intersection* $D = \cap_I A_i$ are defined by

$$\mu_C(x) = \bigsqcup_I \{ \mu_{A_i}(x) \}, \qquad x \in X, \tag{2.13}$$

$$\mu_D(x) = \bigsqcap_I \{ \mu_{A_i}(x) \}, \qquad x \in X. \tag{2.14}$$

Clearly, the fuzzy grades for the *empty fuzzy set ϕ of type 2* and the *universal fuzzy set X of type 2* are $\mu_\phi(x) = 0 (=1/0)$ and $\mu_X(x) = 1 (=1/1)$, for all x in X, respectively.

DEFINITION 2.7. A *fuzzy relation R of type 2* in the product space $X \times Y = \{ (x, y) \}$, $x \in X$, $y \in Y$, is a fuzzy set of type 2 in $X \times Y$ characterized by a fuzzy membership function $\mu_R: X \times Y \to [0, 1]^J$ which associates with each ordered pair (x, y) a fuzzy grade of membership $\mu_R(x, y)$ in R. More generally, an *n-ary fuzzy relation of type 2* in a product space $X = X_1 \times X_2 \times \cdots \times X_n$ is a fuzzy set of type 2 in X characterized by a fuzzy membership function $\mu_R: X_1 \times X_2 \times \cdots \times X_n \to [0, 1]^J$, $x_i \in X_i$, $i = 1, \ldots, n$.

DEFINITION 2.8. Let f be a function from X to Y. Let B be a fuzzy set of type 2 in Y with fuzzy grade $\mu_B(y)$. Then the *inverse* of B, written as $f^{-1}(B)$, is a fuzzy set of type 2 in X, whose fuzzy grade is defined by

$$\mu_{f^{-1}(B)}(x) = \mu_B(f(x)) \qquad \text{for all } x \text{ in } X. \tag{2.15}$$

Conversely, let A be a fuzzy set of type 2 in X with fuzzy grade $\mu_A(x)$. The *image* of A, written as $f(A)$, is a fuzzy set of type 2 in Y whose fuzzy grade is given by

$$\mu_{f(A)}(y) = \bigsqcup_{x \in f^{-1}(y)} \{ \mu_A(x) \} \qquad \text{if } f^{-1}(y) \text{ is not empty}, \tag{2.16}$$

$$= 0 \qquad \text{otherwise},$$

for all y in Y, where the join $\sqcup$ is taken over the set of points $f^{-1}(y)$ in X. $f^{-1}(y) = \{ x \mid f(x) = y \}$.

DEFINITION 2.9. If R is a fuzzy relation of type 2 from X to Y, and S is a fuzzy relation of type 2 from Y to Z, then the *composition* of R and S is a fuzzy relation of type 2 denoted by $R \circ S$ and defined by

$$R \circ S \iff \mu_{R \circ S}(x, z) = \bigsqcup_{y \in Y} \left[\mu_R(x, y) \sqcap \mu_S(y, z) \right] \tag{2.17}$$

for all $(x, z) \in X \times Z$, where $\mu_R: X \times Y \to [0, 1]^J$ and $\mu_S: Y \times Z \to [0, 1]^J$.

DEFINITION 2.10. Let A be a fuzzy set of type 2 of a universe of discourse X, and let B be a fuzzy set of type 2 of a possibly different universe of discourse Y. Then, the *Cartesian product* of A and B is denoted by $A \times B$ and is defined by

$$A \times B \iff \mu_{A \times B}(x, y) = \mu_A(x) \sqcap \mu_B(y), \qquad x \in X, \; y \in Y. \tag{2.18}$$

DEFINITION 2.11. An *ultrafuzzy mapping* f from X to Y is a fuzzy set of type 2 on $X \times Y$ characterized by a fuzzy membership function $\mu_f : X \times Y \to [0, 1]^I$.

An *ultrafuzzy function* $f(x)$ is a fuzzy set of type 2 on Y with fuzzy grade

$$\mu_{f(x)}(y) = \mu_f(x, y). \tag{2.19}$$

Its *inverse* $f^{-1}(y)$ is a fuzzy set of type 2 on X with fuzzy grade

$$\mu_{f^{-1}(y)}(x) = \mu_f(x, y). \tag{2.20}$$

3. DYNAMIC TEAM DECISION PROBLEM

In this section we shall describe a dynamic team decision problem which is based on the concepts given in [1]. We consider a dynamic team composed of $i=1, 2, \ldots, n$ members who choose actions. In a dynamic team, the information and action of member i are dependent on the actions of the other precedent members, for $i=1, 2, \ldots, n$. We index the members in such a way that if member h is a precedent of member i, then $h \langle i$. Environmental conditions beyond the team member's control are termed sates of the environment. Let $X = \{x\}$ be the set of mutually exclusive states of the environment. The uncertainty about x is expressed by a probability distribution on X. Let us assume that the set X is finite, nonempty and does not contain any states that have zero probability.

The information received by a team as a whole, called the team information, has n components. Denote the team information by $y=(y_1, \ldots, y_n)$, where y_i, $i=1, \ldots, n$, is the part of the team information received by the ith team member and is related to the state of the environment x and the action $\bar{a}_{i-1}$ of the other precedent members. Let $Y_i = \{y_i\}$ be the set of all possible alternative information signals that can be received by the ith member, and let $Y = \{y\}$ be the set of all possible alternative information signals that can be received by the team; then $\{y\} = \{y_1\} \times \{y_2\} \times \cdots \times \{y_n\}$, $Y = Y_1 \times Y_2 \times \cdots \times Y_n$. Let η_i be a function from the Cartesian product of X and $\bar{A}_{i-1}$ to set Y_i, i. e., $y_i = \eta_i(x, \bar{a}_{i-1})$, and call it the causal information function or causal information structure of the ith member, where $\bar{A}_0 = \{\bar{a}_0\}$ is an empty set. $y_i \in Y_i$, $\eta_i \in \Pi_i$, $i=1, 2, \ldots, n$. $a_j \in A_j$ for $j=1, 2, \ldots, i-1$, $\bar{a}_{i-1} = (a_1, \ldots, a_{i-1}) \in \bar{A}_{i-1}$, $\bar{A}_{i-1} \equiv A_1 \times \cdots \times A_{i-1}$. a_i is the action taken by the ith member. It is assumed that if the action of member j affect the information of member i, then a_i cannot affect the information of member j for all $i, j \in I = \{1, 2, \ldots, n\}$.

An action taken by a team as a whole, called a team action, also has n components; then $a=(a_1, \ldots, a_n)$, where a_i is the action taken by the ith member, for $i=1, 2, \ldots, n$. Let $A_i = \{a_i\}$ be the set of feasible alternative actions that can be taken by the ith member, for $i=1, 2, \ldots, n$, and let $A = \{a\}$ be the set of feasible alternative team actions that can be taken by the team as a whole; then $\{a\} = \{a_1\} \times \{a_2\} \times \cdots \times \{a_n\}$, $A = A_1 \times A_2 \times \cdots \times A_n$.

The ith member, $i=1, 2, \ldots, n$, takes his action a_i on the basis of his information signal y_i according to a decision function α_i from Y_i to the set A_i of feasible actions, i.e., $a_i = \alpha_i(y_i)$. Thus, given the causal information function η_i and the decision function α_i, if the state of the environment x and the action $\bar{a}_{i-1} = (a_1, \ldots, a_{i-1})$ of other precedent members obtain, then the ith member will take the action $a_i = \alpha_i[\eta_i(x, \bar{a}_{i-1})] = \alpha_{i\circ}\eta_i[x, \bar{a}_{i-1}]$, for $i=1, 2, \ldots, n$.

For any given set of decision functions $\alpha_1, \ldots, \alpha_n$, and causal information structures $\eta_1, \ldots, \eta_n$, We define a set of composite decision functions p_i from X to A_i such that $p_1(x) = \alpha_1[\eta_1(x)]$, $p_i(x) = \alpha_i[\eta_i(x, p_1(x), \ldots, p_{i-1}(x))]$, $i=2, \ldots, n$. Naturally, $a_i = \alpha_i(y_i) = p_i(x)$, $i=1, 2, \ldots, n$. p_i is a composite function which directly maps points of X space into A_i space.

The n-tuple of information structures and decision functions can be denoted by $\eta = (\eta_1, \ldots, \eta_n) \in \Pi = \Pi_1 \times \cdots \times \Pi_n$, and $\alpha = (\alpha_1, \ldots, \alpha_n) \in \Delta_1 \times \cdots \times \Delta_n$, respectively and called the information structure and decision function of the team.

Let us denote the sets of available team information structures and the team decision functions by Π and Δ, respectively, i.e., $\eta \in \Pi$ and $\alpha \in \Delta$.

The expected utility of the team resulting from a given pair (η, α) is

$$\Omega(\eta, \alpha; \omega, \phi) = \sum_{x \in X} \phi(x)\, \omega(x, \alpha[\eta(x)]), \tag{3.1}$$

where $\alpha[\eta(x)]\equiv[\alpha_1[\eta_1(x)],\ \alpha_2[\eta_2(x,p_1(x))], \ldots, \alpha_n[\eta_n(x,\ p_1(x), \ldots, p_{n-1}(x))]]$. $\omega : X\times A\rightarrow R$ denotes a real−valued utility function of the team and $\phi : X\rightarrow[0,\ 1]$ denotes a probability density function. Assume that all members have the same utility function and the same probability function, and further, that both functions are the true ones.

With a given causal information structure and a known probability distribution over the set X of states x of the environment, the team's objective is assumed to be to choose the best decision rules, that is, the $\alpha=\alpha^0$ for which

$$\Phi(\eta^0)=\max_{\alpha\in\Delta}\Omega(\eta^0,\ \alpha). \tag{3.2}$$

Let us now define that an information structure η_1 is not more valuable than an information structure η_2, given ω and ϕ, if

$$\Phi(\eta_2)\geqslant\Phi(\eta_1). \tag{3.3}$$

After each information structure is evaluated in terms of the best decision function that can be used with it, the obvious conclusion is to use the information structure, η^*, that maximizes $\Phi(\eta)$.

4. ULTRAFUZZY DYNAMIC TEAM DECISIONS

This section introduces the concepts of fuzzy sets of type 2, i.e., ultrafuzzy sets, given in section 2 to the dynamic team decision problem defined in section 3 and formulates an ultrafuzzy dynamic team decision problem. For a fixed causal information function η_i of member i, each signal y_i is identified with a subset of the product space of the state space X and the action space $\bar{A}_{i-1}$, namely, the set of states x and the set of the actions $\bar{a}_{i-1}=(a_1, \ldots, a_{i-1})$ of the other precedent members for which $\eta_i(x,\ \bar{a}_{i-1})=y_i$, for $i=1, 2, \ldots, n$. It follows that every causal information structure induces a partition of $X\times\bar{A}_{i-1}$ into an exhaustive family of mutually exclusive subsets, each identified with a particular information signal.

By introducing the concept of fuzzy sets of type 2 proposed by Zadeh [5, 7], let us assume that each member i, $i=1, 2, \ldots, n$, receives an ultrafuzzy information signal y_i about the state of the environment x and the action $\bar{a}_{i-1}$ of other precedent members. Then each ultrafuzzy information signal $\underline{y_i}$ in the information space Y_i is identified with a fuzzy set of type 2 of the product space $X\times\bar{A}_{i-1}$. For each member i, $i=1, 2, \ldots, n$, let C_i be a fuzzy set of type 2 of information signals in the information space Y_i of the ith member. Then, a fuzzy set C_i of type 2 in Y_i is characterized by a fuzzy membership function $\mu_{C_i} : Y_i\rightarrow[0,1]^J$ which associates with each information signal y_i of Y_i a fuzzy grade $\mu_{C_i}(y_i)$ in the finite subset J of $[0,\ 1]$ which represents the fuzzy grade of membership of y_i in C_i.

An information function η_i of the ith member is a function from $X\times\bar{A}_{i-1}$ to Y_i. Then, by applying (2.15), the inverse of C_i, written as $\eta_i^{-1}(C_i)$, is a fuzzy set of type 2 of the product space $X\times\bar{A}_{i-1}$ whose fuzzy grade is defined by

$$\mu_{\eta_i^{-1}(C_i)}(x,\ \bar{a}_{i-1})=\mu_{C_i}(\eta_i(x,\ \bar{a}_{i-1})) \qquad \forall(x,\ \bar{a}_{i-1})\in X\times\bar{A}_{i-1},\ i=1, \ldots, n. \tag{4.1}$$

Conversely, let B_i be an ultrafuzzy set of states in X with fuzzy grade $\mu_{B_i}(x)$, and $\bar{E}_{i-1}=E_1\times\cdots\times E_{i-1}$ be an ultrafuzzy set in $\bar{A}_{i-1}=A_1\times\cdots\times A_{i-1}$. Then $F_i=B_i\times\bar{E}_{i-1}$ is defined by

$$\mu_{F_i}(x,\ \bar{a}_{i-1})=\mu_{B_i}(x)\sqcap\mu_{\bar{E}_{i-1}}(\bar{a}_{i-1}) \qquad \forall x\in X,\ \bar{a}_{i-1}\in\bar{A}_{i-1},\ i=1, \ldots, n. \tag{4.2}$$

Let F_i be an ultrafuzzy set in $X\times\bar{A}_{i-1}$. Then, by applying (2.16), the information function η_i induces a fuzzy set C_i of type 2 of information signals in Y_i whose fuzzy grade is given by

$$\begin{aligned}\mu_{C_i}(y_i)&=\bigsqcup_{(x,\ \bar{a}_{i-1})\in\eta_i^{-1}(y_i)}\{\mu_{F_i}(x,\ \bar{a}_{i-1})\} && \text{if } \eta_i^{-1}(y_i)\neq\phi \\ &=0 && \text{if } \eta_i^{-1}(y_i)=\phi,\quad y_i\in Y_i,\ i=1, \ldots, n\end{aligned} \tag{4.3}$$

where the join $\sqcup$ is taken over the set of points $\eta_i^{-1}(y_i)$ in $X \times \bar{A}_{i-1}$ and $\eta_i^{-1}(y_i)=\{(x, \bar{a}_{i-1}) \mid \eta_i(x, \bar{a}_{i-1})=y_i\}$ for all i.

Let an ultrafuzzy set B be the intersection of ultrafuzzy sets $B_1, \ldots, B_n$. Then, by applying (2.14), an ultrafuzzy set $B=B_1 \cap B_2 \cap \ldots \cap B_n$ in X is defined by

$$\mu_B(x)=\mu_{B_1}(x) \sqcap \mu_{B_2}(x) \sqcap \ldots \sqcap \mu_{B_n}(x), \quad x \in X. \tag{4.4}$$

where the symbol $\sqcap$ means taking the meet of $\mu_{B_i}(x)$, $i=1, \ldots, n$.

In order to define an ultrafuzzy set of information signals to the team, let C be the Cartesian product of n ultrafzzy sets, $C_1, \ldots, C_n$. Then, by applying (2.18), $C=C_1 \times \cdots \times C_n$ is defined by

$$\begin{aligned} \mu_C(y) &= \mu_C(y_1, \ldots, y_n), \quad y_i \in Y_i, \quad i=1, \ldots, n \\ &= \mu_{C_1}(y_1) \sqcap \mu_{C_2}(y_2) \sqcap \ldots \sqcap \mu_{C_n}(y_n). \end{aligned} \tag{4.5}$$

where the symbol $\sqcap$ means taking the meet of $\mu_{C_i}(y_i)$, $i=1, 2, \ldots, n$, and $\{y\}=\{y_1\} \times \cdots \times \{y_n\}$, $Y=Y_1 \times \cdots \times Y_n$.

C denotes an ultrafuzzy set of information signals, $y=(y_1, \ldots, y_n)$, to the team.

In the dynamic team decision problem, a decision function α_i is a function from Y_i to A_i of feasible actions, and a team decision function α is a function from Y to A. Thus, let C_i be an ultrafuzzy set of information signals in the information space Y_i of the ith member. Then, by applying (2.16), for each member i, $i=1, 2, \ldots, n$, the decision function α_i induces an ultrafuzzy set D_i of actions in the action space A_i whose fuzzy grade is given by

$$\begin{aligned} \mu_{D_i}(a_i) &= \bigsqcup_{y_i \in \alpha_i^{-1}(a_i)} \{\mu_{C_i}(y_i)\} && \text{if } \alpha_i^{-1}(a_i) \neq \phi \\ &= 0 && \text{if } \alpha_i^{-1}(a_i)=\phi, \ a_i \in A_i, \ i=1, \ldots, n. \end{aligned} \tag{4.6}$$

where the join $\sqcup$ is taken over the set of points $\alpha_i^{-1}(a_i)$ in Y_i, and $\alpha_i^{-1}(a_i)=\{y_i \mid \alpha_i(y_i)=a_i\}$ for all i.

Let D be the Cartesian product of n ultrafuzzy sets, $D_1, \ldots, D_n$.
Then $D=D_1 \times \cdots \times D_n$ is defined by

$$\begin{aligned} \mu_D(a) &= \mu_D(a_1, \ldots, a_n) \quad a_i \in A_i, \quad i=1, \ldots, n \\ &= \mu_{D_1}(a_1) \sqcap \mu_{D_2}(a_2) \sqcap \ldots \sqcap \mu_{D_n}(a_n) \end{aligned} \tag{4.7}$$

where the symbol $\sqcap$ means taking the meet of $\mu_{D_i}(a_i)$, $i=1, 2, \ldots, n$, and $\{a\}=\{a_1\} \times \{a_2\} \times \cdots \times \{a_n\}$, $A=A_1 \times \cdots \times A_n$.

D denotes an ultrafuzzy set of actions, $a=(a_1, \ldots, a_n)$, in the product space $A = A_1 \times \cdots \times A_n$ which are taken by the team as a whole.

Let us define a set of actions to the team with fuzzy threshold γ as follows:

$$D_\gamma=\{a \mid \mu_D(a) \sqsupseteq \gamma\}, \tag{4.8}$$

where γ is assumed to be a normal convex fuzzy grade and $1/0 \sqsubseteq \gamma \sqsubseteq 1/1$.

Then, the problem becomes one of choosing, for a given set D_γ, the best action, that is, the $a=a^0$ for which

$$\Omega(a^0)=\max_{a \in D_\gamma} \Omega(a) \tag{4.9}$$

$$\Omega(a)=\sum_{x \in X} \phi(x)\, \omega(x, a) \tag{4.10}$$

where $\Omega(a)$ denotes the expected utility. $\omega: X \times A \rightarrow R$ is a real valued utility function of the team and $\phi: X \rightarrow [0, 1]$ is a probability density function. The expected utility $\Omega(a)$ is said to represent the preference ordering $\preccurlyeq_{D_\gamma}$ if for all actions a_i and a_j with $i \neq j$ in D_γ, $a_i \preccurlyeq_{D_\gamma} a_j$ if and only if $\Omega(a_i) \leqslant \Omega(a_j)$.

If the state of the environment, the probability distribution, and the utility are fuzzy and B is an ultrafuzzy set of states in X observed by the team, then the fuzzy expected utility is defined as follows:

$$\Gamma(a)=\{\sum_{x \in X} \varepsilon(x) v(x, a)\, \mu_B(x)\} / \Xi \tag{4.11}$$

where $\Xi = \sum_{x \in X} \varepsilon(x)\mu_B(x)$, $\varepsilon: X \to [0, 1]^P$, $v: X \times A \to [0, 1]^U$, and $\mu_B: X \to [0,1]^J$ denote a fuzzy probability function, a fuzzy utility function, and a fuzzy membership function, respectively, where $P = \{p_1, p_2, ..., p_l\}$, $U = \{u_1, u_2, ..., u_m\}$, and $J = \{w_1, w_2, ..., w_n\}$ are finite subsets of [0, 1], respectively. Here $\varepsilon(x)$, $v(x, a)$ and $\mu_B(x)$ denote a fuzzy probability, a fuzzy utility, and a fuzzy grade which are fuzzy sets, i.e., normal convex fuzzy numbers, in P, U, and J, respectively. $\Gamma(a)$ is a fuzzy subset of the real line, i.e., a normal convex fuzzy number, and can be obtained by using the extension principle. $\varepsilon(x)v(x, a)\mu_B(x)$ and the operation $\sum$ stand for the arithmetic product and arithmetic sum, respectively. The fuzzy expected utility $\Gamma(a)$ represents the preference ordering $\preccurlyeq_A$, such that for any two actions a and b in the finite action space A, $a \preccurlyeq b \Longleftrightarrow \Gamma(a) \sqsubseteq \Gamma(b) \Longleftrightarrow \Gamma(a) \sqcap \Gamma(b) = \Gamma(a)$, or $a \dot{\preccurlyeq} b \Longleftrightarrow \Gamma(a) \dot{\sqsubseteq} \Gamma(b) \Longleftrightarrow \Gamma(a) \sqcup \Gamma(b) = \Gamma(b)$.

If the team takes an ultrafuzzy set D of actions and B is an ultrafuzzy set of states observed by the team, then the maximal fuzzy expected utility of the team, given its information structure η^0, is the fuzzy expected utility maximized with respect to α:

$$\Phi(\eta^0) = \max_{\alpha \in \Delta} \Gamma(\eta^0, \alpha) \tag{4.12}$$

$$\Gamma(\eta^0, \alpha) = \{ \sum_{x \in X} \varepsilon(x) v(x, \alpha[\eta^0(x)])\, \mu_{B \times D}(x, \alpha[\eta^0(x)])\} / \textcircled{H} \tag{4.13}$$

where $\textcircled{H} = \sum_{x \in X} \varepsilon(x)\mu_{B \times D}(x, \alpha[\eta^0[(x)])$. $\Gamma(\eta^0, \alpha)$ denotes the fuzzy expected utility over the Cartesian product of B and D. $\alpha[\eta^0(x)] = \{\alpha_1[\eta_1^0(x)], ..., \alpha_n[\eta_n^0(x, \alpha_1[\eta_1^0(x)], ..., \alpha_{n-1}[\eta_{n-1}^0(x)])]\}$. $\mu_B(x) = \mu_{B_1}(x) \sqcap ... \sqcap \mu_{B_n}(x)$. $\mu_D(a) = \mu_{D_1}(a_1) \sqcap ... \sqcap \mu_{D_n}(a_n)$.

5. ULTRAFUZZY INFORMATION STRUCTURE AND ULTRAFUZZY DECISION RULE

In this section, we shall define the concepts of an ultrafuzzy information structure and an ultrafuzzy decision rule as an extension of the concepts presented in [4].

5.1 Ultrafuzzy information structure

Let us assume that each team member i uses a given vague method of information-gathering, namely, an ultrafuzzy information structure η_i. Then, for each team member i, $i=1, 2, ..., n$, an ultrafuzzy information structure η_i is defined as follows.

Definition 5.1. An ultrafuzzy information structure η_i from $X \times \bar{A}_{i-1}$ to Y_i is an ultrafuzzy set on $X \times \bar{A}_{i-1} \times Y_i$ characterized by a fuzzy membership function $\mu_{\eta_i}: X \times \bar{A}_{i-1} \times Y_i \to [0,1]^J$, for $i=1, 2, ..., n$.

An ultrafuzzy information $\eta_i(x, \bar{a}_{i-1})$ is an ultrafuzzy set on the information space Y_i of team member i with fuzzy grade

$$\mu_{\eta_i(x, \bar{a}_{i-1})}(y_i) = \mu_{\eta_i}[(x, \bar{a}_{i-1}), y_i], \quad i=1, 2, ..., n. \tag{5.1}$$

Its inverse $\overset{-1}{\eta_i}(y_i)$ is an ultrafuzzy set on $X \times \bar{A}_{i-1}$ with fuzzy grade

$$\mu_{\overset{-1}{\eta_i}(y_i)}(x, \bar{a}_{i-1}) = \mu_{\eta_i}[(x, \bar{a}_{i-1}), y_i], \quad i=1, ..., n. \tag{5.2}$$

For each team member i, $i=1, 2, ..., n$, an ultrafuzzy information $\eta_i(x, \bar{a}_{i-1})$ is said to be finer than an ultrafuzzy information $\theta_i(x, \bar{a}_{i-1})$ if

$$\mu_{\eta_i(x, \bar{a}_{i-1})}(y_i) \sqsubseteq \mu_{\theta_i(x, \bar{a}_{i-1})}(y_i) \quad \forall (x, \bar{a}_{i-1}) \in X \times \bar{A}_{i-1}, \quad y_i \in Y_i, \quad i=1, ..., n. \tag{5.3}$$

For each team member $i, i=1, 2, \ldots, n$, let F_i be an ultrafuzzy set on $X \times \bar{A}_{i-1}$ with fuzzy grade $\mu_{F_i}(x, \bar{a}_{i-1})$. Then the ultrafuzzy information structure η_i induces an ultrafuzzy set C_i of information signals in Y_i whose fuzzy grade is defined by

$$\mu_{C_i}(y_i) = \bigsqcup_{(x, \bar{a}_{i-1}) \in X \times \bar{A}_{i-1}} (\mu_{F_i}(x, \bar{a}_{i-1}) \sqcap \mu_{\eta_i}[(x, \bar{a}_{i-1}), y_i]), \quad y_i \in Y_i, \quad i=1, \ldots, n. \tag{5.4}$$

Let C be the Cartesian product of n ultrafuzzy sets $C_1, \ldots, C_n$. Then, $C=C_1 \times C_2 \times \cdots \times C_n$ is defined by

$$\begin{aligned} \mu_C(y) &= \mu_C(y_1, \ldots, y_n), \quad y_i \in Y_i, \quad i=1, \ldots, n \\ &= \mu_{C_1}(y_1) \sqcap \mu_{C_2}(y_2) \sqcap \ldots \sqcap \mu_{C_n}(y_n) \end{aligned} \tag{5.5}$$

where $\mu_{C_i}(y_i)$, $i=1, 2, \ldots, n$, are given by (5.4).

C denotes an ultrafuzzy set of information signals, $y=(y_1, \ldots, y_n)$, to the team.

5.2 Ultrafuzzy decision rule

For each team member i, $i=1, 2, \ldots, n$, an ultrafuzzy decision rule α_i is defined as follows.

Definition 5.2. An ultrafuzzy decision rule α_i from the information space Y_i to the decision space A_i is an ultrafuzzy set on $Y_i \times A_i \to [0, 1]^J$, $i=1, 2, \ldots, n$.

An ultrafuzzy action $\alpha_i(y_i)$ is an ultrafuzzy set on A_i with fuzzy grade

$$\mu_{\alpha_i(y_i)}(a_i) = \mu_{\alpha_i}(y_i, a_i), \quad i=1, \ldots, n. \tag{5.6}$$

Its inverse $\alpha_i^{-1}(a_i)$ is an ultrafuzzy set on Y_i with fuzzy grade

$$\mu_{\alpha_i^{-1}(a_i)}(y_i) = \mu_{\alpha_i}(y_i, a_i), \quad i=1, \ldots, n. \tag{5.7}$$

An ultrafuzzy action $\alpha_i(y_i)$ is said to be finer than an ultrafuzzy action $\beta_i(y_i)$ if

$$\mu_{\alpha_i(y_i)}(a_i) \sqsubseteq \mu_{\beta_i(y_i)}(a_i), \quad \forall\, y_i, a_i \in Y_i, \ A_i, \quad i=1, \ldots, n. \tag{5.8}$$

Let C_i be an ultrafuzzy set of information signals in the information space Y_i with fuzzy grade $\mu_{C_i}(y_i)$. Then the ultrafuzzy decision rule α_i induces an ultrafuzzy set D_i of actions in the action space A_i whose fuzzy grade is defined by

$$\begin{aligned} \mu_{D_i}(a_i) &= \mu_{\alpha_i(C_i)}(a_i) \\ &= \bigsqcup_{y_i \in Y_i} (\mu_{C_i(y_i)} \sqcap \mu_{\alpha_i(y_i, a_i)}), \quad a_i \in A_i, \quad i=1, \ldots, n. \end{aligned} \tag{5.9}$$

Let D be the Cartesian product of n ultrafuzzy sets $D_1, \ldots, D_n$, then $D=D_1 \times D_2 \times \cdots \times D_n$ is defined by

$$\begin{aligned} \mu_D(a) &= \mu_D(a_1, \ldots, a_n) \\ &= \mu_{D_1(a_1)} \sqcap \mu_{D_2(a_2)} \sqcap \ldots \sqcap \mu_{D_n(a_n)}, \quad a_i \in A_i, \quad i=1, \ldots, n, \end{aligned} \tag{5,10}$$

where $\mu_{D_i}(a_i)$, $i=1, 2, \ldots, n$, are given by the eq. (5.9).

D denotes an ultrafuzzy set of actions, $a=(a_1, \ldots, a_n)$, in the product space $A=A_1 \times A_2 \times \cdots \times A_n$ which are taken by the team as a whole.

5.3 Ultrafuzzy constraints

Let us consider the dynamic team decision problem under ultrafuzzy constraints as an extension of the definitions given in [4]. Then the maximization of the fuzzy expected utility of the team, subject to the ultrafuzzy constraints represented by an ultrafuzzy set B in the state space X and an ultrafuzzy set D in the action space A, reduces to a maximization of the function

$$\Gamma^*(a)=\Gamma(a)[\mu_D(a)]^{\lambda}, \tag{5.11}$$

where $\Gamma(a)=\{\sum_{x\in X}\varepsilon(x)\,v(x,a)\,\mu_B(x)\}/(\sum_{x\in X}\varepsilon(x)\,\mu_B(x))$. $\varepsilon : X\rightarrow[0,1]^P$ and $v : X\times A\rightarrow[0,1]^U$ are a fuzzy probabiliity function and a fuzzy utility function, respectively. $\mu_B : X\rightarrow[0,1]^J$ and $\mu_D : A\rightarrow[0,1]^J$ are fuzzy membership functions. λ is any positive number. $\Gamma^*(a)$ is a fuzzy subset of the real line, i.e., a normal convex fuzzy number, and can be obtained by using the extension principle. $\varepsilon(x)\,v(x,a)\,\mu_B(x)$, $\Gamma(a)[\mu_D(a)]^{\lambda}$, and the operation $\sum$ stand for the arithmetic products and arithmetic sum, respectively. The fuzzy expected utility $\Gamma^*(a)$ represents the preference ordering $\preccurlyeq_A$ such that for any two actions a and b in the finite action space A, $a\preccurlyeq b \Longleftrightarrow \Gamma^*(a)\sqsubseteq\Gamma^*(b) \Longleftrightarrow \Gamma^*(a)\sqcap\Gamma^*(b)=\Gamma^*(a)$ or $a\preccurlyeq b \Longleftrightarrow \Gamma^*(a)\sqsubseteq\Gamma^*(b)\Longleftrightarrow\Gamma^*(a)\sqcup\Gamma^*(b)=\Gamma^*(b)$.

The problem then becomes one of choosing the best action that is $a=\hat{a}$ for which

$$\Gamma^*(\hat{a})=\max_{a\in A}\Gamma^*(a). \tag{5.12}$$

The fuzzy expected utility of the team resulting from the use of an ultrafuzzy information structure η and an ultrafuzzy decision rule α is defined by

$$\Gamma(\eta,\alpha;\phi,\varepsilon)=\sum_{x\in X}\varepsilon(x)\,\phi(x,\alpha[\eta(x)]), \tag{5.13}$$

where for $x\in X$ and $a\in A$

$$\phi(x,\alpha[\eta(x)])\equiv\bigsqcup_{a\in A}[\mu_{\alpha\circ\eta}(x,a)\sqcap v(x,a)]$$

$$\mu_{\alpha\circ\eta}(x,a)\equiv\mu_{\alpha_1\circ\eta_1}(x,a_1)\sqcap\ldots\sqcap\mu_{\alpha_n\circ\eta_n}((x,a_1,\ldots,a_{n-1}),a_n)$$

$$\mu_{\alpha_1\circ\eta_1}(x,a_1)\equiv\bigsqcup_{y_1\in Y_1}[\mu_{\eta_1}(x,y_1)\sqcap\mu_{\alpha_1}(y_1,a_1)],$$

$$\mu_{\alpha_i\circ\eta_i}((x,a_1,\ldots,a_{i-1}),a_i)\equiv\bigsqcup_{y_i\in Y_i}[\mu_{\eta_i}((x,a_1,\ldots,a_{i-1}),y_i)\sqcap\mu_{\alpha_i}(y_i,a_i)],\ i=2,\ldots,n.$$

The maximal fuzzy expected utility of the team, given its ultrafuzzy information structure η', is the fuzzy expected utility maximized with respect to α:

$$\Pi(\eta')=\max_{\alpha\in K}\Gamma(\eta',\alpha), \tag{5.14}$$

where $\Gamma(\eta',\alpha)$ is given by eq. (5. 13). K denotes a collection of ultrafuzzy decision rules of the team.

6. CONCLUDING REMARKS

In this paper, we have studied the ultrafuzzy dynamic information structures and have formulated the dynamic team decision processes which contain ultrafuzzy states, ultrafuzzy dynamic information structures, ultrafuzzy information signals, ultrafuzzy decision rules, and ultrafuzzy actions. The dynamic team decision problems under ultrafuzzy constraints have been dealt with and the concepts of fuzzy expected utility of the dynamic team have been given.

REFERENCES

[1] Y.C.Ho and K.C.Chu, Team decision theory and information structures in optimal control problems, Part Ⅰ, Part Ⅱ, IEEE Trans. Auto. Control, AC−17, 1: 15−22(1972), 1: 22−28 (1972).

[2] M.Mizumoto and K.Tanaka, Fuzzy−fuzzy automata, Kybernetes 5: 107−112(1976).

[3] M.Mizumoto and K.Tanaka, Some properties of fuzzy sets of type 2, Information and Control 31: 312−340(1976).

[4] H.Nojiri, Ultrafuzzy team decisions in a dynamic environment, Information Sciences 23: 105−122(1981).

[5] L.A.Zadeh, Quantitative fuzzy semantics, Information Sciences 3: 159−176(1971).

[6] L.A.Zadeh, Outline of a new approach to the analysis of complex systems and decision processes, IEEE Trans. Systems, Man, and Cybernet. SMC−3: 28−44(1973).

[7] L.A.Zadeh, The concept of a linguistic variable and its application to approximate reasoning (Ⅰ), (Ⅱ), (Ⅲ), Information Sciences 8: 199−249(1975); 8: 301−357(1975); 9: 43−80(1976).

Fuzzy Logic in Knowledge-Based Systems,
Decision and Control
M.M. Gupta and T. Yamakawa (Editors)

USING FUZZY LINEAR REGRESSION AS A FORECASTING TOOL IN INTELLIGENT SYSTEMS*

by

Abraham Kandel and Behrooz Heshmathy
Computer Science Department
and The Institute for Expert Systems and Robotics
FLORIDA STATE UNIVERSITY
TALLAHASSEE, FLORIDA 32306-4019
USA

ABSTRACT

In intelligent forecasting, fuzzy linear regression can be used to measure the relationship between the variables, the effectiveness of the model, and the evaluation of the correlation coefficients which can be used to predict the value of the dependent variable from the independent variables.

A fuzzy linear regression model might be more useful in areas of decision-making where there is a great deal of uncertainty as well as vague phenomena. In fuzzy linear regression, the relationship between the dependent variable and the independent variables (the coefficients of regression) is not as sharply defined as in the case of non-fuzzy linear regression. One assumption of fuzzy linear regression is that the residual or deviation of the estimated value from the real value of the dependent variable is due to the fuzziness of the system's parameters. The fuzzy parameters of the linear model are considered to be a "possibility distribution which corresponds to the fuzziness of the system" [23].

The use of fuzzy linear regression as a forecasting tool in intelligent systems is discussed and illustrated in this paper.

Key Words: Fuzzy Linear Regression
Decision Making
Forecasting
Intelligent Systems

*Research partially supported by NSF grant IST 8405953

1. ON FUZZY LINEAR REGRESSION

In fuzzy linear regression, the same techniques of model building apply as in the case of non-fuzzy abstract linear regression models discussed in [1]-[2]. In our fuzzy linear regression model there is more emphasis in objective methods, as opposed to other studies where highly subjective methods were used as far as a priori information was concerned. However, subjective methods are inevitable in forecasting, since heavy reliance on objective methods can result in estimates that are not consistent with the "real world." This inclusion of subjective information [1]-[25] in forecasting is considered in fuzzy linear regression because one cannot appropriately fit the real world into a mathematical model. In environments where human subjective estimation is influential in the regression model, we "must deal with a fuzzy structure of the system" [23]. The deviations between the actual values and the computed values of the dependent variable in the fuzzy linear regression model are due to the "indefiniteness" of the system structure. In other words, fuzziness in the system is indicated in the fuzzy system parameters-- coefficients of regression in the fuzzy model. The fuzzy parameters of the fuzzy linear regression model "mean a possibility distribution which corresponds to the fuzziness of the system" [24].

To define the fuzzy linear regression model, two definitions are given [24]:
Definition : The fuzzy function is denoted by $f : X \rightarrow F(y)$; $Y = f(x,A)$ where $x \epsilon X$, X is a classical set, A is a fuzzy set, Y is the mapping of x from fuzzy set A, and $F(y)$ is the set of all fuzzy subsets on Y.

The fuzzy set Y is defined by the membership function -- the grade of membership of the computed y containing the actual y-- as follows :

$$\chi_Y(y) = \begin{cases} \max\limits_{\{a \;|\; y = f(x,a)\}} \chi_A(a), & a \;|\; y = f(x,a) \neq 0 \\ 0 \quad ; \; otherwise . \end{cases}$$

Fuzzy parameters in the fuzzy linear regression are defined by the fuzzy sets represented as :

$$\chi_A(a) = \min_j[\chi_{Aj}(a_j)]$$

$$\chi_{Aj}(a_j) = \begin{cases} 1 - |\alpha_j - a_j| / c_j \; , \; \alpha_j - c_j \leq a_j \leq \alpha_j + c_j \\ 0 \; ; \; otherwise. \qquad (where \; c_j > 0). \end{cases}$$

The fuzzy parameter of this definition reads " approximate α ," described by the center of the fuzzy parameter α and the width c. To show the fuzzy parameters of the regression center and width in the vector form :

$A=[\alpha,c]$

$\alpha=(\alpha_1, \ldots, \alpha_n)^t$
$c=(c_1, \ldots, c_n)^t$

where A is defined on the vector space R^n, which is a Cartesian product.

To formulate a fuzzy linear regression model, we let the model be of the additive form $y=\alpha^t x$. The estimated values for the dependent variable using the model $y=\alpha^t x$ is denoted by $y_i^*=\alpha^t x_i$, where y_i^* is the computed value in the model, α^t is the transpose of vector $\alpha=(\alpha_1, \ldots, \alpha_n)$, and x_i is the vector of all x's for the independent variables of the $i-th$ sample. The deviation between the observed value y_i (the value of the independent variable in the $i-th$ sample) and the estimated value y_i^*, is $e_i=y_i-y_i^*(i=1,..,N$, where N is the number of samples).

These deviations e_i are related to the fuzziness of the system parameters in fuzzy linear regression as opposed to being related to observation errors which are random variables with zero mean in non-fuzzy linear regression.

If we regard the data as input-output tables, where outputs are the values of the dependent variables for different samples, then the fuzzy linear regression technique can be applied to both non-fuzzy data and fuzzy data (fuzzy output). The latter is another form of input-output relations except that the output (dependent variable) has some fuzziness in it, denoted by e. This is similar to the fuzzy parameters such that there is a center which is the actual value of the dependent variable (observed value) and the width e, which is the fuzziness in the dependent variable. The rules and formulas of fuzzy linear regression apply both to fuzzy and non-fuzzy output data except $e=0$ for the non-fuzzy output data.

The fuzzy linear regression model to represent the data is of the general form: $Y_i^* = A_1^* x_{i1} + \cdots + A_n^* x_{in} = A_i^* x_i$ where Y_i^* is the computed y for the $i-th$ sample, $A_1^*, \ldots, A_n^*$ and the fuzzy parameters of the form (α,c), and the center and width for the independent variables 1 to n, $x_{i1}, \ldots, x_{in}$ are the $x_1, \ldots, x_n$, independent variables for the $i-th$ sample. Given x_i, Y_i^* is computed , and the grade of membership of actual y to the estimated y is [23]:

$$\chi_{Y_i^*}(y) = 1- \frac{| y_i - x_i^t\alpha |}{c^t \, | x_i |}.$$

In order to measure the degree of the fitting of the estimated fuzzy linear regression model $Y_i^* = A_i^* x_i$ to the given (observed) data, $Y_i=(y_i,e_i)$ is measured by h, which maximizes over H, a value chosen by the decision-maker to represent this degree of fitting, such that

$Y_i^{*H} = \{ y \mid \chi_{Y_i^*}(y) \geq H \}$. The vagueness is defined by J, which is the sum of the all the widths of the fuzzy parameters ($J = c_1 + ... + c_n$). This value of J should be as small as possible. The problem of linear regression is then viewed as finding the fuzzy parameters A_i^* such that J is minimized under the condition $\bar{h}_i \geq H$. (J is the sum of the least squares in non-fuzzy linear regression when the least square method is used).

2. THE FORECASTING MODEL

In this section we describe a procedure utilizing fuzzy linear regression to build sales forecasting models for computers and peripheral equipment in the United States. The dependent variable, y, included the sales of electronic computers in addition to peripheral equipment for electronic computing systems. This data was extracted from the Current Industrial Reports, computers and office and accounting machine survey, performed by the Bureau of Census from 1964 to 1981 [6]. The scope of the survey included any computer manufacturer that had more than ten employees, a range which covers all of the industrial manufacturers in the United States. In the data collected from the Census industrial report, the dollar value of sales was considered rather than the units sold.

At the first stage of building a fuzzy linear regression model, data is collected from various sources-- these sources will be specified later -- for the years 1967-1981, a time span of 15 years. Three different measurement models were built, each based on a different data set. In each measurement model there were some common independent variables. In other words, the independent variables were not unique to each of the measurement models. Since most of the data collected was not in a ratio form, an additive model of the form $y=A_1x_1+...+A_nx_n$ was used rather than a multiplicative or a log-log model. The independent variable in all of the three measurement models was the sales of computers and peripheral equipment.

One measurement model, for example, includes the following independent variables:

$Y_i=A_1x_{i1}+ \cdots +A_9x_{i9}$
where $x1$ is the economic ability to buy,
$x2$ is availability (prime rate),
$x3$ is user population expansion,
$x4$ is microsales units,
$x5$ is calculators sales,
$x6$ is micro sales ($),
$x7$ is the price of the microcomputers,
$x8$ is semi-conductor sales ($), and
$x9$ is minicomputer sales ($).

This measurement model was tested with 3 different data sets. The first data set had 15 samples (1967-81), the second data set had 10 samples (1972-81), and the third data set had only 7 samples (1975-1981). These data sets were non-fuzzy output data, $e.g. e_i = 0$, or the actual y's were not fuzzy. Another experiment was performed by running the fuzzy linear regression on the fuzzy output data. The width of the fuzzy output variables was taken as a percentage of actual y, subjectively ranging from 7 percent to 2 percent of the actual (observed) y. The results from running the fuzzy linear regression with the fuzzy output were not as accurate as the results from the runs of non-fuzzy output data. One reason might be that the grade of memberships of actual y to the computed y, $\chi_{Y_i^*}(Y_i)$, will be larger since the actual y according to the linear programming solutions is in the form:

$$\alpha^t x_i - (1-H)[\Sigma_j c_j \ |x_{ij}\ | - e_i] \leq y_i \leq \alpha^t x_i + (1-H)[\Sigma_j c_j \ |x_{ij}\ | - e_i];$$

thus the term $(1-H)[\Sigma_j c_j \ |x_{ij}\ | - e_i]$ is larger when e_i is zero, and as a result the interval where the computed y falls in becomes larger. From this point on, all testing and running of the fuzzy linear regressions are done on non-fuzzy output data (e_i=0). The constant H was chosen by the decision-maker to be 0.5. Thus the support ($\chi_Y{}^*(y) \geq 0.5$) of the fuzzy sets was considered to be a good estimate of the dependent variable in the fuzzy linear regression models. Besides establishing that the centers of the fuzzy parameters not be equal to zero, that the width of the fuzzy parameters be as small as possible, and that the grade of membership be greater than or equal to 0.50, two extra criteria were added to the evaluation of good estimates for the dependent variable. One criterion was the percentage of the deviations between the actual y (y_i) and the computed y (Y_i^*) to the actual y. This percentage is in form of $|\alpha^t x_i - y_i| / y_i$. Another criterion was the percentage of the computed error (width of the fuzzy sets) to the actual y. This percentage is in the form of $\Sigma_j c_j |x_{ij}| / y_i$. The reason for adding these two percentages to the evaluation of the measurement models is that after examining the formula for the grade of memberships :

$$\chi_{Y_i}{}^{*}(y_i) = 1 - \frac{|\alpha^t x_i - y_i|}{\Sigma_j c_j |x_{ij}|},$$

one can see that if $\Sigma_j c_j |x_{ij}|$ is very large, then $\chi_{Y_i}{}^{*}(y_i)$ is going to be a high grade of membership. Also, the smaller the deviation $|\alpha^t x_i - y_i|$, the smaller the fraction will be, and thus the higher the grade of memberships. For the reasons mentioned above, these two percentage criteria were added to avoid the problem of large width in fuzzy sets (estimated y) and to keep the deviation between the actual y and the computed y small. These percentages were subjectively considered to be acceptable outcome by the decision-maker. In this study, 0-20 percent and 0-30 percent were considered to be acceptable for the two percentage criteria mentioned above, respectively. When the measurement Model was tested with the 3 different data sets, the final updated model was found to be acceptable as far as the 3 criteria of $\chi_{Y_i}{}^{*}(y_i) \geq 0.50$ and the percentage $|\alpha^t x_i - y_i| / y_i \leq 20\%$ and percentage $\Sigma_j c_j |x_{ij}| / y_i \leq 30\%$ was concerned. And when the model was run with the data sets, the last data set with 7 samples (N=7, 1975-81) was shown to be the one with more accurate results as far as the three criteria mentioned before are concerned. This is partly due to the fact that the microcomputers contributed to the model, since the data for the micro was not available before 1975.

3 CONCLUSION

In this paper, the concept of fuzzy linear regression and its applications to sales forecasting in an uncertain environment are discussed. The forecasting model was built using fuzzy linear regression on non-fuzzy output data. However, other measurement models were also tested on both fuzzy output data and non-fuzzy output data. It turned out that the model worked better on non-fuzzy output data. Since all of the forecasts for the sales of computers and peripherals are given as fuzzy sets, the decision-maker can choose a sales figure in the interval of the fuzzy set with the triangular membership function. It is noticed that since the width of the fuzzy parameters for the fuzzy linear regression forecasting model are the same for all of the data in the forecasting years, the fuzzy centers (forecasts of sales) with the larger values also have larger fuzzy widths. In fuzzy linear regression, the deviations between the observed value of sales and the estimated value of sales (using the fuzzy linear regression model) are due to the fuzziness in the system parameters of the model. In non-fuzzy linear regression, these deviations are due to the observation error in the data.

By computing the estimated sales forecast and its fuzzy width of the forecast, one gets a fuzzy set of triangular membership function with the computed interval for the fuzzy set. This fuzzy set forecast interval incorporates the possibility of the estimated sales for the given data, which is the grade of membership of the actual y to the computed y. On the other hand, in non-fuzzy linear regression models, the confidence interval "seems to estimate the upper and lower limits of the observation errors" [24]. In conclusion, fuzzy linear regression can be applied to any model that uses abstract non-fuzzy linear regression by fuzzifying the linear regression model using a fuzzy linear function along with the fuzzy parameters of the type of triangular membership functions or any other type of membership functions.

4 REFERENCES

[1] Armstrong, Scott J. "Long-Range Forecasting for a Consumer Durable in an International Market." Diss. MIT 1968.

[2] Armstrong, S. J. **Long-Range Forecasting from Crystal Ball to Computer.** New York: John Wiley & Sons Inc., 1978.

[3] "Artificial Intelligence." **Business Week,** March 8, 1982, pp 66-75.

[4] Bails, D. G. and L. C. Peppers. **Business Fluctuations: Forecasting Techniques and Applications.** New York: Prentice-Hall, 1982.

[5] **BYTE,** July-December, 1977-1981.

[6] **Current Industrial Report: Computer and Office and Accounting Machine.** U.S. Department of Commerce, Bureau of the Census, MA35R(64)-MA35R(81), 1964-1981.

[7] **Datapro Survey of the Microcomputers.** 1982.

[8] Dixon, W. J. **BMDP Statistical Software (Biomedical Programs).** Calif.: University of California Press, 1981.

[9] **Economic Report of the President.** 1966, 1982.

[10] **Electronics.** Vol. 55, No. 1, January 13, 1982, p. 124.

[11] Fiedelman, L. "Minis in Small Business." **Infosystems,** October 1982, pp. 58-64.

[12] Gross, C. W. and R. T. Peterson. **Business Forecasting.** Boston: Houghton Miffin Company, 1976.

[13] Grumman G. S. and Cowen. "Annual Minicomputer Survey." **Datamation.** Vol. 26, No. 11, November 1980, p. 154.

[14] Ingwerson, Marshall. "Mathematics goes soft around the edges." **The Christian Science Monitor,** June 24, 1981, pp. 810-811.

[15] Kandel, A. **Fuzzy Mathematical Techniques with Applications,** Addison Wesley, 1986.

[16] Kandel A, and W. J. Byatt. "Fuzzy Sets, Fuzzy Algebra, and Fuzzy Statistics." Proceeding of the IEEE, Vol. 66, No. 12, December 1978, pp. 1619-1639.

[17] Mead, C. and l. Conway. **Introduction to VLSI Systems,** Addison Wesley, 1978.

[18] Miller, F. W. "Desk-top computers are for real." **Datamation,** Vol. 27, No. 5, 1981, pp. 58-62.

[19] Miller, F. W. "Small Business Computers Mean Business." **Datamation,** Vol. 28, No. 6, 1982, pp. 54-60.

[20] Payne, Joseph. "Minicomputers: Big Bucks from Small Systems." **Datamation,** Vol. 25, No. 5, may 1979, p. 118.

[21] Platt, John R. "Strong Inference." **Science,** 146, October 1964, pp. 347-353.

[22] **Standard and Poors, Industry Report,** Vol. 1 and Vol. 2, 1974-1982.

[23] Tanaka, H., S. Uejima, and k. Asai, "Fuzzy Linear Regression Model." **IEEE Transactions on Systems, Man, and Cybernetics,** Vol. 10, No. 4, 1980, pp. 2933-2938.

[24] Tanaka, H., S. Uejima, and k. Asai, "Linear Regression Analysis with Fuzzy Model." **IEEE Transactions On Systems, Man, and Cybernetics,** Vol. 12, No. 6, November/December 1982, pp. 903-907.

[25] Zadeh, L. A. "Fuzzy sets." **Information Control,** Vol. 8, 1965 pp. 338-353.

Fuzzy Logic in Knowledge-Based Systems,
Decision and Control
M.M. Gupta and T. Yamakawa (Editors)
Elsevier Science Publishers B.V. (North-Holland), 1988

COMP/STAT. FUZZINESS AND RARE ITEMS

F. AZORIN

ABSTRACT

Computational Statistics, or COMP/STAT, is the interface of Computation Science and Statistics: that is, the study of the computational problems and solutions in statistical research and applications (Victor (63)). It should be extended to encompass not only Statistics, but also other sources of variability and vagueness, especially those concerned with fuzziness. In this way it would adapt itself to fuzzy aspects of artificial intelligence and expert systems, and approximate reasoning, (on which there is already an impressive quantity of literature). As an important application we will consider problems concerned with the study of rare items, or more concretely , with statistical sampling of rare items, i.e., units, elements or individuals which have a low, or very low, frequency in a given population. Rarity is, of course, a fuzzy concept. There is fuzziness in the assignment of the predicate "rare" to a given type of item, depending on its frequency or proportion in the population considered.

Examples of rare items are found in very different fields, such as the incidence and prevalence of certain diseases; specimens or couples of endangered species; vacant dwellings in crowded towns; defective industrial or agricultural products; purchasers of an expensive commodity; misprints in books, etc. There are other related problems when detecting if a proportion or class total exceeds a given limit, as with infectious diseases or radiation effects. In these cases fuzzy logical medical diagnosis can be applied. Also in auditing and financial inspection, etc.

In all these cases there is fuzziness in decision and classification of frequencies as low, or very low, and sometimes also in the values or states of certain units in the sample, and in the population or the sample itself.

From here we shall pass to a much debated question: Whether the membership function for rare items should be just assumed, specified, as it is often done in Statistics with the assumption or normality and of regression models in Bayesian sampling theory; or constructed, estimated from observed data, as in statistical non-parametric density estimation(NPDE).

Planning, design, implementation and field work, processing, interpretation and analysis, require the determination of goals, costs, budget, equipment, and different circumstances, some of them fuzzy, and especially of the sampling design, which will be a combination of stratified, cluster, multistage, and multiphase sampling. Once the characteristics or features to be investigated have been decided, in some cases from humanistic or cognitive considerations, as happens with studies of psychological and social motivations, or feelings of happiness and distress, etc. the population proportions must be estimated. This estimation must be examined from statistical, fuzzy, and computational points of view. There is a wide choice of estimators, none having all the desiderata of accuracy, efficiency, consistency, etc. The difficulty of choice is compounded when allowing for fuzziness in characters.

Another aspects to be mentioned are: estimation of spatial patterns and structures, and the connections with search theory and perception.

KEY WORDS: COMP/STAT, Fuzzy sets, Rare items.
(Possible classification: Part III, Applications).

1. INTRODUCTION

Computational Statistics, or Comp/Stat. is a part of the electronic treatment of uncertainty, be it statistical or vague, stochastic or fuzzy. The concept of statistical rarity in populations is itself fuzzy, and originates problems in various fields. Some of these problems arise in estimating the proportions and totals, i.e., the relative and absolute frequency of rare items in a certain population, from which samples have been obtained. And also in estimating the mean and total of a variable or attribute within the class of the items previously defined as rare. The examination of the characters of each item is often followed by decision making and the application of fuzzy logic for diagnosis purposes. In some cases this examination requires the use of lengthy, expensive procedures. If there is a scarcity of available specialists in a given place and time, expert systems can be the only satisfactory instrument to cope with difficult aspects of uncertainty. Then, as said by Zadeh (65): "Fuzzy logic subsumes both predicate logic and probability theory, and makes it possible to deal with different types of uncertainty within a simple conceptual framework", and facilitates "representation and inferences from knowledge bases which are imprecise, incomplete, or not totally reliable".

In cases concerned with rare items, imprecision arises first from statistical variability; secondly, from fuzzy uncertainty in the characters observed; thirdly, because of missing data, nonresponse or unreliability in the responses, observations and measures.

Missing and fuzzy data may arise as well from lost or deteriorated questionnaires, imperfect oral or written expression, or insatisfactory knowledge of a natural language. These can be serious problems in censuses and in sample surveys, and they will be considered in the following Sections.

2. ESTIMATION

2.1. General Aspects

Once the character to be observed or measured is defined, decisions must be taken as to the observational or measuring procedures.

Given $S=(a_1, a_2,\ldots, a_N)$, a finite set of N distinguishable units, it is required to determine the identity of each element $\underline{x}$ of S. If $\underline{x}$ could be directly observed, there would be no problem on this connection. Otherwise (Chakravarti, (14)), one could choose a sequence of functions: $f_1,\ldots,f_m$ from a given family or system F, and take the values of these functions at $\underline{x}$. If in practice $F=(f_1, f_2,\ldots, f_m)$ contains a function which takes on different values for different units or elements of S, a single observation of this function at $\underline{x}$ would reveal its identity. Usually, the number of different values taken on by any f is much smaller than N. F is called a binary search system when each function takes on only two values, 0 or 1.

Now, leaving aside identification problems, we consider the ordinary case of direct observation, in simple sampling, for the estimation of the true or population proportion, $P = \sum_{i=1}^{N} a/N = A/N$, where N is the size of the population, and a_i is the value or state of the character, an attribute which can take only two values, 1 or 0, in the i-th unit.

2.2. Estimators of P.

A substantial part of statistical literature has been devoted to point and interval estimation of a population proportion in samples of size n (see e.g. Azorín and Sánchez-Crespo (5)). The direct, analogic, or "natural" point estimator, is: $\hat{P} = \sum_{i=1}^{n} a/n, = a/n$. This is a maximum likelihood, unbiased estimator, of P, with variance P (1-P)/n.

We will represent $\hat{P}$ by $\hat{P}_1$, for there are several other estimators of P, as the following:

The mid-range estimator: $\hat{P}_2 = 1/2$, which has 0 variance, and a squared bias equal to $(1/2 - P)^2$.

More generally, other guessed estimators, also independent from the sample values, are of the type: $\hat{P}_3 = \tilde{P}$ which have also 0 variance, and a squared bias equal to $(\tilde{P} - P)^2$.

The general weighted estimator is: $\hat{P}_4 = w\hat{P}_1 + (1 - w)\hat{P}_3$.

It is interesting to note that w may represent a "grade of membership", as will be seen afterwards.

The variance of $\hat{P}_4$ is w^2 P (1-P)/n, and its squared bias: $(1 - w)^2 (\tilde{P} - P)^2$.

It is thus seen that the accuracy, expressed as the sum of the variance and the squared bias: $MSE(\hat{P}) = var(\hat{P}) + b(\hat{P})^2$, depends on the true value, P, of the proportion, and that there is no uniformly best or preferred estimator.

Beside those seen above, there are other estimators, worth mentioning. Such as the mean likelihood estimator, $\hat{P}_5$, and the median likelihood estimator $\hat{P}_6$.

Both depend on the prior distribution, F(P) of P. In simple sampling, i.e., with the binomial distribution:

$$\hat{P}_5 = \int_0^1 P\binom{n}{a}P^a (1-P)^{n-a} dF(P) \Big/ \int_0^1 \binom{n}{a} P^a (1-P)^{n-a} dF(P).$$

If F(P) is uniform, i.e.,equal to 1, in (0,1): $\hat{P}_5 = (a+1)/(n+2)$ which is a special case of $\hat{P}_4$ (when $w = n/(n+2)$, $\tilde{P} = 1/2$). (For a=0 see 2.5.).

If F(P) is exponential, $F(P) = (m+1)(1-P)^m$, then $(a+1)/(n+m+2) = \hat{P}_5$,

$$MSE(\hat{P}_5) = \frac{n\ PQ + [1 - (m+1)\ P]^2}{(n + m + 2)^2}.$$

Still more generally, if it is proportional to $P^{\alpha-1}(1-P)^{\beta-1}$, then $\hat{P}_5 = (a+\alpha)/(n+\alpha+\beta)$ (for $w = n/(n+\alpha+\beta)$, $\tilde{P} = \alpha/(\alpha+\beta)$.

The median likelihood estimator $\hat{P}_6$ is obtained from the inequality:

$$\int_0^{\hat{P}_6} \binom{n}{a} P^a (1-P)^{n-a}\ dF(P) = 1/2 \int_0^1 \binom{n}{a} P^a (1-P)^{n-a}\ dF(P).$$

And the minimax estimator (see Kruskal and Tanur (42)):

$$\hat{P}_7 = (\sqrt{n}(a/n) + 1/2)/(\sqrt{n}+1),\ \text{for } w = \sqrt{n}/(\sqrt{n}+1),\ \tilde{P} = 1/2.$$

By averaging some of the above estimators, other expressions can be obtained.

For instance, from $\hat{P}_1$ and $\hat{P}_2$: $\hat{P}_8 = (a + 1/2)/(n+1)$.

This has been justified by Jeffreys and others on certain invariance properties (Good (28)).

Optimization of weighted estimators, minimizing absolute or relative mean squared errors, give values dependent on the true value, P, and respectively equal to:

$$w_1 = (\tilde{P} - P)^2 / [(\tilde{P} - P)^2 + PQ/n]$$
$$w_2 = (\tilde{P} - P)^2 / [(\tilde{P} - P)^2 + \tilde{P}Q/n]$$

(see Azorín and Sánchez-Crespo (5)).

The dependence of w on P does not mean that these expressions are useless. There are other practical situations in which the knowledge of the true value of a certain parameter is needed for optimum estimation. This happens, for instance, in estimating the size of samples required for a given precision. In some occasions a process of successive approximations can be applied, in a fuzzy approach to "good", "better" or "satisfactory" alternatives.

2.3. Estimation of proportions from other sample designs.

(a) Sampling without replacement (hypergeometric distribution).

The maximum likelihood estimator is still: P=a/n, and the mean likelihood estimator is:

$$\hat{P} = \int_0^N A \binom{A}{a}\binom{N-A}{n-a}\ dA / \int_0^N \binom{A}{a}\binom{N-A}{n-a}\ dA$$

for a prior uniform distribution of A.

The median likelihood estimator, $\hat{P} = \hat{A}/N$, is obtained from the equality:

$$\int_0^{\hat{A}} \binom{A}{a}\binom{N-A}{n-a}\ dA = 1/2 \int_0^N \binom{A}{a}\binom{N-A}{n-a}\ dA$$

(b) Stratified sampling (L strata):

Now $\hat{P} = \sum_{h=1}^{L} W_h P_h$, where the strata weights, W_h are known but in some cases only fuzzily.

(c) For tri- and polychotomic attributes, i.e., with 3 or more states or "colours", the maximum likelihood estimators are $\hat{P}_h = a_h/n$, $(h=1,\ldots,m)$, var $(\hat{P}_h) = P_h\,(1 - P_h)/n$, etc., and similary for the other possible estimators.

Fuzzy solutions could also be obtained for fuzzily defined states or colours.

(d) Inverse sampling (negative binomial distribution)

Sampling is continued until a given number a of rare items have emerged. This number a, is supposed to be the minimum number of rare items required, for example, for study, experimentation or analysis. Then the probability is:

$$f(n) = \binom{n-1}{a-1} P^a Q^{n-a}$$

which gives the unbiased estimator (Johnson and Kotz (41)): $\hat{P}=(a-1)/(n-1)$ and its estimated variance: $\hat{V}\,(\hat{P}) \cong \hat{P}\,(1-\hat{P})/(n-2)$

(e) Poisson distribution (infinite populations)

Maximum likelihood estimator: $\hat{\lambda}_1 = a/n$

Mean likelihood estimator: $\hat{\lambda}_2 = (a+1)/n$

Median likelihood estimator: $\hat{\lambda}_3$, obtained from:

$$\int_0^{\hat{\lambda}_3} \lambda^a \; e^{-n\lambda}\, d\lambda = \tfrac{1}{2} \int_0^{\infty} \lambda^a \; e^{-n\lambda}\, d\lambda$$

(f) For cluster and multistage sampling, the unbiased estimator of Horvitz and Thompson, for the total, or population absolute frequency, is: $\hat{A} = \sum_{i=1}^{n} A_i / \Pi_i$ (see, e.g., Azorín and Sánchez-Crespo (6)).

(g) Other desiderata, besides accuracy, can be considered, such as consistency, efficiency, economy, simplicity (some of these desiderata can have two or more definitions), and they may require other sampling designs. Introducing lineal costs (c being the cost per unit), the expression of the loss function is: $w_1^2\, V + w_2^2\, B^2 + (1-w_1^2 - w_2^2)\, cn$, from which the optimum - size of samples can be determined.

On biases, sampling errors and costs in surveys of rare items, see Czaja et alia (18), for multiplicity counting rules.

2.4. Interval estimation and levels.

From a statistical point of view, the P_k confidence interval, i.e., the interval corresponding to a probability P_k before the sample has been selected, is given by: $(\hat{P} - k \sqrt{\hat{V}(\hat{P})}\;\; ; \; \hat{P} + k \sqrt{\hat{V}(\hat{P})}\,)$ where the variance V is estimated by $\hat{P}\,\hat{Q}/\,(n-1)$.

For the sake of simplicity, we consider simple sampling, without finite population corrections. Criteria of departure from normality may be based on the coefficients of skewness and kurtosis, which show that for small values of the proportion P very large sizes of samples may be required. Moreover, it is obvious that the above given confidence intervals cannot be used when P or Q are 0.

A simple classical procedure is to find out lower and upper

confidence limits from the equations:

$$\text{Prob}\ (P \leq \hat{P}_U) = (1 - P_k)/2\ ,\quad \text{Prob}\ (P \geq \hat{P}_1) = (1 - P_k)/2$$

Using now the adequate expressions of simple sampling (binomial distribution).

When the sample gives $a=0$ or $a=n$, the confidence interval would be $(0,\hat{P}_U)$, $(\hat{P}_L, 1)$, respectively.

The accuracy of normal approximation to confidence limits has been investigated by Blyth (8) and others.

Considerations on fuzzy confidence intervals can also be seen in Civanlar and Trussell (17).

2.5. The important special case $a=0$.

In sampling for rare items, it may happen that no rare items appear in the sample (except in inverse sampling, where $\underline{a}$ is established beforehand). Then some of the estimators, such as $\hat{P}_1$, and others, take the value zero. Others, more "conservative", take for instance the value $1/(n+1)$ which can be very small, when n is large, but different from 0. Fuzziness may be introduced when $\underline{a}$ is "practically" zero, or when there are vague situations in deciding if a given unit in the sample is or not rare.

2.6. Ratio estimators.

Once $P=A/N$, for an attribute, and $\overline{X} = X/N$, for a quantitative character have been estimated, there may be necessary to estimate their ratio, e.g., $R = X/A$. As in the case of P, several estimators can be used for R. As a matter of fact, there is now more liberty in choice, and still more if transformations (logarithmic, etc.) are allowed (see Box and Cox(11)).

$\hat{R}_1 = x/a$, and $\hat{R}_4 = w\hat{R}_1 + (1-w)\tilde{R}$, are perhaps the obvious estimators to be chosen. Another type of ratio estimator arise from the fuzzy expression $\sum_{i=1}^{n} a_i \cdot \mu(a_i) / \sum_{i=1}^{n} \mu(a_i)$, and the corresponding formulae defined for other estimators of P.

Another important special case for ratio estimators appear in the estimation of odds, i.e., of the quotient: $A/(N-A)$. The direct estimator is $a/(n-a)$. There is an extensive literature on the estimation of odds (see, e.g., Greenland and Robins (30)), and on its application to Biology, Demography, and related fields.

Silva and Rubi (58), on the estimation of proportions of three or more classes using cluster sampling, suggest the estimator:

$$\hat{P}_{\lambda j} = \hat{P}_j + \lambda_j\ (1 - \Sigma P_j)$$

where $\hat{P}_j = M\,\overline{x}_j/N$, $\overline{x}_j = x_j/m$, $x_j = \sum_{i=1}^{m} X_{ij}$

for M clusters of sizes M_i , j=1,2..., k ; i=1,2,..., M from which a sample is taken of size m. (This is a particular case or Horvitz – Thompson estimator)

Several values can be given to λ_j;

$$\text{For} \quad \lambda_j = \hat{P}_j / \sum_{j=1}^{k} \hat{P}_j \;, \quad \hat{P}_{ij} = \hat{P}_j / \sum_{j=1}^{k} \hat{P}_j = x_j/n$$

which is the direct estimator, in this case not unbiased but with a negligible bias. Knowledge of N is not required. Other advantages are noted in the paper.

3. SAMPLE SURVEYS: PLANNING AND DESIGN

Planning refers now to the conception of all aspects of a certain observational study, or survey, including both crisp and fuzzy considerations, and for large scale sample surveys it requires the use of macro (main frame), mini and micro-computers, which should be integrated in a holistic or systemic approach. The statistical information system could include also censuses, and administrative or institutionalized registers, as happens in the case of rare diseases.

As to sample design in the study of rare items, stratified sampling is useful when a cheap, rapid test is apt to divide the population in a part which is almost free of rare items, and the other heavily populated with them, as was noted by Deming, and other authors. Also when previous knowledge of the population allows a convenient stratification according with the density of rare items and with the cost or difficulty of observation, measurement or response. Otherwise, a selection of large clusters is to be preferred. Also two or multi-phase sampling is applied when filtering or screening is cheap and does not allow too many "false positive" units to pass. A combination of the above designs and procedures should be used in most situations. It must be enphasized that in some cases it is advisable to make use of network surveys, using multiplicity counting rules (see Czaja et alia (18) and previous work by Levy (43), and Sirken and Levy (59)). According to Czaja et alia, when the prevalence rate is $\geq$ 0,05%, a household survey, with reporting on relatives who live outside the household sampled (the child or sibling rule), is to be preferred. When it is below 0,05, other methodologies, such as administrative lists, dual frames (a combination of lists and network surveys), or network surveys with wider counting rules, should be pursued.

It is by no means easy to attain optimum, or satisfactory, or in some cases even reasonable, plans and designs, including estimators and measurement procedures. Specially when there are two or more types of rare items in the investigation, and perhaps multiple objectives and criteria, and possible variations in time to be predicted or guessed. In these cases, approximate reasoning for knowledge representation may be of use, as a complement to the management of coverage errors (Wolter (64)) and imputation procedures (Chiu and Sedransk (16)). These ensure "that responses are within a legal range, are mutually consistent, and conform to the structure of the questionnaires" (see Pullum et alia (51)). On possible causal models for patterns of non response, see Fay (23).

4. THE CONSTRUCTION OF MEMBERSHIP FUNCTIONS.

4.1. General considerations

As said in the Introduction, the consideration of rarity, and other problems connected with the study of rare items, may require the introduction of degrees of membership, and are thus connected with the use of membership functions. Now, there are many authors who still consider the construction of membership functions as an important problem, not quite satisfactorily solved. (See, for instance the report by Dubois, 1984, of the workshop organized by Carlsson in Brussels).

Hisdal (35), has gone so far as to say that the lack of a well defined meaning of the grade of membership concept is the Achilles heel of fuzzy set theory; and Civanlar and Trusdell (17) consider that the determination of membership functions is vital in practical applications of the fuzzy set theory. Hippel (34) quoted S.R. Watson as saying that the membership function is the central concept of fuzzy set theory. Therefore some general aspects on member function estimation will be now discussed.

Three main approaches can be distinguished:

(i) Estimation of membership values.

(ii) Estimation of the form of the membership function.

(iii) Estimation of the parameters of a membership function, given its form, or family, i.e., once its algebraic expression has been specified.

4.2. Membership values.

As it is well known, the membership value of an element, unit, object or individual $\underline{x}$ with respect to a given predicate, character or feature A, can be considered as the degree of compatibility between the predicate, etc. and $\underline{x}$. It can also be said that the membership value expresses how much does $\underline{x}$ belong to A.

Typical examples of fuzzy sets for rare items are:

(i) The set of very tall men, or giants; or of very low men, or dwarfes.

And a typical question:

(ii) What is the first height such that one would say that a man with this height is a giant or a dwarf?.

Still more difficult would be the definition of fuzzy sets for predicates or characters of a more subjective type, such as ugly or beautiful; repulsive or attractive.

Now, for estimating membership, as for estimating frequencies, a typical statistical procedure could be used, giving $\hat{P}$.

As we have seen in Section 2, there are several possible estimators: of a "true" value, P, none of them having all the advantages or satisfying all desiderata.

All these estimations depend very much on the population from which the sample is selected, and on the cost and circumstances of selection. But there are cases in which the membership value could be considered as objective. For instance, when the degree of blackness of a board with squares of equal size is expressed by the ratio or proportion of black squares to the total number of squares. Even in this case the estimation could be difficult, or take too much time. For instance, when there is a very great number of very small squares. Or if there are irregular black and white patches, instead of squares. A very small number of black squares, would serve to exemplify the objective estimation of rare items.

All the objective cases just mentioned are closely related to statistical frequency and the difficulties of estimation satisfying different desiderata, and not with the objectivity of the procedures. On the other hand, if degrees or intensities of blackness, shades of grey are considered, subjective perception may be again introduced, specially in connection with rare, or infrequent shades, when there is a black surface with hardly visible white dots. Dubois and Prade (22) have mentioned "noise removal in grey scale pictures", following Nakagawa and Rosenfeld.

Other examples are rare intermediate cases of hermaphroditism, for the alternative male/female; or in taxonomy, border-line cases between taxa; hybridism, etc. In Medicine practice, difficulties arise due both to fuzziness, and rarity. But coming back to the simpler case of black squares ("good" and "defective"), and as in sampling inspection, the use of the Operating Characteristic Curve (OCC) is often advisable, to portray the statistical performance of the process. The curve is generally the graphical representation of the probability of accepting -or rejecting- a submitted lot as a result of sampling inspection for successive values of the true proportion of defective items.

In Biology and Psychology, yield or results evaluation, and stimulus-response curves, often require the introduction of fuzzy sets considerations. There may be possible but "rare" responses, or results; these problems have a clear connection with the theory and practical treatment of outliers.

Another important case of physical fuzziness is found in experiments of diffusion. The intensity of a fluid spreading from a source may decrease with the distance to the source, and other variables, and be different for different directions. Some considerations on this topic can be seen in Azorín (4), and also on the distinction between diffusion and percolation, depending upon whether it is more natural to ascribe the (usually random) mechanism to the fluid or to the medium. As is has been said, this distinction is not clear-cut, and so we have fuzziness within, or superposed to, fuzziness.

There are other possible realizations of diffusion. For instance the spreading of news, or rumours, which has been studied in Sociology, and Psychology (see Payne (49)). Also, the spreading of economic, or political influence. There could be cases of resistance to this spreading, and rare cases in which a strong influence has not been able to penetrate. This has happened for instance with the persistence of languages, such as Basque, Irish and Welsh, against romanization, or Saxon or Norman influences. Fuzzy rare items are found in studying the geographical distribution and ethnic origin of languages. An interesting case is Shranan language (Surinam).

We pass now to the second aspect mentioned above, i.e., the estimation of the form, or family, of a membership function, or curve.

4.3. Membership curves.

Starting with a "scatter diagram" of successive membership values, an "appropriate" membership curve may be adjusted.

The normal curve could be an acceptable choice only in certain cases. Examples of specification of curves for subsequent estimation of the parameters can be seen in Civanlar and Trussell, and in Dishkant (21). It is important to note in each case whether they have been obtained through physical or through linguistic fuzziness, e.g. cognitive or mechanical procedures.

4.4. Evaluation.

It is desirable that the subjects, judges or evaluators should give reliable conclusions, after proper training, as in statistical appraisal. They should have both knowledge, experience and ability in the substantive field or subject, and in the practice of general evaluation. There is a paper by Hogarth (37), which, with adequate changes, could be applied to fuzzy evaluation. Gaines and Shaw (25) also consider the problem of knowledge elicitation from a skilled person.

Evaluation methods in Statistics and in Fuzzy Sets are far from settled. Box said, in 1980, that "scientific learning is an iterative process employing criticism and estimation. The first can induce model modification when confronting model -or assumption- with available data. New designs would be chosen to explore those shadowy regions where illumination is important". And in the discussion of Box paper, R.B. Miller said: "a model must be viewed as an expression of the scientist thinking at a particular moment and not as an objective entity. The fact that the model is widely accepted among scientists makes it neither objective nor true, only widely accepted. The only objective entity is data, and even data are corrupted by measurement error, selection bias, processing error, etc." These considerations are true for probabilistic and fuzzy modeling.

Payne (49) refers to experts Systems in validation and extension of models, by combining traditional simulation techniques with Artificial Intelligence paradigms which could be used in some cases after a membership function has been estimated, for its possible application to extended situations or conditions. This transportation of parameters values is in some sense analogous to portability of computations operating systems, etc. and also of the "rate of homogeneity" (roh, or rho, $\bar{\rho}$) or intra-cluster correlation coefficient, in sample surveys (Kish (40)), when extending the sample design to a different, but more or less similar population, in order to use the formula of the design effect, DEFF, which is the ratio between the variance in the given design, and the variance in simple random sampling: $1+(\bar{m}-1)\rho$ (Thomsen et alia,(61)).

5. UNCERTAINTY, DIVERSITY, DECISION AND DIAGNOSIS.

As was said in the Introduction, statistical and fuzzy uncertainty arise in connection with rare items, and in decision and diagnostic problems, the simplest case being if a given object or individual should be considered as rare or as non-rare.

Once problems related with membership values and functions have been discused, as indicated in Section 4, attention should be given to measures of fuzzy uncertainty, of which there is a large and growing number.

These measures have appeared in many fields, some of them widely separated. Therefore results attained in one domain often remain unknown to research workers in other areas.

As it is well-known,many measures of uncertainty and information are related with the notion of entropy, which pervades many disciplines after have been originated in Physics.

It is also connected with other concepts, such as diversity, (see e.g. C.R. Rao (52), Patil and Taillie (48), Azorín (3)). One of the basic formulae are logarithmic entropy (Shannon, Gibbs, Wiener, Weaver), and the exponential expression derived from the sum of the squared probabilities (Gini, Gibson). Kaufmann (1973) and De Luca and Termini (1979) provided useful fuzzy set entropy measures.

Hopf and Brown (1985), use expressions related to the above mentioned measures of diversity. Rare items, belonging to rare species, tend to diminish the value of diversity indexes. These items may even be absent in small samples. To increase expectations, larger samples, purposely designed, should not be used, only proper probabilistic designs. Sometimes combined with some reuse techniques, such as jackknife or bootstrap method, in order to better estimate their precision. In any case, adequate computerization would be required, and in some cases also proper statistical packages (Comp/Stat).

Czogala et alia (1982) show a general fuzziness measure which gives also an entropy measure in the sense De Luca and Termini. It is important also to consider the notion of energy (Teodorescu, 1977) which has been developed by Pardo and Menéndez, and others. Czogala uses the degree of fuzziness $\Sigma\mu_A(x)$, normalized by $\Sigma\mu_A(x)$ / $\mu_A(x)$, and $\Sigma\overline{\mu_A(x)}^n$ is a particular case of an energy measure from the point of view of a decision maker.

Also Gil et alia (27) start from a general measure of information, based on entropy, and suggest criteria to compare information systems, in different spaces.

Salicrú, and Salicrú and Calvo (57), have presented properties of uncertainty indexes and divergence, starting from uncertainty functions derived from entropy, in the line of Lindley, Renyi, Havrada-Charvat, and others (see the previously mentioned papers on the diversity concept). A fundamental role is played by the notion of divergence, introduced by Matusita, as shown by Burbea and Rao (13).

García-Carrasco (26) has also analysed uncertainty functions concerning information purposes. and suggested their use in the Bayesian line of De Groot and others.

The impact of the above mentioned concepts in the study of rare items stems from the above emphasized fact that rarity itself is a fuzzy notion. But there are other aspects to be considered, especially when there are more than one simple character to be measured. This happens, for instance, in the above mentioned studies of biological diversity. For instance, when not "small specimens" are looked for as rare, but rather a type of ethnic dwarfs; or not just "giants", but "yetis", or remnants of Gigantopithecus, which may be not only rare, but even non existent.

It should be mentioned that Gaines and Shaw (1986), have used an information theoretical measure of uncertainty reduction to determine an optimum in the trade-off between fuzzy logical and probabilistic hypotheses. They start from repertory grids as suggested by Kelly, which are a representation of personal constructs. The theory, say Gaines and Shaw, was developed as a model that has strong systemic and psychologic foundations. Here again appears the problem of eliciting the experts vocabulary, and of structuring some aspects of the domain knowledge. Constructs are dichotomous distinctions that a person creates to fit over the realities of which the world is composed. The repertory grid is a way of representing these personal constructs in the problem domain, which may consist of personal relations. Thus it could be related to survey designs such as snowball schemes and networks which are useful in sampling for rare items.

The high cost of these surveys makes it convenient to look for several traits in one sample, and thus divide the cost among the cooperating investigations (Kish (40)). On this connection it is important to explore methods and results of multipurpose operations. Borisov (9) evaluated multicriteria alternatives, using linguistic scales to measure the degree of satisfaction attained, in order to be able to choose the best alternative, making use of an expert system. Buckley (1985) investigates the use of fuzzy data in fuzzy decision problems. He arrives at the conclusion that further research is required to identify various types of optimization procedures and their properties. Ríos (55) surveys the present situation in multi-criteria decision theory, and indicates some aspects in which theory and techniques are connected with fuzzy theory and practice. On the applications of expert systems, Buckley et alia (12) describe a fuzzy rule based expert system which has as an input a fuzzy vector, and produces as output a fuzzy set of conclusions. He emphasizes the need of using a fuzzy expert system, when conventional statistical methods of representing uncertainty appear to be inadequate. And how the use of a concept of reality is in a sense wider than the purely statistical, to cope with fuzzy aspects, especially in the application or upper and lower confidence levels in decision making. In this decision, they assign a "personality" to the expert system,according with its willingness to take risks.

Problems of diagnostic can be seen as a locus of convergence of the two main purposes of scientific activity, i. e. decision and understanding (Anderson, 1985). Ben-Bassat (7) describes his experience in the research and development of large scale medical expert systems. A fundamental problem is the integration and summary of evidence, and a key point, trying not to get the computer involved in tasks where its contribution would be insignificant, or questionable. The problem of abduction, a syllogism with the major premise certain and the minor only probable, is treated by Reggia et alia (53) as a theoretical framework for diagnostic expert systems and approximate reasoning. They emphasize the necessity of explanation of the findings of a set of manifestations (symptoms signs, test results), of determining the whys and hows. They call abductive inference the general reasoning processes involved in diagnostic problem solving. For instance, when the physician postulates the presence of a set of possible diseases that could explain the patient's manifestations. His hypotheses are based on his medical "knowledge base", and looks for a resolution of two conflicting goals; coverage (explanation of all the manifestations) and parsimony (minimizing the complexity of the model). This is a trade-off akin to what we have called "Occam's razor" versus "Menger's palette" (Azorín (2)).

6. RARE ITEMS IN SPATIAL PATTERNS

In some cases, not only the relative frequency (proportion) and the absolute frequency (total) of a rare item is to be estimated, but also a certain knowledge of the location in a given area or domain is required, to be used in the "découpage" of the population, for the determination of strata and clusters or other sampling units, and for a deeper understanding of physical structure.

Plane or, in general, spatial sampling, can be used for the estimation of structures, configurations, patterns, and also nexus and snowball sampling. In the next Section some comments are made on different meanings of search in their connection with rare units, Comp/Stat and fuzzy sets. It should be noted that the study of rare items is strongly connected with Poisson processes. Points or units may be distributed as a definite random process on a surface or plane, but not necessarily so. A first examination of randomness in distribution is often useful, as a basic fact tobe introduced in the sampling plans and designs. In Ecological applications, Hopf and Brown (38) examine quantitative ecological data to determine whether they can be regarded as being random in character. Their work concerns a very important general problem: how to distinguish signal from noise, or order from chaos, or "interesting" patterns from a messy background. Other authors have suggested that apparently organized structure can be due to chance alone. They add: "how do we know that examples that appear to prove some rules are 'statistically typical'; and not culled from a set of random data". As a matter of fact, there are two problems in this question. First, the decision about the randomness of a disposition of elements. Secondly the "typicality" of statistical data. On the first, it should be noted that there may be a fuzzy region between non-randomness and randomness, Hofstadter (36) mentions some patterns which are "delicately balanced between order and chaos". For instance, the behaviour of the sequence x, f(x), f[f(x)], ..., where f(x), is some function with interesting properties, and x, the initial value, is called a "seed". Starting from the function: $y = 4\lambda x(1-x)$ with $\lambda < 1$, and $x \in (0,1)$, there are impressive differences in the degree of regularity of the patterns or sequences ("orbits"), according to the values of the parameter or "knob". Beyond a critical value of λ, the sequences cease to be regular or "patterned", and convergent, and appear essentially chaotic, with aperiodic sequences, or orbits. There are intermediate situations of orbits which are still periodical, but with a periodicity difficult to discover, because the period is very large, and also because the extreme oscillations make the behaviour hardly distinguishable from total chaos. Another example is a recursive function defined as $Q(n) = Q[n - Q(n-1)] + Q[n - Q(n-2)]$ for any natural number $n > 2$, and $Q(1) = Q(2) = 1$. The result is an "erratic" sequence, with a chaotic appearance.

These situations, of predetermined order and apparent disorder, arise also in the construction of pseudo-random numbers, and they exemplify cases of fuzziness between randomness and non-randomness. It should also be observed that apparently chaotic configurations or white noise, can in some cases reproduce themselves point to point (or "almost"), after very long periods of time, giving raise to patterns of red noise (Anscombe (1)), or pink noise (May (45)).

Sometimes randomness depends on the level of observation. Thus, a random table is expected to contain some "non random" sequences of digits, with a probability which grows with the extension of the table. On the other hand, a random part of the table or even the whole table, if repeated, originates a regularity or periodic sequence, composed of random sequences, which in turn contains "non random" sequences, etc. and in this way a hierarchy of successive random and non random components can be established. So we can find regularity or chaos, depending on which level we stand.

Now, conceiving pattern in a more general context, which also encompasses patterns of growth, and patterns in the balance of nature, we could deal with the dynamical behaviour of population. May (45) shows how simple non linearities, when compounded with environmental unpredictability and/or hetereogeneities in individual behaviour or spatial distribution, can cause problems in analyzing data and in predicting events. Then he surveys possible effects of competition, mutualism, and predator-prey associations upon community structure. And lastly, he discusses work on community level patterns having to do with food web structure, the relative abundance of species or of individuals in different physical size classes.

It would be worth exploring the possible impact on patterns of fuzzy uncertainty, besides probabilistic un-predictability, and variability due to heterogeneity in behaviour or in spatial distribution and also on spatial distributions of community structure.

To the considerations which we have made on the influence of rare items in the estimation of configurations through plane and spatial sampling, it should be added that this estimation or reconstruction of pattern can refer as well to special cases, such as facial restoration (see Mieno and Ishikawa (46). The sampling design is often probabilistic, systematic sampling, and connected issues are theoretical point processes. (See the 1986 ASA meeting on Statistics of Spatial Data).

A more complex situation consists of data structures involving pairwise relationships: for instance, friendships, transactions between bank accounts and diffusion in the plane or in space. See, e.g., Frank and Strauss (24).

7. SEARCH

Search is used with several meanings in scientific theory and practice. In some cases it is applied to purposive, non probabilistic sampling. In a general sense it may involve research, and also looking for the solution of certain problems. And reduction of the size of the interval which contains an optimum by evaluation of a function at suitable search points. Stone considered a class of resource allocation, when the search assumes an a priori probability distribution, for the target's location, in a search plane or space. Pregibon (50) equates data analysis with search.

Haley and Stone (32) refer to objects to be found, and to the exploration of natural resources. Applications include rescue surveillance, and a distinction between stationary and moving targets (non evasive and evasive). Koopman, B.O., a pioneer of Search theory, suggested classifications such as: Reactive and non reactive pursue; simple and repetitive search effort, etc. Within the domain of search procedures Comp/Stat can help in the estimation of probability distributions, the target's location and motion, and in the estimation of a detection function, representing the effort devoted to searching and finding, and evaluation of results. Use of prior information from past surveys or from other sources, can be seen in Chiu and Sedransk(16).

The connection between search and sampling of rare items has special relevance when each sample unit is the subject of further scrutiny and analysis. In Section 3, network sampling and multiplicity counting rules were mentioned, which are related with search of infrequent, or rare items, or events.

In Czaja et alia (18), the multiplicity of event I_j by rule k, in a population of M households, N individuals or units at risk, and C events or items of interest (linking the event j to the household i by rule k) is given by:

$$S_{jk} = \sum_{i=1}^{M} \delta_{ijk}, \text{ where } \delta_{ijk} = 1$$

if event I_j is linked to household i by rule k, and 0 otherwise. In the model of Casady et alia (see Czaja), $a_{ijk} = 1$ if event I is reported at household i by rule k, and 0 otherwise. Introducing fuzziness, μ_{ijk} should be substituted for δ_{ijk}, ν_i for δ_i which takes the value 1 when the household i appears in the sample) and α_{ijk} for a_{ijk}, and the formulae of the multiplicity estimator $\hat{\theta}$ for θ, and their correspondent mean squared error as sum of its squared bias, and variance.

8. SOME FIELDS OF APPLICATIONS OF FUZZY SETS WITH RESPECT TO RARE ITEMS AND COMP/STAT.

As was said in the Introduction, Comp/Stat is the interface of Computation and Statistics. We have examined their interface with fuzzy sets, and with rare items. It is convenient to see now the interface with several substantive fields, which in their turn may have also fuzzy frontiers between themselves, as happens between Medicine and Biology, etc.

Health and Medicine

This field is connected with the incidence and prevalence of rare diseases, the care and hospitalization of diminished or handicapped persons, and the protection of populations against infectious diseases and accidents. There may be both probabilistic and fuzzy uncertainties in decision making about mass vaccination, isolation of individuals, or confinement of contagious cases. And also on budgeting for further research and evaluation. We have already mentioned medical diagnosis as an area of application of artificial intelligence and expert systems, with approximate reasoning. In some cases diagnosis of some diseases, as happens with rare types of cancer, etc. seems to be exceedingly difficult.

Applications of medical decision making and diagnosis, which were presented in the works edited by Gupta and Sánchez (1982) and Gupta et alia (31), show a wide experience and a useful panoply of procedures. The accumulation of initial findings, and their integration (Ben-Bassat (7)), have special importance in the sampling for rare items. See also Gaines and Shaw (25) on different approaches to the combination of fuzzy and probabilistic inference.

Biology, Ecology

The protection of environment and natural resources, especially of endangered species, of rare specimens and couples, are of capital importance. The connections between uncertainty measures and measures of diversity which were considered in Section 5, seem to point to synthesis and further comparisons. On the other hand, the search for rare and shy specimens may require very specialized techniques and some previous knowledge of distributions on plane or spatial regions, in three or more dimensions. This may be the case in the detection of sources, or potential sources of contaminants or pollution, to be erradicated or at least controlled.

There are as well classification and taxonomic problems, not only with respect to the existence of rare individuals or specimens, but also to the appearance of whole species, which may be still unknown, due to their scarcity, inaccessibility, or other circumstances.

May (45) says that "it is appalling that we do not know how many species of plants and animals we share the world with. Fewer than two millions are currently classified, and estimates for the total number range from 5 to 5o million". The sheer volume of the required surveys for the discovery of new species imply an intensive use of Comp/Stat procedures, and probabilistic and fuzzy measures, and decision making. If the problem is extended to already extinguished species, which according to some authors could reach millions, it is to be feared that 80% or more species would remain for ever unknown to scientists. This points to the urgence of applying a well planned and coordinated effort, including sample surveys of rare items, uncertainty techniques and Comp/Stat procedures.

Sociology, Psychology, Jurismetrics.

Many problems of social, psychological and criminological interest are associated with rare items, fuzziness and Comp/Stat. On the use of statistical packages, an important aspect is the application to large scale sample surveys and censuses, when looking for small ethnic or social groups; and the study of centenarians, albinos, etc. Psychological studies could refer to fuzzy situations of ambivalence, and plurivalent feelings. Jurismetrics, to uncertainty problems in testimonies, and to grouped or isolated, actual or potential criminals, delinquents or outlaws. A related topic is concerned with security matters (see, for instance, Rine (54)); fuzzy terms are used, and support and management of security systems using fuzzy logic should be considered.

Economics, Finance, Cost/Benefit analysis, Business Administration, etc.

An example of sampling of rare items, which is often mentioned, is the demand and supply of very expensive commodities, and of potential purchasers, in market research. The typically fuzzy terms, "rich", and "very rich"; "poor", and "very poor", have an impact in studies of income and wealth distribution. Auditing problems should also be mentioned, being related with sampling of rare items and intensive computerization, and with possible fuzziness in the conclusions about the financial status of certain firms or institutions.

Operation research and management methods are often needed in connection with optimization and planning. Fuzzy considerations and techniques have been introduced by Zimmermann and other authors in mathematical programming (see e.g. Delgado et alia (20)). The costs and effects of looking for rare items, which may be needed or convenient to enhance the quality of a certain product, could be introduced in a global approach.

9. Concluding Remarks.

Computational Statistics, or Comp/Stat, has been detaily discussed by Victor (63) and others. As has been said above, these methods and techniques can be extended to other forms of uncertainty, especially to fuzziness. An important field to be further explored is concerned with rare items, or events. One of the first examples mentioned by Hansen, Hurwitz, Madow(33) and others, was the proportion of vacant dwellings in crowded towns, this being a topic of current interest to planners, economists, sociologists, etc. In a more experimental context, the proportion, number, and distribution of misprints in books has been considered useful for comparison purposes between methods and procedures of sampling design and estimation. Many other could be examined, beside those mentioned in Section 8, with respect to Statistics and Fuzzy Sets and Systems and in connection with decision theory, when estimating if a frequency exceeds an established safety bound. And on the adequacy of estimations and predictions, when deciding what is "low enough to be negligible", and other aspects of approximation and asymptotic theory and practice. Tukey (62), dealing with the strengthening of graphic displays, says that there are two ways to make better plots: "1. By flattening the picture so that we can make the intervals of uncertainty bigger... and 2. By shifting emphasis from 'what might be', inevitably truly fuzzy, to 'what we know cannot be', which only has fuzzy edges". And he emphasizes some roles for expert systems, the necessity of thinking through a strategy, and of continuing education.

In conclusion, there is still a long way to attain a deeper understanding of phenomena connected with uncertainty and scarcity, and thus a challenge to an efficient use of expert systems and artificial intelligence in those problems for which a growing panoply of instruments is contending for search, synthesis and selection.

KEY WORDS. COMP/STAT, Fuzzy sets, Rare items.
(Possible classification: Part III, Applications).

REFERENCES

(1) Anscombe, F.J. "Computing in Statistical Science Through APL". (Springer-Verlag, 1981).

(2) Azorín, F. (a). "Uncertainty and Evaluation", in "Proceedings of the 3rd Prague Symposium on Asymptotic Statistic". (Mandl and Hušková, eds., Elsevier, 1983)

(3) Azorín, F. (b). "Diversity and Uncertainty", Fall International Seminar on Applied Logic. (Universidad de Palma de Mallorca, Spain, 1983).

(4) Azorín, F. "On Fuzzy Aspects of Size and Shape" in Gupta, M.M. et alia (eds.), "Approximate Reasoning in Expert Systems" (North Holland, 1985).

(5) Azorín, F. and Sánchez-Crespo, J.L. (a) "On Survey Sampling for Rare Items", in Second Catalán Symp. on Statistics (CIDC, CIRIT, Barcelona, 1986).

(6) Azorín, F. and Sánchez-Crespo, J.L. (b) "Métodos y Aplicaciones del Muestreo", (Alianza Editorial, Madrid, 1986).

(7) Ben-Bassat, M. "Expert Systems for Clinical Diagnosis", in Gupta, M.M. et alia (eds) "Approximate Reasoning in Expert Systems" (North Holland, 1985).

(8) Blyth, C.R. "Approximate Binomial Confidence Limits", JASA, 81, 843-855, 1986.

(9) Borisov, A. and Naglis, L. "Multicriteria Choice of Alternatives in an Expert System for Computer-Aided Design of Industrial Robot Installation", FSS, 16, 93-102, 1985 .

(10) Box, G. E. P. "Sampling and Bayes Inference" JRSS, A, 143, 383-430,1980).

(11) Box, G.E.P. and Cox, D.R. "An Analysis of Transformations", JRSS, B, 26, 211-246, 1964 .

(12) Buckley, J.J. et alia, "A Fuzzy Expert Systems". FSS, 20, 1, 1-16,1986 .

(13) Burbea, J. and Rao, C.R. "Entropy Differential Metric, Distance and Divergence Measures in Probability Spaces", J. of Multivariate Analysis, 12, 575-596, 1982.

(14) Chakravarti, I.M. "Search in the Presence of Noise", Sankhya, 38, A, 381-393, 1976.

(15) Carlsson Ch. See Dubois, D. "Report on the Workshop on the Membership Function", FSS, 13, 312-319, 1984.

(16) Chiu, H. Y.and Sedransk, J. "A Bayesian Procedure for Missing Imputing Values in Sampling Surveys", JASA, 81, 667-676, 1986.

(17) Civanlar, M. R. and Trussell, H.J. "Constructing Membership Functions Using Statistical Dat", FSS, 18, 1-14, 1986.

(18) Czaja, R.F. et alia "Reporting Bias and Sampling Errors in a Survey of a Rare Population Using Multiplicity Counting Rules", JASA, 81, 411-419, 1986.

(19) Czogala, E. et alia "Contribution to Application of Energy Measure of Fuzzy Sets", FSS, 8, 205-214, 1986.

(20) Delgado, M. et alia "Solving the Biobjective Linear Programming: A Fuzzy Approach", in Gupta, M.M. et alia (eds), "Approximate Reasoning in Expert Systems" (North Holland, 1985).

(21) Dishkant, H. "About Membership Functions Estimation", FSS, 5, 141-148, 1981.

(22) Dubois, D. and Prade, H. "Fuzzy Sets and Systems: Theory and Applications" (Academic Press, 1980)

(23) Fay, R.E. "Causal Models for Patterns of Non-Response", JASA, 81, 354-365, 1986.

(24) Frank, O. and Strauss, D. "Markov Graphs", JASA, 81, 832-842, 1986.

(25) Gaines, B.R. and Shaw, M.L.G. "Systemic Foundations for Reasoning in Expert Systems", in Gupta, M.M. et alia (eds),"Approximate Reasoning in Expert Systems" (North Holland, 1985).

(26) García-Carrasco, M.P. "Incertidumbre Simétrica Generalizada", XVI Reunión SEIO, Málaga, Spain, 1986.

(27) Gil, M.A. et alia "Quantity of Information: Comparison between Information Systems", 1. Non-Fuzzy States, and 2. Fuzzy States, FSS, 15, 65-78 and 129-146, 1985.

(28) Good, I.J. "The Estimation of Probabilities: An Essay on Modern Bayesian Methods", Research Monograph, 30, M.I.T., MA, 1965.

(29) Good, I.J. "What is the Use of a Distribution", in Krishnaiah, ed., Multivariate Analysis II, Academic Press, 1969.

(30) Greenland, S. and Robins, J.M. "Estimation of a Common-Effect Parameter

from Sparse Follow-Up Data", Biometrics, 41, 55-68, 1985.
(31) Gupta, M.M. et alia (eds) "Approximate Reasoning in Expert Systems" (North Holland, 1985).
(32) Haley, K.B. and Stone, L.D.(eds) "Search Theory and Applications" (Plenum Press, 1980).
(33) Hansen, Hurwitz, Madow, "Sampling Methods and Theory" (Wiley, 1953).
(34) Hippel, K.W. "Fuzzy Set Methodologies in Multicriteria Modelling", in Gupta, M.M. and Sánchez, E. (eds) "Approximate Reasoning in Decision Analysis" (North Holland, Amsterdam, 1982).
(35) Hisdal, E. "Reconciliation of the Yes-No Versus Grade of Membership Dualism in Human Thinking", in Gupta, M.M. et alia (eds), "Approximate Reasoning in Expert Systems" (North Holland, 1985).
(36) Hofstadter, D. "Gödel, Escher, Bach- An Eternal Golden Braid", (Basic Books, 1979).
(37) Hogarth, R.M. "Cognitive Processes and the Assessment of Subjective Probability Distributions", JASA, 70, 271-289, 1975.
(38) Hopf, F.A. and Brown, J.H. "The Bull's-Eye Method for Testing Randomness in Ecological Communities", Ecology, 67, 1139-1155, 1986.
(39) Kaufmann, A. "Théorie des Ensembles Flous", Masson, París, 1975.
(40) Kish, L. "Survey Sampling" (Wiley, 1965).
(41) Kotz, S. and Johnson, N.L. "Encyclopedia of Statistical Sciences" (Wiley, 1981).
(42) Kruskal, E. and Tanur, J. "International Encyclopedia of Statistics" (The Free Press, 1978).
(43) Levy, P.S. "Optimum Allocation in Stratified Random Sampling Network for Estimating the Prevalence of Attributes in Rare Populations": JASA, 72, 758-763, 1977.
(44) Luca, A. De and Termini, S. "A Definition of Non Probabilistic Entropy in the Setting of Fuzzy Sets Theory", Information and Control, 20, 301-312, 1972 .
(45) May, R.M. "The Search for Patterns in the Balance of Nature: Advances and Retreats", Ecology, 67, 1115-1126, 1986.
(46) Mieno, H., and Ishikawa, A. "An Expert System for Facial Restoration" in Gupta, M.M. et alia (eds) "Approximate Reasoning in Expert Systems" (North Holland, 1985).
(47) Pardo, L. and Menéndez, M.L. "The Information Energy as a Bound in the Discrimation Problem with Fuzzy Classes and Fuzzy Information", First IFSA Congress, Vol. 1, 1985.
(48) Patil, G.P. and Taillie, C. "Diversity as a Concept and its Measurement", JASA, 77, 548-561, 1982.
(49) Payne, E.C. "the Integration of Traditional Simulation with Expert Systems and A.I. Knowledge Representation", in Gupta, M.M. et alia (eds), "Approximate Reasoning in Expert Systems" (North Holland, 1985).
(50) Pregibon, D. "Data Analysis and Search" in Second Catalan Symposium on Statistics(CIDS, CIRIT, Barcelona, 1986).
(51) Pullum, T.W. et alia "The Machine Editing of Largo Sample Surveys: The Experience of the World Fertility Survey", Intern. Statistical Review, 54, 311-326, 1986.
(52) Rao, C.R. "Diversity: Its Treatment, Decomposition, Apportionment and Analysis", Sankhya A, 44, 1-22, 1982.
(53) Reggia, J.A. et alia "A Theoretical Foundation for Abductive Expert Systems", in Gupta, M.M. et alia (eds), "Approximate Reasoning in Expert Systems" (North Holland, 1985).
(54) Rine, D.C. "Some Applications of Multiple-Valued Logic and Fuzzy Logic to Expert Systems", in Gupta, M.M. et alia (eds) "Approximate Reasoning in Expert Systems" (North Holland, 1985).
(55) Ríos, S. "Enfoque Actual de la Decisión Multicriterio" , in Primera Reunión Nacional de Investigación Militar Operativa, Madrid, 1985.
(56) Salicrú, M. "Sobre la Convergencia y Simetría de la J-Divergencia Generalizada", XVI Reunión SEIO, Málaga, Spain, 1986.
(57) Salicrú, M. and Calvo, M. "Un Indice de Incertidumbre asociado a Divergen-

cias". XVI Reunión SEIO. Málaga, Spain, 1986.
(58) Silva and Rubi, A. "On the Estimation of Proportions Using Complete Cluster Sampling". The American Statistician, 37, 53-54, 1983.
(59) Sirken, M.G. and Levy, S. "Multiplicity Estimation of Proportions Based on Ratios of Random Variables". JASA, 69, 68-74, 1974.
(60) Stone, L.D. "Theory of Optimal Search". Academic Press, 1975.
(61) Thomsen, I. et alia "Estimation of Design Effects and Intraclass Correlations Using Outdated Measures of Size". International Statistical Review, 54-349, 1986.
(62) Tukey, J.W. "Sunset Salvo". The American Statistician. 40, 72-76, 1986.
(63) Victor, N. "Computational Statistic-Tool or Science" (with discussion), Statistical Software Newsletter, 10, 105-125, 1984.
(64) Wolter, K.M. "Some Coverage Error Models for Census Data". JASA, 81, 1986.
(65) Zadeh, L.A. "The Role of Fuzzy Logic in the Management of Unvertainty in Expert Systems", in Gupta et alia, "Approximate Reasoning in Expert Systems". (North Holland, 1985)

FSS = Fuzzy Sets and Systems.
JASA = Journal of the American Statistical Association.
JRSS = Journal of the Royal Statistical Society.

APPENDICES

Fuzzy Logic in Knowledge-Based Systems,
Decision and Control
M.M. Gupta and T. Yamakawa (Editors)

Appendix A
Major Current Bibliographical Sources on Fuzzy Sets and Systems

Since the inception of the theory in 1965 over 7000 research papers, reports, monographs and books have been published on fuzzy set theory and its applications to many branches of soft sciences. Many symposia, meetings and international conferences have been organized in the field. Over 100 research institutions around the world are engaged in the development of the theory and its applications to engineering, medical, management, social and other soft sciences.

Readers may find the following additional bibliographical sources very useful in their own studies and research work.

1. **Fuzzy Sets and Systems:** An IFSA (International Fuzzy Systems Association) publication, North-Holland, Amsterdam. (This journal, the first one completely devoted to Fuzzy Sets and Systems, publishes theoretical and applied papers, and presents reviews on current literature, current conferences and meetings on the subject.) Also, IFSA sponsors international conferences, the first one being held at Palma de Mallorca, Spain (1985), and the second one at Tokyo (1987).
2. **Busefal:** Bulletins for studies and exchanges in fuzziness and its applications, University of Paul Sabatier, Toulouse, France. (As the name indicates, this is a very useful bulletin which publishes current research results (without review) and gives information on current literature, conference and meetings).
3. **Journal of Fuzzy Mathematics:** Wuham, Peoples Republic of China, Huazhong, University of Science and Technology. (This is a research journal published in the Peoples Republic of China. Most of the papers appear in Chinese with occasional papers appearing in English).
4. **NAFIPS** (North American Fuzzy Information Processing Society) (NAFIPS sponsors conferences and publishes occasional newsletters). International Journal of Approximate Reasoning: North Holland, Official Publication of NAFIPS.
5. **Japanese Working Group in Fuzzy Systems.**
6. **Chinese Working Group in Fuzzy Systems.**
7. **European Working Group in Fuzzy Sets.**
8. **Indian Working Group in Fuzzy Sets.**
9. Other conferences and meetings devoted to fuzzy set theory and related subjects are:

   IEEE - Systems Man and Cybernetics

   IEEE - Automatic Control Systems

   IEEE - Multivalued Logics

   Applied General Systems Research etc.
10. In addition to sources 1, 2, 3 and 4 (IFSA Journal of Fuzzy Sets and Systems, BUSEFAL, J. of Fuzzy Mathematics and Int. J. of Approximate Reasoning), cited above, several Bibliographies on the subject have appeared. Most of these bibliographies may be found in the books listed in Appendix B.

Fuzzy Logic in Knowledge-Based Systems,
Decision and Control
M.M. Gupta and T. Yamakawa (Editors)

Appendix B
Selected Books on Fuzzy Theory and Applications

[1] Bezdek J.C., [1981], "Pattern Recognition with Fuzzy Objective Function Algorithms," **Plenum Press,** New York.

[2] Dubois D. and Prade H., [1980], "Fuzzy Sets and Systems: Theory and Applications," **Academic Press,** New York.

[3] Goodman I.R. and Nguyen H.T., [1985], "Uncertainty Models for Knowledge-Based Systems", **North-Holland,** Amsterdam.

[4] Gupta M.M. (Ed.), Saridis G. and Gaines B.R. (Assoc. Eds.), [1977], "Fuzzy Automata and Decision Processes," **North-Holland,** Amsterdam.

[5] Gupta M.M. (Ed.), Ragade R.K. and Yager R.R. (Assoc. Eds.), [1979], "Advances in Fuzzy Set Theory and Applications," **North-Holland,** Amsterdam.

[6] Gupta M.M. and Sanchez E. (Eds.), [1982], "Approximate Reasoning in Decision Analysis," **North-Holland,** Amsterdam.

[7] Gupta M.M. and Sanchez E. (Eds.), [1982], "Fuzzy Information and Decision Process," **North-Holland,** Amsterdam.

[8] Gupta M.M., Kandel A., Bandler W. and Kiszka J.B. (Eds.), [1985], "Approximate Reasoning in Expert Systems," **North-Holland,** Amsterdam.

[9] Gupta M.M. and Yamakawa T. [1988], "Fuzzy Computing: Theory, Hardware and Applications", **North Holland,** Amsterdam.

[10] Gupta M.M. and Yamakawa T. [1988], "Fuzzy Logic in Knowledge-Based Systems, Decision and Control," **North Holland,** Amsterdam.

[11] Kacprzyk J. and Yager R.R. (Eds.), [1985], "Management Decision Support Systems Using Fuzzy Sets and Possibility Theory," **Verlag TUV, Rheinland,** Koln, Germany.

[12] Kandel A. and Lee S.C., [1979], "Fuzzy Switching and Automata," **Crane and Russak,** New York.

[13] Kandel A., [1982], "Fuzzy Techniques in Pattern Recognition", **John Wiley and Sons,** New York.

[14] Kandel A., [1986], "Fuzzy Mathematical Techniques with Applications", **Addison-Wesley Publishing Company, Reading, Mass.**

[15] Kaufmann A., [1973], "Introduction a la Theorie des Sous-Ensembles Flous, Vol. 1, Elements Theoriques de Base", **Masson et cie,** Paris, France.

[16] Kaufmann A., [1975], "Introduction at la Theorie des Sous-Ensembles Flous, Vol. 2, Applications a la Lingustique et al Semantique, **Masson et cie,** Paris, France.

[17] Kaufmann A., [1975], "Introduction a la Theorie des Sous-Ensembles Flous, Vol. 3, Applications a la Classification et al Reconnaissance des Formes, aux Automates et aux Systems, aux Choix des Criteres," **Masson et cie,** Paris, France.

[18] Kaufmann A., [1975], "Introduction to the Theory of Fuzzy Subsets," Vol. 1, **Academic Press,** New York.

[19] Kaufmann A., Dubois T. and Cools M., [1975], "Exercices avec Solutions sur la Theorie des Sous, Ensembles Flous, **Masson,** Paris.

[20] Kaufmann A., [1976], "Introduction a la Theorie des Sous-Ensembles Flous, Vol. 4," Ed. **Masson,** Paris.

[21] Kaufmann A., [1979], "Modeles Mathematiques pour la Stimulation Inventive," Ed. **Albin Michel,** Paris.

[22] Kaufmann A., [1979], "Introduction to the Theory of Imprecise Sets," (in Russian), **Radio Moscow.**

[23] Kaufmann A., [1982], "Introduccion a la teoria de los subconjuntos borrosos, Elementos teorieos de base" **C.E.C.S.A.,** Mexico.

[24] Kaufmann A., and Gupta M.M., [1985], "Introduction to Fuzzy Arithmetic: Theory and Applications," **Van Nostrand Reinhold, Electrical/Computer Science and Engineering Series, New York.**

[25] Kaufmann A. and Gupta M.M. [1988], "Fuzzy Mathematical Models with Applications to Engineering and Mangement Science", **North Holland,** Amsterdam.

[26] Kickert W.J.M., [1979], "Fuzzy Theories on Decision Making," **Hortinus Nijhoff, The Hague,** Boston, London.

[27] Mamdani E.H. and Gaines B.R., [1981], "Fuzzy Reasoning and Its Applications," **Academic Press,** New York.

[28] Moisil G.C., [1975], "Lectures on the Logic of Fuzzy Reasoning," **Scientific Editions,** Bucharest.

[29] Negoita C.V. and Ralescu D.A., [1975], "Applications to Fuzzy Sets to Systems Analysis," **Birkhauser, Bale and Stuttgart.**

[30] Negoita C.V., [1979], "Managements Applications of System Theory," **Birkhauser, Bale and Stuttgart.**

[31] Negoita C.V., [1981], "Fuzzy Systems," **Tunbridge Wells,** England.

[32] Negoita C.V., [1985], "Expert Systems and Fuzzy Systems," **Benjamin/Cummings, Menlo Park,** California.

[33] Nishida T. and Takeda E., [1978], "Fuzzy Sets and Its Applications," **Morikita,** Tokyo (in Japanese).

[34] Prevot M., [1980], "Sous ensembles Flous, une Approche Theorique," Ed. Sircy, E. Trillas, 1977, Conjuntos Borrosos, Ed. **Vincenso Universitad.**

[35] Skala H.J., Termini S. and Trillas E. (Eds.), (1984), "Aspects of Vagueness," **D. Reidel, Dordnecht.**

[36] Wang P.P. and Chang S.K. (Eds.), [1980], "Fuzzy Sets - Theory and Applications to Policy Analysis and Information Systems," **Plenum Press,** New York.

[37] Yager, R.R. (Ed.), [1982], "Recent Advances in Fuzzy-Set and Possibility Theory," **Pergamon Press,** Elmsford, NY.

[38] Yager R.R., [1983], "Fuzzy Sets - A Bibliography," **Intersystems Publications,** Seaside, CA

[39] Zadeh L.A., Fu K.S., Tanaka K. and Shimura M. (Eds.), [1974], "Fuzzy Sets and Their Applications to Cognitive and Decision Processes," **Academic Press,** New York.

[40] Zemankova-Leech M. and Kandel A., [1984], "Fuzzy Relational Data Bases - A Key to Expert Systems," **Verlag TUV Rheinland,** Koln, Germany.

[41] Zimmermann H.J., [1985], "Fuzzy Set Theory - and Its Applications," **Kluwer, Nijhoff,** Publishing, Dordrecht.

[42] Zimmermann H.J., Zadeh L.A. and Gaines B.R. (Eds.), [1984], "Fuzzy Sets and Decision Analysis," **North Holland, Published by Elsevier Science Publishers B.V.** as Vol. 20 in TIMS/Studies in the Management Sciences.

BIOGRAPHICAL INFORMATION

Fuzzy Logic in Knowledge-Based Systems,
Decision and Control
M.M. Gupta and T. Yamakawa (Editors)

BIOGRAPHICAL INFORMATION ABOUT THE EDITORS AND CONTRIBUTING AUTHORS

Francisco Azorin, [23].

Universidad Autonoma de Madrid, Spain.

Francisco Azorin received the degrees Lic. in 1943 and Doctor en Ciencias Matematicas in 1953 from the University of Madrid and Doctor Honoris Causa in 1985 from the University of Malaga, Spain. He is a professor (retired) of mathematics, statistics and probability at the University Autonoma of Madrid, University of Santiago de Compostela, Spain, and University Central de Venszuela. He is the author or co-author of over 80 articles and four books on applied and mathematical statistics, sampling and fuzzy sets. He is a former president of the National Institute of Statistics, Spain and member of the ISI and of the Spanish Royal Academy of Sciences.

Bernadette Bouchon, [5].

LAFORIA, Universite Paris VI, Tour 45, 4 Place, Jussieu, 75252 Pairs Cedex 05, France.

Bernadette Bouchon is a researcher at the National Center for Scientific Research (CNRS) in France. She is also Vice-Director of the LAFORIA (Form and Artificial Intelligence Laboratory) at Paris VI University. She received her M.S. degree in mathematics and computer science from the University of Paris in 1970 and 1971, as well as her Dr. S. degree in applied mathematics in 1978. She is a member of the Institute of Electrical and Electronic Engineers and the International Fuzzy Systems Association. She is also the President of the Systems Science Section of the French Association for Economical and Technical Cybernetics (AFCET). She was a co-chairperson of the International Conference on Information Processing and Management of Uncertainty in Knowledge-based Systems, held in Paris in 1986. Her research interests include information theory, including questionnaire theory, and possibility theory, particularly its applications to the management of uncertainty in automated process.

James J. Buckley, [9].

Mathematics Department, University of Alabama at Birmingham, Birmingham, AL 35294, USA.

James Buckley received the Ph.D. degree in applied mathematics from the Georgia Institute of Technology in 1970. From 1970 to 1976 Dr. Buckley was an assistant professor of mathematics at the University of South Carolina. In 1976 Dr. Buckley joined the faculty at the University of Alabama at Birmingham and is currently an associate professor of mathematics. His research interests include decision theory, fuzzy sets, risk analysis, game theory, expert systems, operations research models, finance and economics, statistics, probability, mathematical programming and analysis.

Dan Butnariu, [13].

Department of Mathematics, University of Halifa, 31999 Haifa, Israel.

Dan Butnarius received his M.S. and Ph.D. degrees in mathematics from the University of Iassy, Romania in 1975 and 1980 respectively. From 1979 to 1983 he was with the Department of Mathematics of the Polytechnic Institute of Iassy and from 1984 to 1986 he was a fellow in the Department of Theoretical Mathematics of the Weizmann Institute of Science, Rehovot, Israel. Dr. Butnariu is currently a senior lecturer in the Department of Mathematics, University of Haifa, Israel. His research interests are fuzzy topologies, fixed points, random sets, point- to-fuzzy set mapping, fuzzy measures, and their applications in game theory and system analysis.

Mihir K. Chakraborty, [14].

Department of Pure Mathematics, Calcutta University, 35, Ballygunge Circular Road, Calcutta 700 019, India.

Mihir K. Chakraborty was awarded the M.Sc. degree in pure mathematics by Calcutta University and the Ph.D. degree by Kalyan University, India. In the early 1970's he published papers on fixed point theory. He has had visiting appointments in the Department of Mathematics and Statistics, University of Regina, Canada and RWTH, Aachen, West Germany. His current research interests are fuzzy set theory, mathematical logic and topology. Dr. Chakraborty is one of the organisers of ISFUMIP (The Indian Society of Fuzzy Mathematics and Information Processing).

Tu Chengyuan, [18].

Department of Radioelectronics, Beijing Polytechnic University, the Eastern Suburb of Beijing, China.

Xu Chunxi, [20].

Editorial Office Journal of Fuzzy Mathematics, Huzhan University of Science and Technology, Wuhan, China.

Xu Chunxi graduated in linguistics from the Xiangiang Normal College, China in 1981. Since 1981, he has been the editor of the journal "Fuzzy Mathematics." He is also a special adviser to the Guiyang micro-computer research association. His research interests are linguistics and logistic.

Siegfried Gottwald, [11].

Logic Group, Department of Philosophy, Karl Mark University, Leipzig, G.D.R.

Siegfried Gottwald is an associate professor of mathematical logic. He received his Ph.D. and Habilitation degrees from the Karl Marx University in 1969 and 1977 respectively. His current research interests include many-valued logics, fuzzy set theory, approximate reasoning and other applications of fuzziness in systems modelling and expert systems. He is a member of the Editorial Boards of Fuzzy Sets and Systems and the Journal of Non-Classical Logics, a Correspondant Etranger of BUSEFAL, and co-author of a four-volume *First Course in Analysis* (in German, Teubner, Leipzig, 1972-74) as well as of an *Introduction into Non-Classical Logics* (in German, Akademie-Verlag, Berlin, 1987)

Madan M. Gupta (Editor), [1, 2, 4].

Intelligent Research Laboratory, College of Engineering, University of Saskatchewan, Saskatoon, S7N OWO, Canada.

Dr. M.M. Gupta received the B. Eng. (Honors) in 1961, and the M.Sc. in 1962, both in electronics-communications engineering, from BITS Pilani, India. He received the Ph.D. degree for his studies in adaptive control systems in 1967 from the University of Warwick, U.K.

From 1962 to 1964, he served as a lecturer in electrical-communication engineering at the University of Roorkee, India. He joined the faculty of the College of Engineering at the University of Saskatchewan in 1967 as a sessional lecturer, becoming a full professor in July 1978. He has received numerous grants and research contracts since 1969. He has acted as a consultant to various agencies such as Atomic Energy of Canada Limited, Saskatchewan Power Corp., MacMillan Bloedal Research Limited, the Defense Research Board, and Honeywell of Canada. He was a visiting professor at the Florida State University during 1984. He has initiated research in the areas of optimization and adaptive control systems, fuzzy automata and decision processes, and signal processing. His present research interests are in the areas of non-invasive methods in medical diagnosis, medical imaging, intelligent robotic systems, cognitive information and neural networks.

He is a co-author, with A. Kaufmann, of the books *Introduction to Fuzzy Arithmetic: Theory and Applications* (Van Nostrand Reinhold, 1985) and *Fuzzy Mathematical Models with Applications to Engineering and Management Science* (North Holland, in press) and is the editor of the books *Fuzzy Automata and Decision Processes* (1977), *Advances in Fuzzy Set Theory and Applications* (1979), *Approximate Reasoning in Decision Analysis* (1982), *Fuzzy Information, Knowledge Representation, and Decision Analysis* (1983), *Approximate Reasoning in Expert Systems* (1985), *Fuzzy Computing* (1988), *Fuzzy Logic in Knowledge Based Systems, Decision and Control* (1988), all with North Holland and *Adaptive Methods for Control Systems Design* (IEEE Press, 1986). He is the subject editor for the *Encyclopedia of Systems and Control* (Pergamon Press, Oxford, 1987) and has authored or coauthored over 200 research papers. He is an advisory editor for the *International Journal of Fuzzy Sets and Systems* (IFSA) and other journals in the field.

Dr. Gupta was a program chairman of the special symposium on Fuzzy Set Theory and Applications held during the 1977 IEEE-CDC at New Orleans. He was also a co-chairman of the IFAC symposium on Fuzzy Information, Knowledge Representation and Decision Analysis held at Marseille (France) in July 1983.

Behrooz Heshmathy, [22].

University of Florida State, Tallahassee, Florida, 32306-4019 USA.

Behrooz Heshmathy received his M.Sc. degree from the Department of Mathematics and Computer Science at Florida State University in 1983. Presently he is involved in research and development work in the fields of artificial intelligence and software engineering.

Ellen Hisdal, [3].

Institute of Informatics, University of Oslo, Box 1080 Blindern, 0316 Oslo 3, Norway.

Ellen Hisdal received the B.A. and M.A. degrees in physics from the University of California, Berkeley, and the Ph.D. degree, also in physics, from the University of Oslo. Her current research interests include fuzzy and probability logic adapted to the use of logic in natural language, and a natural language computer system with emphasis on the semantic side. Dr. Hisdal is an associate professor of informatics at the University of Oslo.

Li Hongxing, [12].

Section of Mathematics, Tianjin Institute of Textile Engineering, Tianjin, China.

Li Hongxing graduated from the Department of Mathematics, Nankai University, Tianjin, in 1978. In 1978, he joined the faculty at the Tianjin Institute of Textile Engineering, Tianjin, as an assistant professor of mathematics. He became a lecturer of mathematics at the Institute in 1984 and is currently an associate professor and director of the mathematics section. He is also a counsellor of the Chinese Fuzzy Systems Association, and Associate President of the Tianjin Council of the Fuzzy Mathematics Association. Since 1982 he has published several papers and books on mathematics and fuzzy mathematics. His research interests include lattic topology, upper space theory, upper algebra structure, ordered structure, fuzzy decision-making, fuzzy clustering analysis, fuzzy relation equations and fuzzy pattern recognition.

Abraham Kandel, [22].

Computer Science Department, Florida State University, Tallahassee, Florida, USA.

Abraham Kandel is a professor in and the Chairman of the Computer Science Department at Florida State University in Tallahassee, Florida. He is also the Director of the Institute for Expert Systems and Robotics at Florida State University. He received his Ph.D. degree in electrical engineering computer science from the University of New Mexico, his M.S. degree in electrical engineering from the University of California, and his B.Sc. in electrical engineering from the Technion (Israel Institute of Technology). Dr. Kandel is a senior member of the Institute of Electrical and Electronics Engineering and a member of NAFIPS the Pattern Recognition Society, and the Association for Computing Machinery, as well as an advisory editor to the international journals Fuzzy Sets and Systems, Information Sciences and Expert Systems. He is also co-author of *Fuzzy Switching and Automata: Theory and Applications* (1979), author of *Fuzzy Techniques in Pattern Recognition* (1982), co-author of *Discrete Mathematics for Computer Scientists* (1983), co-author of *Fuzzy Relational Databases - A Key to Expert Systems* (1985), co-editor of *Approximate Reasoning in Expert Systems* (1985), co-author of *Designing Fuzzy Expert Systems* (1986), and author of *Fuzzy Mathematical Techniques with Applications* (1986). He has written more than 150 research papers for numerous national and international professional publications in computer science.

Etienne E. Kerre, [10].

Seminar for Mathematical Analysis, State University of Gent, B-9000 Gent, Belgium.

Etienne E. Kerre is a member of the academic staff of the Seminar for Mathematical Analysis and Nuclear Power Technology at the State University of Gent. He received his doctoral degree in mathematics from the State University of Gent in 1970 for research concerning a theoretical approach to low energy electron diffraction. He is co-author of three books on basic mathematics at the undergraduate level. Since 1977 he has worked on fuzzy sets theory. He has published 23 papers on fuzzy topology, medical diagnosis, information retrieval and databases, expert systems. Since 1985 he has been giving a new course on fuzzy sets theory at the graduate level in mathematics at the State University of Gent.

George K. Knopf, [4].

Intelligent Systems Research Laboratory, University of Saskatchewan, Saskatoon, Sask., S7N OWO, Canada.

George K. Knopf received his B.A., B.Sc. and M.Sc. from the University of Saskatchewan. He is presently pursuing his Ph.D. in the area of vision perception. His research interests include intelligent systems, vision systems, cognitive processes, neural networks and fuzzy systems.

Javier Montero, [15].

Universidad Complutense, 28040, Madrid, Spain.

Javier Montero received the B.Sc. degree in mathematics in 1978, and the Ph.D. degree in statistics and operational research in 1982, both from the Complutense University of Madrid. In 1979 Dr. Montero joined the faculty at the Complutense University as an assistant professor of mathematics. Dr. Montero is currently Titular Professor in the Department of Statistics and Operational Research at the same university. His research interests include decision making, fuzzy systems and modelling in biology.

Peter N. Nikiforuk, [4].

College of Engineering, University of Saskatchewan, Sask., S7N OWO, Canada.

Peter N. Nikiforuk received the B.Sc. degree in engineering physics from Queen's University, Kingston, Ontario, Canada in 1952 and the Ph.D. degree in electrical engineering from Manchester University, England in 1955. In 1970 he received the D.Sc. degree from Manchester University for research on control systems. Prior to joining the University of Saskatchewan as an assistant professor in 1960, he worked in the aircraft and missile industry in Canada and the USA. Dr. Nikiforuk became a professor in 1965, was head of the Department of Mechanical Engineering from 1966 to 1973 and since 1973 has been Dean of Engineering. He has served as a member of a number of national research councils and is presently a fellow of six engineering and scientific institutions. His research is mainly in the areas of adaptive control systems and hydraulic control systems.

Hideyuki Nojiri, [21].

Kumamoto University of Commerce, 2-5-1 Oe, Kumamoto 862, Japan.

Hideyuki Nojiri received a M.Sc. degree in mechanical engineering from Waseda University in 1969 and a doctoral degree in system science from the Tokyo Institute of Technology in 1981. Dr. Nojiri is currently a professor of management science and also the Director of the Computer Center at KUC. His research interests are in the areas of team theory, theory of group decision-making and expert systems.

Takehisa Onisawa, [19].

Department of Basic Engineering, Faculty of Engineering, Kumamoto University, 2-39-1, Kurokami, Kumamoto, 860 Japan.

Takehisa Onisawa received the B. Eng. and the M. Eng. degrees in control engineering and the D.Eng. in systems science from the Tokyo Institute of Technology in 1975, 1977 and 1986, respectively. He was a research associate in the Department of Systems Science at the Tokyo Institute of Technology from 1977 to 1986. He is currently a lecturer in the Faculty of Engineering at Kumamoto University. His research interests include the applications of fuzzy theory, especially reliability and ergonomics.

Walenty Ostasiewicz, [16].

Academy of Economics, Komandorska 118/120, 53-345, Wroclaw, Poland.

Witold Pedrycz, [11].

Department of Automatic Control and Computer Science, Silesian Technical University, Gliwice, Poland.

Witold Pedrycz received his Ph.D. and Habilitation degrees in computer engineering from the Silesian Technical University, where he is now an associate professor. Currently, he is also with the Research and Development Centre of Electronic Medical Instruments, Zabrze, Poland. His research interests include fuzzy systems, computer science, knowledge based systems, pattern recognition and computer control. He is an associate editor of Pattern Recognition Letters, correspondant etranger of BUSEFAL, and co-author of an *Introduction into Fuzzy Set Theory* (in Polish, PWN Warsaw, 1985).

Armando Freitas da Rocha, [17].

Department of Physiology, State University of Compinas, UNICAMP, 13100, Campinas, Brazil.

Armando F. da Rocha is an associate professor in the Department of Physiology and Biophysics at the Institute of Biology, State University of Campinas, Brazil. He received his M.D. from the Medical School at Sao Paulo in 1970 and his Ph.D. from the Institute of Biology, State University of Campinas in 1973. His interests are neurophysiology and informatics. He has published several papers in journals and specialized books about the formalization of neural processing and the application of this knowledge in creating fuzzy expert systems. He is presently working on two main projects on the techniques for knowledge acquisition and representation for expert systems, and on the development of an expert system having some heuristic learning capability.

Chen Shi-quan, [20].

Department of Mathematics, Guiszhou Normal University, Guiyang, Guiszhou, China.

Chen Shi-quan was at the Guiyan Health School from 1961 to 1964 studying medicine and at Guizhou University from 1973 to 1975 studying physics. Before 1986 he was at the Guizhou Institute for the Prevention and Control of Industrial Hygene and Occupational Diseases. He is now an associate professor in the Department of Mathematics at the Guizhou Normal University, and the Director of the Research Section of the Soft Sciences. His research interests are pattern recognition, cluster analysis, fuzzy statistics, fuzzy reasoning, decision making and the application of fuzzy sets in medicine. He is a member of the China Chapter of the IFSA, and an executive committee member of the Chinese Society of Fuzzy Mathematics and Systems (CSFMS).

William Siler, [9].

Mote Marine Laboratory, SARASOTA, Florida, 33577, USA.

William Siler received the M.S. degree in fluid dynamics from the Steven Institute of Technology, in 1949, and the Ph.D. degree in biology from the City University of New York in 1972. From 1959 to 1965, he served in the Radiological Physics Section of the Memorial Sloan Kettering Cancer Center and from 1965 to 1972 he was an associate professor of biomedical computer science at the Downstate Medical Center in Brooklyn, N.Y. From 1972 to 1980, he was professor of biomathematics at the University of Alabama at Birmingham and from 1972 to 1980 he was the Director of Clinical Computing at Carraway Methodist Medical Center in Birmingham, Alabama. He is now a senior research associate at the Mote Marine Laboratory in Sarasota, Florida. His research interests include system simulation models and artficial intelligence.

Heinz J. Skala, [7].

University of Paderborn, Paderborn, FRG.

Heinz J. Skala received the M.S. and Ph.D. degrees in economics from the University of Vienna. In 1973 Dr. Skala received the Habilitation degree in econometrics and statistics from the University of Heidelberg where he joined the Faculty of Economics as Privatdozent. Since 1974 he is a professor of econometrics and statistics at the University of Paderborn. His research interests include decision making and fuzzy set theory. He is editor of the Theory and Decision Library, Series: Mathematical and Statistical Method (Reidel) and co-editor of Statistical Papers (Springer).

Marly Theoto, [17].

Nursing School, State University of Sao Paulo, Av. Dr. Eneas C., Aguiar, 419-05403, Sao Paulo, Brazil.

Marly Theoto received her B.S. in nursing and her B.S. and M.S. degrees in public health from the State University of Sao Paulo in 1971, 1971 and 1978 respectively. She is currently an assistant professor and responsible for the Laboratory of Informatics of the Nursing School of Sao Paulo. Her research interests include decision making systems and goal oriented systems applied to nursing and public health programs.

Pietro Torasso, [17].

Departimento di Informatica, Universita di Torino, Corso Svizzera, 185-10149, Torino, Italy.

Piero Torasso is an associate professor in informatics at the University of Torino. He is actively engage in projects dealing with decision making systems in medicine and is the author and co-author of several papers related to this subject. He participated in the development of LITO2, an expert system that is used for hepatic diseases. He was also involved with a project to establish the guidelines of an expert system in hypertension during his visit to Brazil as an invited professor at the State University of Campinas.

Antoine A.C. Van Schooton, [10].

Seminar and Laboratory for Computer Science, State University of Gent, B-9710 Gent, Belgium.

Antoine Van Schooten was born in Belgium in 1962. He graduated from the Department of Electronic Engineering at the State University of Gent in 1985, with a thesis called "Design of a Fuzzy Expert System". In 1986 he received a M.S. degree in computer science from the same university with a thesis called "The Modelling of Uncertainty and Imprecision in Expert Systems." At present he is working as a researcher at the Belgian Fund for Scientific Research at the Computer Science Laboratory of the State University of Gent, where he is working towards a Ph.D. degree in computer science. His current research interests are fuzzy sets and possibility theory, artificial intelligence, expert systems and databases.

S.T. Wierzchon, [8].

Institute of Computer Science, Polish Academy of Sciences, 00-901 Warsaw PKN, P.O. Box 22, Poland.

S.T. Wierzchon is an assistant professor in the Institute of Computer Science and a consultant in the Institute for Management in Machine Industry, Warsaw. He received a M.S. degree in automation and a Ph.D. degree on the modelling of imprecise systems, both from the Technical University of Warsaw. His research interests include knowledge-based systems, decision-making theory and the theory and applications of fuzzy set.

Tu Xiangchu, [18].

Department of Radioelectronics, Beijing Polytechnic University, the Eastern Suburb of Beijing, China.

Tu Xiangchu graduated from Qinghua University, Beijing, China, in July 1952, and joined the faculty in 1957. He was involved in research on the automatic control of industrial processes. Since 1980 he has been on the faculty at Beijing Polytechnic University. Professor Tu Xiangchu is also a research-professor at the Institute of Fuzzy Systems and Knowledge Engineering, Guangzhou University, China, and a senior adviser of the Chinese Institute of Electronics, Beijing, and the Jiangxi Institute of Computing Technique, Jiangxi Province, China.

Takeshi Yamakawa (Editor).

Department of Electrical Engineering and Computer Science, Kumamoto University, Kumamoto 860, Japan.

Takeshi Yamakawa received the B.E. degree in electronic engineering from the Kyushu Institute of Technology in 1969, and the M.E. and Ph.D. degrees in electrochemistry from Tohoku University in 1971 and 1974, respectively. In 1974 Dr. Yamakawa joined the Faculty of Engineering at Tohoku University as a research assistant in electronics. In 1977 Dr. Yamakawa joined the Faculty of Engineering at Kumamoto University as a research assistant in electrical engineering. Since 1981, he has been an associate professor of electrical engineering and computer science at Kumamoto University. He is also a member of the editorial association of the International Journal of Fuzzy Sets and Systems. His research interests include the development of fuzzy logic hardware systems and multiple-valued logic circuits (both in integrated circuits) and also medical sensors. He is a pioneer of fuzzy chips and fuzzy computer hardware, and has applied for 53 patents in Japan, 7 in the USA and 5 in Europe.

Milan Zeleny, [6].

Department of Management Science, Graduate School of Business, Fordham University, New York, NY, USA.

Milan Zeleny is a professor of management systems at the Graduate School of Business of Fordham University. He holds the Dipl. Ing. (Prague) in political economics, and the M.S. degree (Rochester) in operations research. His previous appointments have been at Columbia University, Copenhagen School of Economics, and the Institute for Advanced Studies in Management, University of South Carolina SUNY at Binghamton. He has published about 200 papers, articles and books and serves on the editorial boards of six journals. His current research interests are concentrated on human systems.

THE COMPANION VOLUME "FUZZY COMPUTING: THEORY, HARDWARE, AND APPLICATIONS"

CONTENTS LIST OF FUZZY COMPUTING: THEORY, HARDWARE, AND APPLICATIONS

M.M. Gupta and T. Yamakawa, eds. Published by Elsevier Science Publishers B.V. (North-Holland), 1988. Pp. xx + 500. ISBN: 0 444 70449 3.

BIOGRAPHICAL INFORMATION